T0311948

Laboratory Assessment
of Vitamin Status

CHAPTER 7   Methods for Assessment of Vitamin B₂ 165

Roy A. Sherwood

CHAPTER 8   Methods for Assessment of Pantothenic Acid (Vitamin B₅) 173

Rick Huisjes and David J. Card

Contents

Academic Press is an imprint of Elsevier
125 London Wall, London EC2Y 5AS, United Kingdom
525 B Street, Suite 1650, San Diego, CA 92101, United States
50 Hampshire Street, 5th Floor, Cambridge, MA 02139, United States
The Boulevard, Langford Lane, Kidlington, Oxford OX5 1GB, United Kingdom

Library of Congress Cataloging-in-Publication Data
A catalog record for this book is available from the Library of Congress

British Library Cataloguing-in-Publication Data
A catalogue record for this book is available from the British Library

ISBN 978-0-12-813050-6

For information on all Academic Press publications
visit our website at https://www.elsevier.com/books-and-journals

Working together
to grow libraries in
developing countries

www.elsevier.com • www.bookaid.org

Publisher: Stacy Masucci
Acquisition Editor: Tari K. Broderick
Editorial Project Manager: Carlos Rodriguez
Production Project Manager: Maria Bernard
Designer: Miles Hitchen

Typeset by SPi Global, India

Laboratory Assessment of Vitamin Status

Edited by

Dominic Harrington

ACADEMIC PRESS

An imprint of Elsevier

Contributors

David J. Card
Nutristasis Unit, Viapath, St. Thomas' Hospital, London, United Kingdom

Neal E. Craft
Eurofins Craft Technologies, Inc., Wilson, NC, United States

Rachel S. Carling
Biochemical Sciences, Viapath, St Thomas' Hospital, London, United Kingdom

Graham Carter
North West London Pathology, Imperial College Healthcare NHS Trust, Charing Cross Hospital, London, United Kingdom

Martin A Crook
Department of Clinical Biochemistry and Metabolic Medicine, Guys and St Thomas' Hospital; Department of Clinical Biochemistry and Metabolic Medicine, University Hospital Lewisham, London, United Kingdom

Krutika Deuchande
Nutristasis Unit, Viapath, St. Thomas' Hospital, London, United Kingdom

Harold C. Furr
Department of Nutritional Sciences (retired), University of Connecticut, Storrs, CT, United States

Renata M. Górska
Nutristasis Unit, Viapath, St. Thomas' Hospital, London, United Kingdom

Dominic J. Harrington
Nutristasis Unit, Viapath, St. Thomas' Hospital; Faculty of Life Sciences and Medicine, King's College London, London, United Kingdom

Rick Huisjes
Nutristasis Unit, Viapath, St. Thomas' Hospital, London, United Kingdom

Scott W. Leonard
Linus Pauling Institute, Oregon State University, Corvallis, OR, United States

Roy A. Sherwood
King's College London, London, United Kingdom

Agata Sobczyńska-Malefora
Nutristasis Unit, Viapath, St. Thomas' Hospital; Faculty of Life Sciences and Medicine, King's College London, London, United Kingdom

Maret G. Traber
Linus Pauling Institute; College of Public Health and Human Sciences, Oregon State University, Corvallis, OR, United States

Charles Turner
WellChild Laboratory, Evelina London Children's Hospital, Guy's & St Thomas' NHS Foundation Trust, London, United Kingdom

Preface

This book had its origin in some of the collaborations that I have enjoyed since unexpectedly gaining an interest in the laboratory assessment of vitamin status 30 years ago. As a student, I had the good fortune to study in the laboratory of Dr. Martin Shearer at Guy's Hospital in London. Dr. Shearer had pioneered the laboratory assessment of vitamin K during the late 1960s and successfully measured the vitamin in plasma at endogenous concentrations for the first time during the 1970s. I soon recognized that his keen attention detail was one key to his considerable success and I am grateful that he recognized my curiosity. I spent happy holidays trying to measure vitamin K properly and mastering the idiosyncrasies of temperamental early generation chromatography systems. With hindsight, overcoming these challenges provided invaluable experiences that I still call on today—little is learnt in a laboratory when all is functioning as it should.

Several decades earlier Guy's Hospital happened to play another role in the vitamin story through the training of the mature student Frederick Hopkins. Sir Frederick Gowland Hopkins went on to be awarded the Nobel Prize in Physiology or Medicine in 1929, with Christiaan Eijkman for the discovery vitamins, although Casimir Funk, while working in Chelsea is widely credited with the discovery. Casimir Funk's most notable prodigy was Jack Drummond. It was Professor Drummond who coined the word "vitamin" much to the discontentment of his mentor. Tragically Jack, his wife Anne, and their young daughter Elizabeth, will always be remembered as the victims of the infamous "Dominici affair."

After relocating to St. Thomas' Hospital in 1994 I went on to found the Nutristasis Unit and benefitted greatly from quickly making the acquaintance of talented (and now longstanding) colleagues. The Unit is focused on the development and application of novel markers of vitamin status. My colleagues subsequently developed their own areas of interest and I am delighted that several of them have contributed chapters to this book. Other chapters have been contributed by collaborators whose unrelenting generosity in sharing their expertise and time are much appreciated and are representative of the characteristics of so many who work tirelessly to unravel the roles of vitamins.

Since vitamins share little structural similarity with each other a variety of approaches have been required to facilitate their study. Present-day methods for the assessment of vitamin status are products of sequential advances which have been achieved since the beginning of the vitamin story. Many of the available techniques and instrumentation would be unrecognizable to our predecessors.

Laboratory workers spend a great deal of their time investigating the routine. It is to those who recognize and find the time to properly investigate the abnormal that particular credit is due. Recognition of the abnormal is made more probable when those performing laboratory investigations understand the significance and limitations of the tests that they carry out. If this book assists in supporting an improved understanding then it has been a worthwhile endeavor. If it also promotes a curiosity that furthers the future development of methods for the laboratory assessment of vitamin status then all the better.

I should like to thank all of the authors for their contribution and camaraderie. A particular thank you to Denise Oblein for her assistance throughout this project, and to Gordon Avery and Barbara Maniglia for providing the cover image of vitamin B_{12}. I would like to acknowledge Kieran Voong for his contribution to the Nutristasis Unit, and Professor Roy Sherwood for his valuable advice while preparing content for this book. Thanks of course are also due to Pip, Anoushka, Dhruv, and Juno.

Discovery to diagnosis

Krutika Deuchande*, Dominic J. Harrington*,†

Nutristasis Unit, Viapath, St. Thomas' Hospital, London, United Kingdom
Faculty of Life Sciences and Medicine, King's College London, London, United Kingdom†

Chapter Outline

Essential Nutrients for Life

The idea that food may contain small quantities of life-important substances is an old one. Yet, the substances that play such a vital role in health and disease eluded discovery until the 20th century. Their elusiveness appears wholly disproportionate to the importance of their function, and reflects not only the minute quantities in which they are required, but also the degree of analytical sophistication necessary for their detection and characterization.

Early scientific studies of living things were dependent on observation using the human eye. This approach proved sufficient for ancient Greek, Roman, and Arab physicians to conclude that diet plays a role in the prevention and cure of some diseases. Ultimately it was attempts to better understand the causes of beriberi and scurvy, and the innovative interpretation of experimental findings, that eventually

Laboratory Assessment of Vitamin Status. https://doi.org/10.1016/B978-0-12-813050-6.00001-2

revealed our dependence on an exogenous supply of micronutrients to support development and maintain health.

Beriberi (a disorder now known to be caused by vitamin B_1 deficiency) was endemic during the 19th century in Japan.[1] In 1872, Takaki Kanehiro joined the Imperial Japanese Navy and observed that the disease was uncommon among the crewmen of Western and Japanese navies whose diet consisted of various vegetables and meat—yet common in those crewmen whose diet consisted almost exclusively of white rice. Low-ranking crewman ate white rice because it was provided free of charge.

While working in Malay, Englishman W Leonard Braddon reported in 1907 that more than 150,000 cases of beriberi had been treated in Government hospitals and infirmaries during the preceding twenty years from a population of one million people. Of these, 30,000 people had died.[2] Since approximately only one-third of the deaths took place in hospitals, Braddon estimated that the total deaths from beriberi were 100,000. No drug showed any promise in alleviating the disease. Braddon noted that patients remaining under the conditions in which they became ill rarely recovered, yet removal to a different environment or a change of food remedied the disorder.

For the poor, scurvy (now known to be caused by vitamin C deficiency) had been an annual winter affliction in England for hundreds of years. It was not until the devastating impact of the disease on the productivity of crewman, and the subsequent negative commercial implications for trade by sea were recognized, that advances were made towards a cure. Scurvy had successfully been treated by teas, brews and beer made from spruce needles,[3,4] and by giving oranges and lemons.[5] In 1601 Sir James Lancaster introduced the regular use of oranges and lemons into the ships of the East India Company.[1] Others in the 17th and 18th centuries also confirmed that fresh fruit and vegetables were effective in preventing or curing scurvy: Woodall (in 1639),[6,7] Kramer (1739),[8] Lind (1757),[8] and Captain Cook (1772).[9]

The Advent of Biochemistry

The discovery and study of vitamins marks the advent of biochemistry. Until the second half of the 19th century there had been very little collaboration between exponents of biology and chemistry or physics. Scientists were strongly biased in favor of one science or the other until some physiologists began to realize that it should be possible to think about organisms in terms of chemical mechanisms and started to adapt methods from chemistry and physics.

Chemists had learnt how to analyze foods, and ascertained that they were composed of proteins, carbohydrates, and fats, together with certain mineral elements and water. Together these components accounted for nearly 100% of the chemical analysis. One of the foremost investigators of nutrition, Carl von Voit (1831–1908), mentioned that the outcome of feeding pure foodstuffs—preparations of protein, fat, sugar, starch, and inorganic compounds would be no different from that achieved by feeding naturally occurring food mixtures. Indeed, the accomplished investigator Röhmann claimed to have been able to satisfy the nutritive requirements of mice with rations prepared by mixing a number of isolated and purified food components.

Not only did the mice put on weight, but they were sufficiently nourished to produce young. Later, scrutiny of this work suggests that insufficient care was taken in ensuring the purity of the components of his food mixture.

In 1881, the Russian student N. Lunin published a paper concerning the significance of inorganic salts for animal nutrition.[10] Working in Prof Gustav von Bunge's laboratory in Basle, he was the first to produce experimental evidence, and came very near to discovering the vitamins, when he failed in his attempt to rear young mice on a mixture of purified proteins, carbohydrates, fats and minerals, he compounded to resemble the composition of milk, and which according to contemporary theories, should have provided all that was required for the mice to thrive. Others in Central Europe reported similar results. Among them was Carl Socin, also from Bunge's laboratory, who concluded that unknown substances were "present in egg yolk and milk" and which it was "the first task of the future to discover." The question was approached in a different way in Germany by Wilhelm Stepp. Rather than use synthetic diets, he explored the effect of subjecting a natural diet such as bread and milk to an extraction process using alcohol and ether. The mice died when fed the diet but others flourished when the extracts were put back. To some extent these findings were not novel. As early as 1873 Forster had reported that washed meat is not an adequate diet for dogs. He had also observed that pigeons fed an artificial diet food mixture developed symptoms similar to that later to be described by Eijkman.

Germ Theory and the Great Diversion

In 1883 the Dutch Government sent a Commission to Java (part of the Dutch East Indies) to investigate the worryingly high prevalence of beriberi.[11] So shortly after the wide acceptance of Pasteur's "germ theory" and the disproving of the "spontaneous generation doctrine" it was natural for the leaders of the Commission, Professors Winkler and Pekelharing, to think of beriberi in terms of germs and infectivity. They were assisted by the young Army doctor Christian Eijkman. Having discovered a micrococcus, which they suspected to be the cause of beriberi, Winkler and Pekelharing returned to the Netherlands, leaving Eijkman behind as the institute's director. Eijkman concentrated all of his attention on the infective agent and the possibility of it causing beriberi, until a chance observation was made. Financial constraints at the institute led his laboratory assistant to feed the experimental chickens surplus food from the hospital kitchen. Eijkman observed that some of the chickens developed an inability to walk and showed symptoms of beriberi.[11] He initially thought that the birds had become infected by the responsible germ, but when the diet was changed the chickens recovered. Eijkman observed that while the chicken feed used in the laboratory had been unpolished rice, the hospital kitchen rice was polished. In 1897 Eijkman concluded that a polyneuritis gallinarum in fowls and pigeons is analogous to the human beriberi and induced by a diet of polished rice.

Eijkman's colleague, Gerrit Grijns, discovered that the addition of the outer layer of rice (the pericarp, removed from brown rice polishing) to the diet prevented the occurrence of polyneuritis gallinarum in fowls and pigeons.[11] Eijkman stated

"There is present in rice polishing a substance of a different nature from proteins, fats or salts which is indispensable to health and the lack of which causes nutritional polyneuritis."

Nutrition at the Beginning of the 20th Century

At the beginning of the 20th century nutrition was viewed primarily from the standpoint of energy requirement. Public health initiatives were focused on improving housing, and on the provision of a pure water supply and efficient drainage.[12] In London, the capital city of a nation with vast national wealth, poor families lived practically entirely on white bread.[13] A memorandum sent from Sir William Taylor, Director General of the Army Medical Service to the Government reported in 1907 that the Inspector of Recruiting was having the greatest difficulty in obtaining sufficient men of satisfactory physique for service in the South African War. The rejection rate in some areas was as high as 60%, and over the whole country nearly 40%. The chief grounds for rejection were bad teeth, heart affections, poor sight or hearing, and deformities.[14] The shortage of men was so serious that it had been found necessary in 1902 to reduce the minimum height of recruits for the infantry to 5 ft (150 cm); it had already, in 1883, been lowered from 165 cm to 160 cm.

Armies were to provide other useful incites. During the siege of Kut-el-Amara (December 1915 to April 1916), the difference in the type of ration issued to the British and Indian troops formed the basis of an unconscious experiment. In an account from a British military surgeon Patrick Hehir's diary he describes the appearance of beriberi and scurvy in the troops. He writes that British troops were receiving white wheaten flour until February 5th 1916, after which date they were compelled to take part of their flour ration in the form either of barley flour or of *atta*. Beriberi occurred in the British troops while they were enjoying white wheaten bread and cleared up when they were required to share the coarsely milled, germ-containing flour of their Indian comrades. Though during the siege, the incidence of scurvy showed an entirely opposite distribution; the British troops were protected by benefitting from a large ration of meat, while the Indian troops, largely vegetarian in habit, suffered terribly from the disease.

Biological Assays and the Path to Discovery

The experimental "beriberi" observed in birds by Eijkman provided the first biological assay for accessory food factors and set the path towards identifying the anti-beriberi factor. Yet, it remained difficult to implant the idea of disease as a dietary deficiency because the collective bias of thought remained strongly towards that of germ theory. Eijkman and Grijns had incorrectly concluded that rice endosperm contained a toxin and that the antitoxin was present in the pericarp which was removed during rice polishing. Eijkman attempted to isolate the antitoxin and showed that the aqueous extract from rice polishing would cure polyneuritis gallinarum in fowls.

He also showed that when foods were heated they lost some of their effectiveness in preventing and curing the disease.

In Christiania (now Oslo) 1907, Axel Holst and Theodor Frölich attempted to produce experimental beriberi in guinea pigs, their interest stemmed from "ship" beriberi which afflicted Norwegian seaman. To their surprise, the guinea pigs fed on the unbalanced cereal diets did not develop beriberi but a disorder which was quickly recognized as scurvy. Having discovered by chance one of the few animal species in which scurvy could be induced, they had produced experimental scurvy, and the second biological assay to support the investigation of what would later be known as the vitamins.[1,11,15]

The idea that the cause of a third disease may also be linked to diet began to emerge. "The fault is of quality not quantity. A child may be reduced to the last stage of atrophy and yet not be rickety. Conversely it may be overfed, fat and gross yet extremely rickety. Rickets is produced as certainly by a rachitic diet as scurvy by a scorbutic diet."[16] It was also said that "Deficiency of fat is the prime cause of the disease and all observers are agreed upon the extremely beneficial effects of cod-liver oil."[16] However other theories were also advanced: infective process, condition of hypothyroidism, confinement and lack of exercise, lack of lime salts in the food, excessive production of lactic acid.

A group of scientists at Cambridge lead by Dr Frederick Gowland Hopkins (later Sir Gowland Hopkins, President of the Royal Society) concluded that natural foods such as bread and milk must contain minute amounts of some hitherto unknown substance necessary for life that were different to those already recognized. Hopkins' conception of unknown indispensable food substances was first formulated in an address delivered to the Society of Public Analysts in November 1906. Hopkins suggested the term "accessory food factors." Simultaneously, but independently, Casimir Funk was exploring the link between diet and disease at the Lister Institute of Preventative Medicine in London. Sir Charles James Martin, the first Director of the Institute, spoke to Funk of his friend Leonard Braddon, the medical officer working in British Malay, who was interested in the disease beriberi. Funk met Braddon when he next visited London. Martin suggested that the disease was caused by the lack of an amino acid in polished rice and set Funk to tackle this problem. Before beginning practical work, Funk immersed himself in the subject and read Braddon's book.[2] He worked alone except for Robbins, the institutes' diener and laboratory helper. Since no chemical reaction for the manifestation of beriberi had yet been developed, all work was reliant on pigeon biological assays. Roosters were to have been used, but complaints from a sculptor living next door to the laboratory forced a change of plan. After several weeks of intensive work Funk concluded that the cause of beriberi was not a protein deficiency, but the lack of an unknown agent present in rice polishings.

Funk's experimental approach was to feed a pigeon polished rice until polyneuritis developed—signified by the appearance of contractions to the neck, wings, and legs.[17] Rice polishings or ground yeast (Funk found it easier to work with yeast) were then fed to the bird and polyneuritis would disappear. If the polishings were withheld the pigeons died within 12 h. The method of extraction used by Funk was adopted from Fraser and Stanton. The rice polishings (or yeast) were divided by chemical reaction into fractions A and B. In short, Funk would shake the rice polishings with alcohol saturated up to a certain point with gaseous hydrochloric acid. The alcoholic solution was then concentrated by vacuum, yielding a fatty residue. The residue was melted using a water bath, extracted with hot water, and the fractions A and B separated in a funnel while still hot. The watery extract was treated with sulfuric acid until a five percent solution was obtained and then precipitated with a 50 percent phosphotungstic acid solution. The precipitate was decomposed with baryta (barium hydroxide) and the resulting filtrate, after removal of excess baryta, tested for its curative power on pigeons with "beriberi." One bird would recover (e.g., the one given fraction B) and the other would die (faction A). Fraction A would be discarded and fraction B known to contain the active substance retained. Fraction B was then fractionated once again and the experiment repeated with two more birds. Experiments were complicated because some extracts contained large amounts of choline which is toxic to pigeons. Further fractionation of fraction B resulted in the discovery of a trace element that could cure polyneuritis in the pigeons. Ultimately Funk used 200 pounds of yeast and extracted from it one-twelfth of an ounce of the active agent. The activity of the agent was such that one fifteen-thousandth of an ounce cured paralyzed pigeons within a few hours.

Funk obtained a crystalline crude fraction which was adequate both in maintaining the health of pigeons and curing beriberi. The compound belonged to a group of compounds known as the pyrimidine, and the substance became known as thiamine. Funk was convinced that more than one agent existed. By dividing the crude fraction into three well-crystallized substances, of which one was later identified as nicotinic acid, he began the basis of what is now referred to as the vitamin B complex. He showed at the time that none of the three fractions alone could maintain good health in pigeons.

For the substances Funk isolated, he coined the term vitamine. "Vita" meaning life and "amine" meaning nitrogen-containing compound. Funk struggled to get the term vitamine accepted. The first paper on the vitamines was published in 1911.[18] The word was not approved by the journal or by Funk's employers to feature in the manuscript "Experiments on the causation of Beri-Beri." The word appears in print for the first time one year later[19] thanks to a fellow Pole Dr Rajchman, a bacteriologist at the Royal Institute of Health and one of the editors of the Institute's publication, who invited Funk to write a review on the subject of vitamines (the publication of reviews did not require prior approval by Funk's employers). This was a revolutionary publication which combined beriberi, scurvy, pellagra, and rickets all under one group 'nutritional deficiency' for the first time. In the review Funk writes, "..*the deficiency substances, which are of the nature of organic bases, we will call 'vitamines,' and we will speak of a beriberi or scurvy vitamine, which means a substance preventing the special disease* ..." The almost simultaneous publication of the experiments of Hopkins on "purified" diets and those of Funk on the "anti-beriberi vitamine" raised much interest in the "vitamin question." Funk soon left the Lister Institute and began working at the Cancer Hospital, where he was given an assistant, Jack Drummond who was to become professor of Biochemistry at University of London (later Sir Jack Drummond—and the victim, along with his family, of the notorious "Vitamin Murders"). With Drummond's help a more vigorous approach to the vitamin problem began. Rice was imported from Malay States and a microanalytical laboratory was equipped. The laboratory was modeled on that of Preggl's—an Austrian who had revolutionized methods used for analyzing compounds in microproportions.

Somewhat overlooked at the time because his papers were mostly published in Japanese, Umetaro Suzuki was undoubtedly (according to accounts by Funk during the 1930s) the first to tackle the problem of "accessory food factors" using chemical methods. From 1910 to 1912, Suzuki had found that the curative agent for beriberi could be precipitated by phosphotungstic acid from rice bran. Using picric acid, he thought that he had isolated the curative substance to which he gave the name "Oryzanin." The agent however was largely an impurity and could not be placed into any group of known chemical structure.[20]

The First World War gave impetus to vitamine research. Prices increased and disarrangement of nutritional elements became more common. The British War Office directed Captain Plimmer, a physiological chemist attached to the Directorate of Hygiene (later Professor of Medicine at St. Thomas' Hospital, London), to undertake analyses of common British foods—producing the first database for food in Britain. The resulting tables were published in 1921[21]. He grouped foods into those containing no vitamines, or any combination of Fat-soluble A (or antirachitic); Water-Soluble B (or antineuritic); or water-soluble C (or antiscorbutic).[21]

The discovery of the fat soluble and water soluble vitamins is shown in Fig. 1.1.

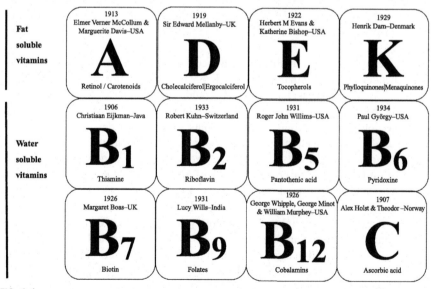

FIG. 1.1

Vitamins and their discovery. With the proposed discovery year country and discoverer.

Fat- and Water-Soluble Vitamins

Vitamins A to D

American biochemists, Elmer Verner McCollum and Marguerite Davis, were the first to confirm that there were at least two accessory food factors when they discovered that a fat-soluble substance in butter and egg yolk was required for rat growth.[22,23] It was suggested that this was a type of "growth factor" mice required to survive; however, from 1912 to 1915 there was confusion in that some thought that the growth factor was found only in butter whereas others thought it was found in yeast. This confusion was clarified in 1915 when McCollum and Davies showed that there were at least two growth factors.

McCollum and Davies discovered that a substance in whole cereals that prevents polyneuritis in chickens and pigeons was also needed in rats.[24] However, even when the diet contained whole cereals, young rats needed something else, which was found in butterfat, but not lard or olive oil.[22] McCollum and Davis called it the "fat-soluble A" factor. The other was found in certain watery food extracts; it was soluble in water but not in fats, so they called it "water-soluble B." Soon after it was found that "water-soluble B" material not only helped the rats to grow in experiments with these synthetic diets but it also acted as though it contained Funk's antiberiberi vitamin. "Fat-soluble A" was later found to have certain vitamine-like properties too.

McCollum and Cornelia Kennedy sought to introduce the classification "fat-soluble A" for the antirachitic vitamine and "water-soluble B" for the antiberiberi vitamine. Drummond proposed "water-soluble C" for the antiscorbutic vitamine. Funk resisted these terms and stated that they were incorrect, chemically and logically and suggested the terms vitamine A, vitamine B, and vitamine C. Drummond went on to propose the word vitamin rather than vitamine which Funk would not agree too. He remained wedded to the idea that all vitamines were nitrogenous in nature into the 1920s.

Osborne and Mendel were the first to notice that laboratory rats without vitamins in their diet develop xerophthalmia and are liable to lung infection.[25,26] It was thought that the antirickets vitamin (fat-soluble A) could also cure these afflictions since it was found that those foods that prevent rickets were also generally effective at preventing xerophthalmia and lung infection in rats. However, evidence began to accumulate that they were different: American workers[27] reported that certain food stuffs which were potent as sources of "vitamin A" were little good at preventing rickets.

Sir Edward Mellanby, taken by the work of McCollum and the discovery of fat-soluble A, decided to investigate the cause of rickets further. In 1919 he succeeded in producing a bone disease in puppies fed a diet of low-fat milk and oatmeal (also kept indoors and away from sunlight). Even adding yeast to the dogs' diet (to provide the water-soluble B-vitamins) and orange juice (to prevent scurvy) did not prevent the appearance of rickets within three to four months. Rickets was prevented by the addition of butterfat to their diet or, most effectively, of cod-liver oil. He wrote: "*Rickets is a deficiency disease which develops in consequence of the absence of some accessory food factor or factors. It therefore seems probable that the cause of rickets is a diminished intake of an anti-rachitic factor, which is either [McCollum's] fat-soluble factor A, or has a similar distribution to it.*[28,29]

In 1920 Hopkins[30] found that the fat-soluble factor A in butterfat could be destroyed by heating and aeration. Butterfat that had been heat treated no longer had growth-promoting activity, and rats fed the treated butterfat developed xerophthalmia and died within 40–50 days. In 1922 American Chemist Theodore Zucker developed a laboratory method capable of separating the antirickets factor from cod-liver oil by an extraction process, leaving fat-soluble A behind.[31] The key experiment was performed by McCollum and his coworkers in 1922,[32] when they observed that heated, oxidized cod-liver oil could not prevent xerophthalmia but could cure rickets in the rats. They showed that oxidation destroys fat-soluble A without destroying another substance which plays an important role in bone growth. They concluded that fat-soluble factor A consisted of two entities, one later called "vitamin A," the other being the newly discovered antirickets factor. Because the water-soluble factors then discovered were termed vitamin B and the known antiscurvy factor was called vitamin C, they named the new factor vitamin D.

Vitamin E

In 1922 Prof H.M. Evans and Katherine Bishop at the University of California discovered a new factor, without which rats could not reproduce. Evans and Bishop reported that the addition of small amounts of yeast or fresh lettuce to the purified diet would restore fertility of both sexes.[33] At first it was provisionally assigned the letter X or the antisterility factor. Evans and Bishop found factor X activity in dried alfalfa, wheat germ, oats, meats, and in milk fat, which was extractable with organic solvents. They distinguished the new fat-soluble factor from the known fat-soluble vitamins by showing that a single droplet of wheat germ oil administered daily completely prevented gestation resorption, where cod-liver oil known to be rich source of vitamins A and D failed to do so. In 1924 Barnett Sure, in an independent study, concluded that this fat-soluble factor was a new vitamin and assigned it the available letter in the Roman alphabet—"E."[34]

Vitamins B$_1$ and B$_2$

Funk had initially suggested that pellagra is the consequence of a vitamin deficiency[20]; however, for a decade Dr. Joseph Goldberger the pioneering pellagra investigator and others thought the disease was caused by a inferior protein.[35] Until 1929 Goldberger tested his theories with human experiments and animal models ("black tongue" in dogs) using well-designed epidemiologic investigations, all of which rejected toxic and infectious theories while strongly supporting a dietary deficiency explanation for pellagra.[36]

Experiments showed vitamin B contained the "old" beriberi vitamin and a new antipellagra vitamin. Fresh yeast, for example, was an effective treatment for both beriberi and pellagra but after being heated in an autoclave, it was of no longer any use in beriberi although retained antipellagra potency. Goldberger and his associates established that a small amount of dried brewer's yeast could cure or prevent pellagra less expensively than fresh meat, milk, and vegetables. A heat-stable component of yeast was shown to prevent the development of black tongue.[37]

Goldberger called the two factors "A-N" or antineuritic, and "P-P" or pellagra preventive[37]. Certain American workers, however, preferred F (which recalls Funk and his original "vitamine") and G (for "Goldberger")—these also being the next two vacant letters in the alphabet. In Britain, the Accessory Food Factors Committee of the Medical Research Council in 1927 recommended the use of the symbols B_1 and B_2, respectively, for the antiberiberi vitamin and its newly discovered companion. However, it was soon recognized that there were "several" B_2 vitamins.

Biotin

The discovery of biotin was a result of several independent experiments that appear to have passed almost unnoticed. Steinitz in 1898 was the first to notice that the consumption of raw egg white led to vomiting and diarrhea in dogs. Similar findings were observed in 1913 by Mendel and Lewis and in 1916, Bateman reported[38] that raw egg white produced dermatitis and characteristic "spectacle-eye" hair loss in rats.

In 1927 Margaret Averil Boa at the Lister Institute in London[39] found that young rats fed a diet with dried egg white as the protein source soon developed a condition labeled "egg white injury"—characterized by severe dermatitis, alopecia, an abnormal kangaroo-like posture attributed to a spastic gait, and ultimately death. Boas also found the curative substance ("protective factor X") in several food stuffs like dried yeast, raw liver, raw potato, crude lactalbumin, and spinach. She stated that the "protective factor X" showed a similar distribution to the previously described water-soluble B vitamins, however was not identical with either the antineuritic factor or Goldberger's pellagra-preventive factor. Later Paul György isolated the curative factor (which he named vitamin H) from liver,[40] later he demonstrated that vitamin H and biotin were the same.

Vitamin B_{12}

The discovery of vitamin B_{12} was only possible by the endeavor to find the cure for pernicious anemia (PA). Pernicious anemia was first described by Comb in 1822, Edinburgh, Scotland. He described a male who suffered a rapidly progressing severe anemia, with autopsy showing megaloblastic anemia. Thomas Addison of Guy's hospital in London was the first to establish the use of anemia as a clinical entity and referred to "idiopathic anemia" in his publications from 1849 and 1855. Addison observed 11 patients admitted into Guy's Hospital and states "a very remarkable form of general anemia occurring without any discernible cause whatever." In 1872 a German physician Anton Biermer gave a comprehensive description of this disorder and used the term "progressive pernicious anemia" for the first time.

The First World War resulted in innumerable deaths due to massive blood loss and inadequate transfusion service, consequently stimulating research into blood substitutes and improving hematopoiesis.[41] This could have also encouraged George Whipple to examine the role of the liver in hematopoiesis. In 1920, at the University of California, Whipple conducted a series of experiments in dogs that were made anemic through venesection and investigated the effects of various dietary treatments.[42] Whipple continued his research on the effect of dietary regimes for the treatment of PA at the University of Rochester. He examined the effects of various dietary factors including arsenic, iron pills, and germanium dioxide,[43] but found that only liver and especially raw, uncooked liver turned out to be effective in treatment of anemia.[44] Realizing raw liver's superior potency at treating PA than cooked liver was serendipity. Without Whipple's permission, laboratory technician Frieda Robscheit-Robbin fed the anemic dogs raw liver instead of cooked and a profound effect on blood regeneration was observed.[41]

Two physicians from Boston; George Minot and William Murphy, learned about Whipple's work with dogs and attempted to use raw liver for treating patients with PA.[41] Forty five had been given a high protein diet that included 100–240 g of liver and 20 g of meat for between 6 weeks and 2 years.[45] Minot had learnt a method of counting reticulocytes that allowed him to study early hematological responses to liver treatment. He observed raised reticulocyte counts within 2 weeks of starting the diet.[45] In 1926 Minot and Murphey presented their results to the meeting of the Association of American Physicians in Boston. Murphy presented a motion picture of the graveness of the disease in historic PA patients before and under treatment with liver extracts. The motion picture was also presented during Murphy's Nobel lecture December 12th, 1934.[45] The anti-PA factor was isolated from liver[46] and kidney and named vitamin B_{12} in 1948.[47]

Vitamin K

While investigating reports that newborn chicks failed to thrive when fed diets deficient of sterols Henrik Dam noticed that these animals often developed subdural or muscular hemorrhages at about the age of three weeks which were not prevented by dietary supplementation with cholesterol, lemon juice, yeast, cod-liver oil, or vitamins A, C, or D.[48,49] The diathesis was also reported in chicks fed ether-extracted fish meal (prepared from white nonoily fish) or meat meal (from which the fat was partially extracted during preparation) in an attempt to establish their requirement for vitamins A and D.[50] Initially this hemorrhagic disease was incorrectly attributed to vitamin C insufficiency due to the similarities of the symptoms with scurvy and the protective properties of cabbage.[51] At about this time synthetic ascorbic acid became available enabling Dam to clearly demonstrate that neither oral nor subcutaneous administration of the vitamin prevented the development of this syndrome.[52]

The high potency of hog liver in preventing this disease when compared with wheat-germ oil enabled Dam to exclude the recently discovered vitamin E as the protective agent. Chicks fed a diet consisting exclusively of cereals or seeds plus salts did not bleed, signifying that the cause of the disease was a deficiency of an unknown antihemorrhagic factor normally present in cereals and seeds.[52] When further studies revealed that this factor was fat soluble, Dam postulated the existence of a previously unidentified fat-soluble vitamin. He named this substance vitamin K. Not only was *K* the first letter of the alphabet not used to describe an existing or a suspected vitamin activity at the time, but it was also the first letter of the Danish and German word Koagulation.[53,54] Later, Dam elucidated that vitamin K is abundant in liver and the green leaves of plants, and noted that the poor coagulability of plasma from bleeding chicks could not be reversed by the in vitro addition of vitamin K.[55,56] Almquist and coworkers found that while dry fish meal did not prevent the development of hemorrhagic disease, ingestion of ether extracts of putrefied fish meal was protective.[57,58] These experiments led to the discovery of the biosynthesis of vitamin K by certain species of bacteria and the observation that despite the severe bleeding problem in vitamin K-deficient chicks, their feces contained significant amounts of the vitamin, suggesting that the bioavailability of this source of the vitamin is poor[59,60]

Vitamin B₅

Pantothenic acid (also known as vitamin B_5) was initially discovered in 1931 by chemist Roger J. Williams during his studies on the vitamin B complex and yeast *Saccharomyces cerevisiae* growth.[61] In 1933 he named the substance pantothenic acid from the Greek word panthos, meaning "from all sides" and its widespread presence in food. A few years later in 1937, a specific dermatitis in chicks which was misleadingly called 'chick pellagra[62] was identified as was an endemic disease in southern India known as "burning-feet" syndrome. Extracts of liver were able to cure this disease and this triggered further studies identifying this "chick antidermatitis" factor. Similarly, Thomas H Jukes also identified a factor required by chicks for growth and prevention of dermatitis, which he called the filtrate factor.[63] In 1939 Jukes recognized that his filtrate factor, Norris's chick antidermatitis factor and the pantothenic acid required for yeast growth were all identical.[64]

Folate

Lucy Wills went to Bombay to investigate macrocytic anemia in pregnancy in female textile workers. The fact that anemia was most frequent in poorer populations with diets deficient in protein, fruit, and vegetables led Wills to study the effects of changes in diet on the macrocytic anemia of albino rats produced by deficient diet and bartonella infection. The anemia was prevented by yeast added to a diet otherwise lacking B vitamins.[65,66] Yeast or a yeast extract (Marmite) was then found to correct macrocytic anemia in the pregnant Bombay patients.[67]

Wills and Evans reported that patients with tropical macrocytic anemia were cured by injections of a crude liver extract or by feeding autolyzed yeast extract, after failing to respond to a purified liver extract in patients with PA.[68] Others also confirmed the existence of a second nutritional factor in liver other than the anti-PA factor.[69–71] Folic acid was isolated in 1941.[72]

Vitamin B₂ leading to vitamin B₆

Interestingly, it was not a nutritional deficiency that led to the discovery of vitamin B_2 (riboflavin). The discovery of vitamin B_2 can be traced back to 1872 when British chemist Alexander Wynter Blyth isolated a yellow-green fluorescence pigment from milk, he called it lactochrome, "lacto" from milk and "chrome" because of the fluoresce color.[73] Though it was not until the 20th century when research on accessory factors began that scientist started to investigate the yellow pigment. Near the end of 1932, Paul György noted that the vitamin B_2 complex could be further divided into two factors, one which was the antipellagra factor and the second was shown to possess growth-promoting activity for rats. It was Richard Kuhn in 1933 who showed that the growth component of vitamin B_2 complex had the same characteristics as the factor isolated by Blyth,[74] riboflavin was the first component of the "vitamin B_2 complex" to be isolated and identified. It was later found that a deficiency of riboflavin is the cause of cheilosis in man.

The second component of the "vitamin B_2" complex to be identified was vitamin B_6 in 1934 by Paul György and his colleagues.[75] They carried out a series of experiments which showed that the using rat acrodynia as a model, in which they showed that when the factor was added to the semisynthetic rat diet it cured rat acrodynia. This consequently led to the isolation and chemical characterization of vitamin B_6.[75,76] The following year Birch,[77] György and Harris,[78] showed that the third component of the "vitamin B_2" complex was the pellagra-preventing factor (P-P). In 1937[79] it was identified as nicotinic acid or nicotinic amide (also known as niacin and niacin amide, respectively).

Standardization of the Novel Methods

As the number of investigators with an interest in vitamins began to increase dramatically so did the analytical methods for evaluation. New methods fell into four broad types: biological, microbiological, chemical, and physical. It was estimated in 1935 that three separate papers on vitamins were published every day[80]—collating the immense output had become laborious. It became clear that the meaningful comparison of findings from different groups required standardization.

In 1931, an International Vitamin Conference was held in London, under the Permanent Commission on Biological Standardisation of the League of Nations Health Organisation. The aim of the meeting was to consider accepting an international stable standard for each vitamin (at that time vitamins A, B_1, C, and D had been described) and to define international units for vitamin activity.[81]

The conference report suggested international standards and units, which were provisional for two years. However, certain of the standard preparations recommended were not available until 1932, and so a proposed second conference was postponed until 1934, at which date experience of the practical application of the provisional standards was available.

At the second conference standards for vitamin B_2 and vitamin E were recommended; however, it was felt that there was insufficient knowledge to justify the adoption of standards and units. It was necessary for the 1931 standards for vitamin A and C to be altered because they were found to have certain defects which impaired their usefulness. The standard for vitamin B_1 had proved to be particularly satisfactory and a large stock was available at the National Institute for Medical Research, London. The vitamin D standard also remained unaltered, with the provision that it may be replaced when exhausted by crystalline vitamin D in a suitable solution. Large quantities of the original standard solution of irradiated ergosterol were still available.

Historic Methods 1912–80—A Journey Toward Clinical Utility

Biological Methods

Biological assays (bioassays) require live animals such as rats or chicks. Bioassays require the recording of physical properties of animals such as weight, observation, and behavior. Largely, two forms of bioassays were used—the curative and the prophylactic method. The curative method involved animals placed on a diet lacking the vitamin concerned until the deficiency symptoms appear. Once deficiency developed, the animals were given a daily or biweekly dose of the test substance or of a known standard material for a specific period. The response to the vitamin was measured by the animal's behavior from the day they were given the first dose of the test substance or standard. In the prophylactic method, animals were given the vitamin-free diet plus doses of the test substance or standard from the beginning of the experiment. The response to the vitamin was measured by the animals' behavior from the beginning of the experiment.

Using a spectrophotometer, the activity of a vitamin-dependent enzyme could be determined in blood samples. The animal would be placed on the vitamin-free diet for a specific period of time. Once the deficiency symptoms had started to develop, a blood sample would be drawn and the enzyme activity measured to provide a baseline value. By gradually increasing doses of the test vitamin given to the animal at specific time points, deficiency symptoms would start to resolve and this would be reflected by a corresponding increase in enzyme activity. This procedure allowed the generation of vitamin concentration vs enzyme activity standard curves, and the estimation of vitamin concentration. Although bioassays led to breakthroughs in vitamin discovery they had little clinical utility.

Microbiological Methods

Microbiological methods have been used extensively for the determination of many vitamins although primarily vitamins B_5, B_6, B_9, and B_{12}. In this approach microorganisms are grown in broth or culture agar that have been prepared to be void of a particular vitamin in the presence of a sample containing an unknown concentration of the vitamin of interest. Calibration curves are constructed by making known amounts of the vitamin available to the microorganism and measuring growth either by turbidity or bacterial suspension. A negative control that has no vitamin at all is also used. Lactic acid formation can also be quantified, which is directly proportional to vitamin concentration in the sample.

Physiochemical Methods

Physicochemical methods in vitamin analysis include colorimetric, spectrophotometric chromatographic, fluorometric/chemiluminescence, and immunological techniques.

Colorimetric assays involve the development of a color after a chemical reaction between a vitamin and a chromogen occurs and this can be used as a qualitative method to show the presence of the vitamin. Likewise, a spectrophotometer and Lambert-Beer law can be used to measure the intensity of the color which is proportional to the vitamin concentration and therefore allowing quantitative analysis.

Fluorescence assays depend on the ability of certain vitamins, or their lysis products, to produce fluoresce when reacted with a fluorophore. The fluorescence is directly proportional to the vitamin concentration. Vitamins of interest can also react with a luminol (a pale crystalline compound that is insoluble in water and soluble in polar organic solvent), which is the basis of chemiluminescence assays. The vitamin-luminol mixture is injected through a capillary where "kinetic reaction" is measured which is directly proportional to the vitamin concentration.

Chemical and fluorescence assays can be automated, require a small volume of sample, and are easier to perform when compared to microbiological and biological assays. However, they cannot be used for all vitamins and other substances may also interfere with the chemical reaction. There are various types of immunoassays that have been used in vitamin analysis including radioimmunoassays which utilize radioisotopes, Enzyme Linked Immuno-Sorbant Assays (ELISA) that utilizes enzyme linked antibodies, and Fluorescent immunoassays that utilize fluorophore-labeled antibodies.

The Arrival of High Performance Liquid Chromatography

The arrival of widely available HPLC systems from the late 1970s transformed the analysis of all vitamins with the exception of vitamin B_{12}. Superior selectivity, and the capability of coquantification of clinically significant vitamin isoforms in biological matrixes, that is, α-tocopherol or individual carotenoid determination made this technique attractive to analysts.

For vitamin A and E analysis, traditional, well-established chemical methods were quickly replaced by superior specificity, lower detection limits, and automation possibilities. The analysis of vitamin D by HPLC revolutionized the resolution and purification of molecular forms. Serum vitamin K determination became a reality for the very first time.

The origins of HPLC can be dated to the same time as vitamin discovery, through to the introduction of partition and paper chromatography in the 1940s, to the introduction of liquid chromatography in the early 1960s. Cecil Tarbet introduced the first variable wavelength monitor for HPLC in the late 1970s. Reversed phase chromatography had also been developed which allowed for improved separation between very similar compounds. Innovation continued with ESA Biosciences, Inc. producing a new type of electrochemical detector in 1982 known as the Coulochem. Detectors capable of detecting fluorescence also became commercially available. Following its first appearance in 1974, LC-MS was more commonly available in laboratories by the 1990s.

For vitamin analysis, HPLC is the preferred method because it requires small sample volume and is rapid and highly sensitive, even though sample extraction and/or filtration is required before analysis. Today, HPLC is used as a reference technique to quantitatively and qualitatively analyze water- and fat-soluble vitamins in biological matrices. A reversed phase HPLC chromatographic procedure has been developed for the determination of water-soluble vitamins. Nowadays, coupling mass spectrometer (MS) to HPLC has added a new dimension to vitamin analysis since it enables the analysis of nonvolatile compounds. In general, LC-MS/MS methods are now available for each fat- and water-soluble vitamin. The LC-MS/MS assays are characterized by high sensitivity and high specificity.

Sample Type

The gradual move away from biological assays toward the direct measurement of the vitamins was a feature of the first one hundred years of vitamin research—a transition that moved as fast as the available technology would allow. In many cases the abundance of a vitamin in blood or urine does reflect dietary exposure, but it may not necessarily be representative of target tissue utilization. Functional biomarkers can prove useful additions to the laboratory repertoire with abnormal metabolite formation or impaired enzyme function.

Sample availability is also a key factor with blood or urine most commonly available. Whole blood, plasma, serum, or red blood cells provide different indications of status for different vitamins. Blood samples collected after a period of fasting can reduce the impact of recent dietary intake on the generated results.

Complete urine samples may be collected during a 24-hour period for the estimation of daily excretion and to eliminate the influence of circadian rhythms; however, the risk of incomplete collections introduces error. Alternatives include a first-voided morning urine specimen or a randomly voided urine sample. The interpretation of results from this type of collection is reliant on correcting for several factors including physical activity, size of sample, liquid consumption prior to collection, and time of day sample collected—correction by urinary creatinine is a standard approach.

The technology that is available today provides opportunities for international comparison surveys. It has become possible to estimate an individual's present, recent, or long-range nutritional status for many of the vitamins.

References

1. McDowell L. Ancient diseases related to diet. In: *Vitamin history, the early years*. Sarasota, FL: First Edition Publishing; 2013.
2. Braddon WL. *The cause and prevention of beri-beri*. New York: Rebman Limited; 1907. p. 544.

3. Cartier J, Biggar HP, Cook R. *The voyages of jacques cartier*. Toronto: University of Toronto Press; 1993. p. 177.

4. Martini E. Jacques Cartier witnesses a treatment for scurvy. *Vesalius Acta Int Hist Med* 2002;**8**:2–6.

5. Hawkins R. *Observations of Sir Richard Hawkins in his voyage into the south sea in the year 1593*. London: Hakluyt Society; 1847.

6. Woodall J. *The surgeons mate: or, military and domestique surgery*; 1639.

7. Woodall J. 1570–1643. *Br J Surg* 1916;**4**:369–72.

8. Lind J. *A treatise on the scurvy*. London: Printed for A. Millar, in the Strand; 1757.

9. Cook J. The method taken for preserving the health of the crew of his Majesty's ship the resolution during her late voyage round. *Philos Trans R Soc Lond* 1776;**66**:402–6.

10. Lunin N. Ueber die Bedeutung der anorganischen Salze für die Ernährung des Thieres. *Zeitschrift Für Physiol Chemie* 1881;**5**:31–9.

11. Carter KC. The germ theory, beriberi, and the deficiency theory of disease. *Med Hist* 1977;**21**:119–36.

12. Drummond JC, Wilbraham A, Hollingsworth D. *The Englishman's food: a history of five centuries of English diet*. London: Pimlico; 1991. p. 482.

13. Rowntree BS. *Poverty a study of town life*. 1st ed. London: Macmillan; 1901. p. 437.

14. Watt Smyth A. *Physical deterioration, its causes and the cure*. London: John Murray; 1904.

15. Combs GF. *The vitamins*. 3rd ed. San Diego, CA: Elsevier/Academic Press; 2012. p. 3–33.

16. Vincent RH. *The nutrition of the infant*. New York: William Wood & Company; 1904.

17. Funk C. On the chemical nature of the substance which cures polyneuritis in birds induced by a diet of polished rice. *J Physiol* 1911;**43**:395–400.

18. Cooper EA, Funk C. Experiments on the causation of beri-beri. *Lancet* 1911;**178**:1266–7.

19. Funk C. The etiology of the deficiency diseases, Beri-beri, polyneuritis in birds, epidemic dropsy, scurvy, experimental scurvy in animals, infantile scurvy, ship beri-beri, pellagra. *J State Med* 1912;**XX**:341–68.

20. Funk C, Dubin HE, Fraser RJ. *Vitamin and mineral therapy*. New York: U.S. Vitamin Corporation; 1936. p. 2–5.

21. Plimmer RHA. *Analyses and energy values of foods*. London: H. M. Stationery Off; 1921.

22. McCollum EV, Davis M. The necessity of certain lipins in the diet during growth. *J Biol Chem* 1913;**15**:167–75.

23. Semba RD. On the "discovery" of vitamin A. *Ann Nutr Metab* 2012;**61**:192–8.

24. McCollum EV, Davis M. The essential factors in the diet during growth. *J Biol Chem* 1915;**23**:231–346.

25. Osborne TB, Mendel LB. The influence of butter-fat on growth. *Exp Biol Med* 1913;**11**:14–5.

26. Osborne TB, Mendel LB. The relation of growth to the chemical constituents of the diet. *J Biol Chem* 1913;**15**:311–26.

27. Shipley PG, Park EA, McCollum EV, Simmonds N. Studies on experimental rickets. VII. The relative effectiveness of cod liver oil as contrasted with butter fat for protecting the body against insufficient calcium in the presence of a normal phosphorus supply. *Am J Epidemiol* 1921;**1**:512–25.

28. Mellanby E. Nutrition classics. *Lancet* 1919;**1**:407–12. An experimental investigation of rickets. Nutr Rev 1976;**34**:338–40.

29. Wolf G. The discovery of vitamin D: the contribution of Adolf Windaus. *J Nutr* 2004;**134**:1299–302.

30. Hopkins FG. The effects of heat and aeration upon the fat-soluble vitamine. *Biochem J* 1920;**14**:725–33.

31. Zucker TF, Pappenheimer AM, Barnett M. Observations on cod-liver oil and rickets. *Exp Biol Med* 1922;**19**:167–9.

32. McCollum EV, Simmonds N, Becker JE, Shipley PG. Studies on experimental rickets. An experimental demonstration of the existence of a vitamin which promotes calcium deposition. *J Biol Chem* 1922;**53**:293–312.

33. Evans HM, Bishop KS. On the existence of a hitherto unrecognized dietary factor essential for reproduction. *Science* 1922;**56**:650–1.

34. Sure B. Dietary requirements for reproduction. II. The existence of a specific vitamin for reproduction. *J Biol Chem* 1924;**58**:593–709.

35. Goldberger J, Tanner WF. Amino-acid deficiency probably the primary etiological factor in pellagra. *Public Health Rep* 1922;**37**:462–86.

36. Lanska DJ. Chapter 30: Historical aspects of the major neurological vitamin deficiency disorders: the water-soluble B vitamins. In: *Handbook of clinical neurology*. Amsterdam: Elsevier; 2009. p. 445–76.

37. Goldberger J, Wheeler GA, Lillie RD, Rogers LM. A further study of butter, fresh beef, and yeast as pellagra preventives, with consideration of the relation of factor P-P of pellagra (and black tongue of dogs) to vitamin B. *Public Health Rep* 1926;**41**:297–367.

38. Bateman WG. The digestibility and utilization of egg proteins. *J Biol Chem* 1916;**26**:263–91.

39. Boas MA. The effect of desiccation upon the nutritive properties of egg-white. *Biochem J* 1927;**21**:712–24.

40. György P, Kuhn R, Lederert E. Attempts to isolate the factor (vitamin H) curative of egg white injury. *J Biol Chem* 1939;**131**:745–59.

41. Sinclair L. Recognizing, treating and understanding pernicious anaemia. *J R Soc Med* 2008;**101**:262–4.

42. Whipple GH, Robscheit FS, Hooper CW. Blood regeneration following simple anemia. IV. Influence of meat, liver and various extractives, alone or combined with standard diets. *Am J Phys* 1920;**3**:236–62.

43. Whipple GH, Robscheit-Robbins FS. Blood regeneration in severe anemia. III. Iron reaction favorable- arsenic and germanium dioxide almost inert. *Am J Phys* 1925;**72**:419–30.

44. Robscheit-Robbins FS, Whipple GH. Blood regeneration in severe anemia. II. Favorable influence of liver, heart and skeletal muscle in diet. *Am J Phys* 1925;**72**:408–18.

45. Minot GR, Murphy WP. Treatment of pernicious anemia by a special diet. *J Am Med Assoc* 1926;**87**:470–6.

46. Smith EL. Purification of anti-pernicious anaemia factors from liver. *Nature* 1948;**161**:638.

47. Rickes EL, Brink NG, Koniuszy FR, Wood TR, Folkers K. Crystalline vitamin B12. *Science* 1948;**107**:396–7.

48. Dam H. Cholesterinositoffwechsel in huhnereiern und huhnchen. *Biochem Z* 1929;**215**:475–92.

49. Dam H. Über die Cholesterinsynthese in Tierkörper. *Biochem Z* 1930;**220**:158–63.

50. Mcfarlane WD, Graham WR, Richardson F. The fat-soluble vitamin requirements of the chick: the vitamin A and vitamin D content of fish meal and meat meal. *Biochem J* 1931;**25**:358–66.

51. Holst WF, Halbrook ER. A "scurvy-like" disease in chicks. *Science* 1933;**77**:354.

52. Dam H, Schonheyder F. A deficiency disease in chicks resembling scurvy. *Biochem J* 1934;**28**:1355–9.

53. Dam H. The antihaemorrhagic vitamin of the chick. *Biochem J* 1935;**29**:1273–85.

54. Dam H. The antihaemorrhagic vitamin of the Chick. Occurrence and chemical nature. *Nature* 1935;**135**:652–3.

55. Dam H, Schønheyder F. The occurrence and chemical nature of vitamin K. *Biochem J* 1936;**30**:897–901.

56. Dam H, Glavind J. Determination of vitamin K by the curative blood-clotting method. *Biochem J* 1938;**32**:1018–23.

57. Almquist HJ, Stokstad ELR. Dietary haemorrhagic disease in chicks. *Nature* 1935;**136**:31.

58. Almquist HJ, Stokstad ELR. Hemorrhagic chick disease of dietary origin. *J Biol Chem* 1935;**111**:105–13.

59. Almquist HJ, Stokstad ELR. Factors influencing the incidence of dietary hemorrhagic disease in chicks. *J Nutr* 1936;**12**:329–35.

60. Almquist HJ, Pentler CF, Mecchi E. Synthesis of the antihemorrhagic vitamin by bacteria. *Exp Biol Med* 1938;**38**:336–8.

61. Williams RJ, Lyman CM, Goodyear GH, Truesdail JH, Holaday D. "Pantothenic acid," a growth determinant of universal biological occurrence. *J Am Chem Soc* 1933;**55**:2912–27.

62. Ringrose AT, Norris LC, Heuser GP. The occurrence of a pellagra-like syndrome in chicks. *Poult Sci* 1931;**10**:166–77.

63. Jukes TH. Further observations on the assay, distribution, and properties of the filtrate factor. *J Biol Chem* 1937;**117**:11–20.

64. Jukes TH. Pantothenic acid and the filtrate (chick anti-dermatitis) factor. *J Am Chem Soc* 1939;**61**:975–6.

65. Wills L, Mehta MM. Studies in pernicious anaemia of pregnancy. Part I. Preliminary report. *Indian J Med Res* 1930;**17**:777–92.

66. Wills L, Mehta MM. Studies in "pernicious anemia" of pregnancy. Part IV. The production of pernicious anemia (bartonella anemia) in intact albino rats by deficient feeding. *Indian J Med Res* 1931;**18**:663–83.

67. Wills L. Treatment of "pernicious anaemia of pregnancy" and "tropical anaemia". *Br Med J* 1931;**1**:1059–64.

68. Wills L, Evans BDF. Tropical macrocytic anaemia: its relation to pernicious anaemia. *Lancet* 1938;**232**:416–21.

69. Davidson LS, Davis LJ, Innes J. Megaloblastic anaemia of pregnancy and the puerperium. *Br Med J* 1942;**2**:31–4.

70. Fullerton HW. Macrocytic anaemia of pregnancy and the puerperium. *Br Med J* 1943;**1**:158–60.

71. Watson J, Castle WB. Nutritional macrocytic anemia: response to a substance other than the anti-pernicious anemia principle. *Proc Soc Exp Biol Med* 1943;**58**:84–6.

72. Mitchell HK, Snell EE, Williams RJ. The concentration of "folic acid". *J Am Chem Soc* 1941;**63**:2284.

73. Clewes C, Thurnham D. The discovery and characterization of riboflavin. *Ann Nutr Metab* 2012;**61**:224–30.

74. Kuhn R, György P, Wagner-Jauregg T. Über Lactoflavin, den Farbstoff der Molke. *Ber Dtsch Chem Ges* 1933;**66**:1034–8.

75. György P. Vitamin B2 and the pellagra-like dermatitis in rats. *Nature* 1934;**133**:498–9.

76. Rosenberg IH. A history of the isolation and identification of vitamin B6. *Ann Nutr Metab* 2012;**61**:236–8.

77. Birch TW, Chick H, Martin CJ. Experiments with pigs on a pellagra-producing diet. *Biochem J* 1937;**31**:2065–79.

78. Harris LJ. The vitamin B2 complex further notes on "monkey pellagra" and its cure with nicotinic acid. *Biochem J* 1938;**32**:1479–81.

79. Koehn CJ, Elvehjem CA. Further studies on the concentration of the anti-pellagra factor. *J Biol Chem* 1937;**118**:693–9.
80. Harris LJ, Hopkins FG. *Vitamins and vitamin deficiencies*. 1st ed. London: J & A Churchill; 193835.
81. Aykroyd WR. International vitamin standards and units. In: Weichardt W, editor. *Ergebnisse der Hygiene Bakteriologie Immunitätsforschung und Experimentellen Therapie: Fortsetzung des Jahresberichts Über die Ergebnisse der Immunitätsforschung Unter Mitwirkung Hervorragender Fachleute Vierzehnter Band*. Berlin, Heidelberg: Springer Berlin Heidelberg; 1933. p. 376–81.

Methods for assessment of Vitamin A (Retinoids) and carotenoids

2

Neal E. Craft*, **Harold C. Furr**[†]

Eurofins Craft Technologies, Inc., Wilson, NC, United States[*], *Department of Nutritional Sciences (retired), University of Connecticut, Storrs, CT, United States*[†]

Chapter Outline

Introduction

Vitamin A was one of the first substances to be recognized as a vitamin. Preformed vitamin A is found only in foods of animal origin; some carotenoids (from plants) can be converted to vitamin A in the human body. Its isolation and the determination of its modes of action were triumphs of organic chemistry and biochemistry. Nevertheless, human vitamin A deficiency remains a significant public health problem in parts of the world, particularly in Central America, parts of Africa, and parts of Southeast Asia, and so determination of human vitamin A status has major importance in public health. In addition to their role in vision, retinoids and carotenoids can also have roles in cancer prevention and treatment as they control gene expression. Because significant amounts of vitamin A can be stored in the human liver and other tissues and mobilized when needed, simply determining serum or plasma concentrations of vitamin A is helpful but does not give a fully accurate measure of vitamin A status except in frank deficiency; assessing human vitamin A status remains a major challenge.

Background

"Fat-soluble factor A" was first recognized as an essential micronutrient by E.V. McCollum and Margurite Davis at the University of Wisconsin and independently by T.B. Osborne and L. Mendel at Yale.[1] Animal studies found that vitamin A deficiency was characterized by xerophthalmia, squamous metaplasia of epithelial tissues, susceptibility to infections, and problems with reproduction.[2] By 1928 vitamin A had been termed an "antiinfective agent".[3] Liver oils and butter and their ether extracts were found to be rich sources of vitamin A activity. T. Moore showed that, whereas vitamin A is colorless, yellow extracts of butter could be converted to vitamin A as found in liver extracts thus explaining the provitamin A role of some carotenoids. This transformation was explained by the determination of the structures of vitamin A and β-carotene by P. Karrer's research group in the 1930s: butter contains the yellow provitamin A carotenoid β-carotene whereas liver contains colorless vitamin A. The similarity of chemical structure between the two compounds confirmed that β-carotene is a nutritional precursor of vitamin A. G. Wald clarified the role of retinal in vision.[4] Arens and van Dorp demonstrated that retinoic acid could not support vision but is adequate for many of the other functions of vitamin A.[5]

Structure and Function

"Vitamin A" refers to retinol, retinyl esters, retinal, and retinoic acid. Structurally, these 20-carbon compounds all contain a cyclohexene ring (with three methyl groups substituted on it), connected to an isoprenoid chain ($H_2C=C\,(CH_3)-CH=CH_2$) which terminates in a polar carbon-oxygen functional group: $-CH_2OH$ for retinol, $-CHO$ for retinal, and $-COOH$ for retinoic acid (Fig 2.1). The conventional numbering system

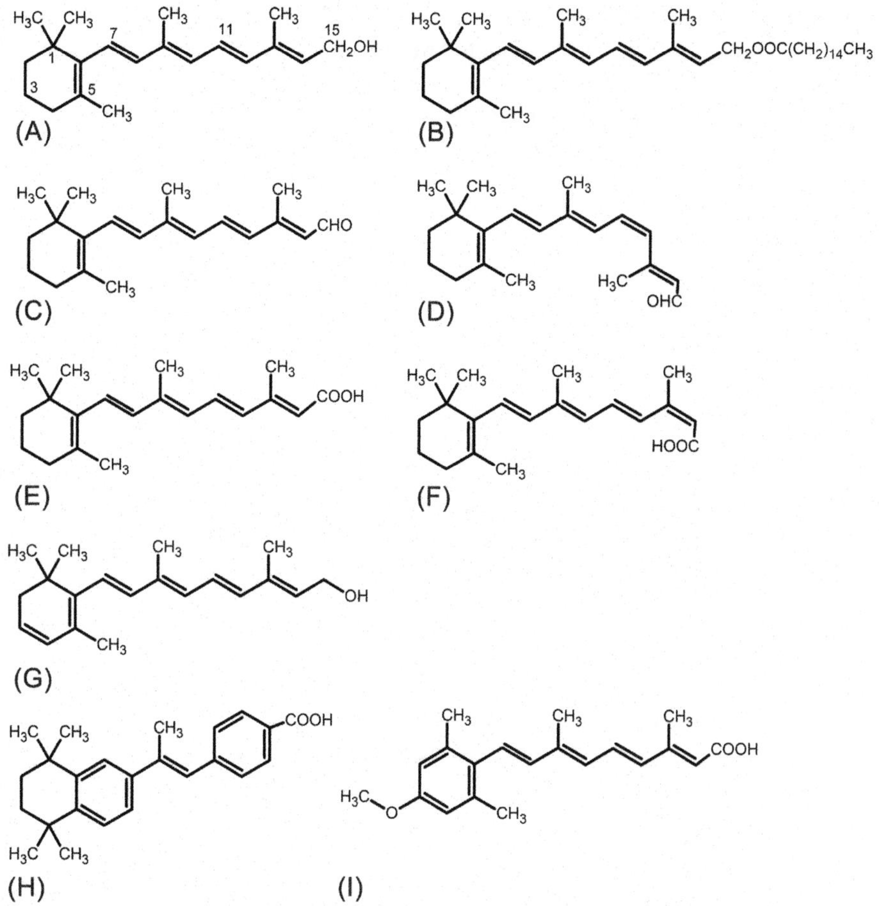

FIG. 2.1

Some representative retinoids. (A) all-*trans* retinol (vitamin A alcohol); numbers represent IUPAC numbering system. (B) all-*trans* retinyl palmitate (vitamin A palmitate). (C) all-*trans* retinal (all-*trans* retinaldehyde; vitamin A aldehyde). (D) 11-*cis* retinal. (E) all-*trans* retinoic acid (vitamin A acid, tretinoin). (F) 13-*cis* retinoic acid (isotretinoin, accutane). (G) all-*trans* 3,4-didehydroretinol (dehydroretinol, vitamin A₂). (H) Tetrahydrotetramethylnaphthalenylpropenylbenzoic acid (TTNPB), a synthetic retinoid. (I) Acitretin, a synthetic retinoid.

is shown in the structure for retinol (Fig 2.1A). For retinoic acid (Fig 2.1E), IUPAC numbering begins at the polar end of the molecule; the systematic name for all-*trans* retinoic acid is (all-*trans*) 3,7-dimethyl-9-(2,6,6-trimethylcyclohex-1-en-1-yl)-nona-2,4,6,8-tetraen-1-oic acid. The discovery of the role of some of these compounds in control of gene expression (see later) has led to the development of a number of synthetic compounds which do not bear close resemblance to the traditional vitamin A compounds but which can bind to and activate nuclear retinoid receptor proteins;

any compounds which can fulfill these functions (including the natural vitamin A compounds) are referred to as "retinoids." Most of these synthetic retinoids cannot be converted to visual pigment and cannot serve in vision, however.

"Carotenoids" (usually) have a trimethylcyclohexenyl ring on each end of an isoprenoid chain; most (but not all) carotenoids contain 40 carbons (Fig 2.2). More than 600 carotenoids have been reported in nature. Compounds having fewer than

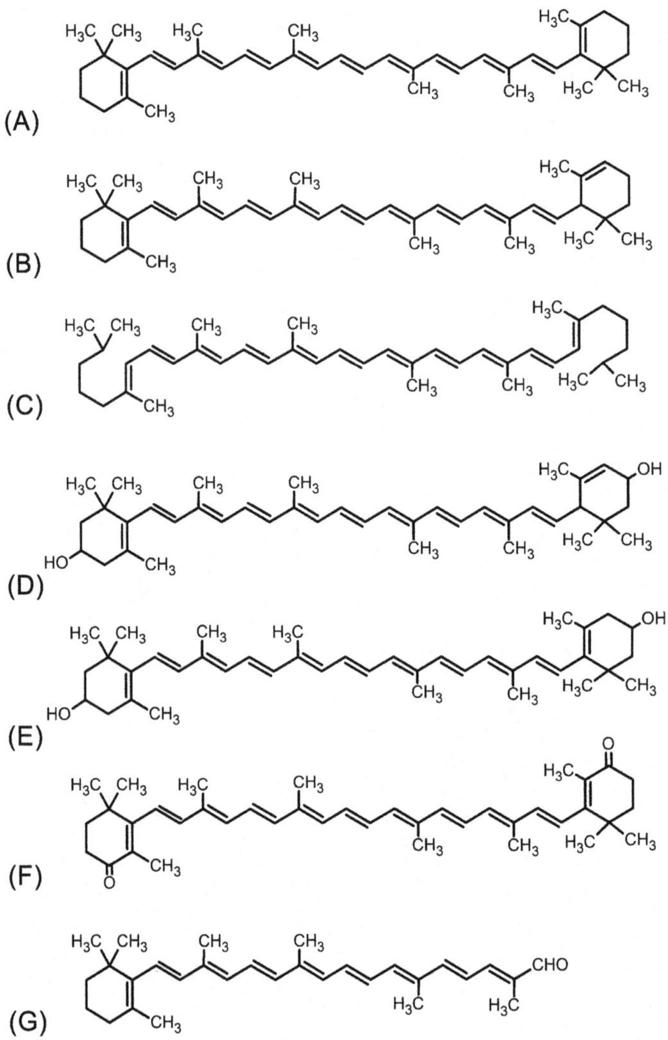

FIG. 2.2

Some representative carotenoids. Carotenes: (A) all-*trans* β-carotene, (B) all-*trans* α-carotene, (C) all-*trans* lycopene. Xanthophylls: (D) all-*trans* lutein, (E) all-*trans* zeaxanthin, (F) all-*trans* canthaxanthin. An apocarotenoid: (G) all-*trans* β-apo-8'-carotenal.

40 carbon atoms and with only one terminal cyclohexenyl ring are called "apo-carotenoids" (e.g., β-apo-8′-carotenal). Carotenoids are divided into two classes, "carotenes" which contain only carbon and hydrogen atoms (e.g., β-carotene, α-carotene, and lycopene), and "xanthophylls" which contain oxygen atoms in addition to carbon and hydrogen (e.g., lutein, zeaxanthin, and canthaxanthin as well as β-apo-8′-carotenal). β-Carotene and α-carotene and β-cryptoxanthin can be cleaved to yield vitamin A; other carotenoids, such as lycopene or lutein or lycopene, cannot.

Human Requirements and Recommendations

The Dietary Recommended Intake (DRI) in the United States for vitamin A for those 14 years and older is 900 μg Retinol Activity Equivalents (RAE) (3000 International Units, IU) for males and 700 μg RAE (2310 IU) for females. The recommendations are different for other groups such as infants, children, and pregnant and lactating women.[6] Different countries and international organizations may have different dietary recommendations but all are similar; there is general agreement on human dietary requirements.

The International Unit of vitamin A activity historically was defined as 0.30 μg of all-*trans* retinyl acetate or 0.6 μg of all-*trans* β-carotene or 1.2 μg of all-*trans* α-carotene. However, this definition does not account for the variable intestinal absorption of provitamin A carotenoids. The lower nutritional value of provitamin A carotenoids was acknowledged by introduction of the term Retinyl Equivalents (RE): 1 RE = 1 μg all-*trans* retinol = 6 μg all-*trans* β-carotene = 12 μg other provitamin A carotenoids. Another correction was provided by the introduction of Retinol Activity Equivalents (RAE), in which 1 μg RAE = 1 μg all-*trans* retinol = 2 μg all-*trans* β-carotene in oil (highly absorbable) = 12 μg all-*trans* β-carotene in other foodstuffs (poorer intestinal absorption) = 24 μg other provitamin A carotenoids in foods. Use of molar units instead of mass units helps to reduce confusion.

Chemical and Physical Properties

The immediately noteworthy feature of the chemical structures of retinoids and carotenoids is their polyene structure (pattern of alternating double- and single-bonds). The resulting delocalization of the electron cloud results in particularly efficient absorption of light, such that these compounds have recognizable colors in the visible spectrum. Furthermore, different compounds have different wavelengths of maximum absorbance and different shapes of absorbance curves, which are useful in qualitative identification. Some representative spectra are presented in Fig. 2.3. The molar absorption coefficient ε (theoretical absorbance of a solution of 1 mole/L at the wavelength of maximum absorption) or alternatively the $E^{1\%}$ (absorbance of a solution of 1 g/100 mL) provides a very useful means of determining concentrations of retinoids and carotenoids in solution (Tables 2.1 and 2.2). As indicated in the

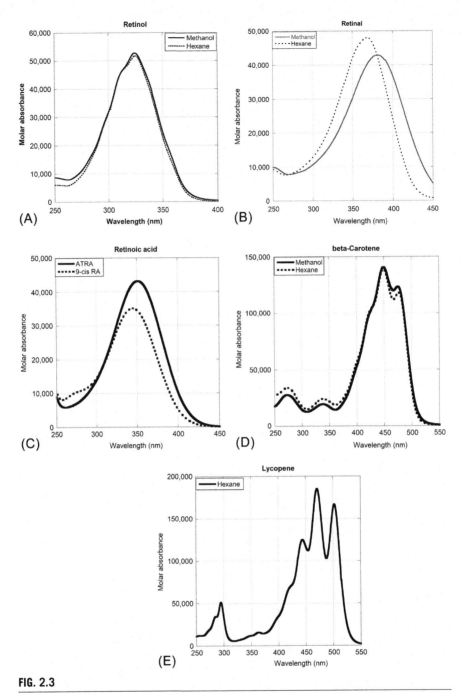

FIG. 2.3

Representative absorption spectra of some retinoids and carotenoids. (A) all-*trans* retinol, in methanol and in hexane. Note slight differences in absorbance maxima in the two solvents. (B) all-*trans* retinal (retinaldehyde), in methanol and in hexane. (C) Retinoic acid, all-*trans (ATRA)* and 9-*cis (9-cis RA)*, in methanol. *Cis*-isomers of retinoids and carotenoids usually (but not always) have shifted λ_{max} and reduced molar absorbance compared with the all-*trans* isomers. (D) β-Carotene, in methanol and hexane. Note slight effects on fine-structure of the spectra. (E) Lycopene, in hexane. Lycopene has much more pronounced fine structure compared with β-carotene.

Table 2.1 Ultraviolet/Visible Spectral Properties of Selected Retinoids[a]

Retinoid	Solvent	λ_{max} (nm)	ϵ	$E^{1\%}_{1cm}$
All-*trans* retinol	Ethanol	325	52770	1845
	Hexane	325	51770	1810
All-*trans* retinyl acetate	Ethanol	325	51180	1560
	Hexane	325	52150	1590
All-*trans* retinyl palmitate	Ethanol	325	49260	940
All-*trans* retinal	Ethanol	383	42880	1510
	Hexane	368	48000	1690
11-*cis* retinal	Ethanol	380	24935	878
	Hexane	365	26360	928
9-*cis* retinal	Ethanol	373	36100	1270
All-*trans* retinoic acid	Ethanol	350	45300	1510
13-*cis* retinoic acid	Ethanol	354	39750	1325
9-*cis* retinoic acid	Ethanol	345	36900	1230
All-*trans* retinoyl β-glucuronide	Methanol	360	50700	1065
All-trans retinyl β-glucuronide	Methanol	325	44950	973
All-*trans* 3,4-didehydroretinol (vitamin A$_2$)	Ethanol	350	43120	1455
All-*trans* 4-oxoretinoic acid	Ethanol	360	58220	1854
	Hexane	350	54010	1720
Anhydroretinol	Ethanol	371	97820	3650
Anhydrovitamin A$_2$	Ethanol	370	79270	2980
All-*trans* acitretin (TMMP-retinoic acid)	Ethanol	361	41400	1270
13-*cis* acitretin	Ethanol	361	40450	1241
TMMP-retinol	Ethanol	325	49800	1596
α-Retinol	Ethanol	311	47190	1650

[a]*More complete listings are given in Refs. 7–9.*

tables, the light absorption spectrum of a given retinoid or carotenoid is dependent on the solvent used, with slight differences in wavelength of maximum absorbance (λ_{max}) and molar extinction coefficient; solutions in hexane typically show greater "fine structure" (separation between peaks and valleys) than solutions in methanol or ethanol. Cis-isomers (*Zusammen, Z*-isomers) usually have lower wavelength of maximum absorbance (hypsochromic shift) and lower molar extinction coefficient than their *trans* (*Entgegen*) analogs (Fig 2.3).

Retinol and retinyl esters are also quite fluorescent (excitation ~ 325 nm, emission ~ 485 nm), a property which is very useful in their analysis.

In general the hydrocarbon nature of retinoids and carotenoids assures that these compounds are not soluble in water but are very soluble in organic solvents such as alcohols (methanol and ethanol), ethers (diethyl ether and tetrahydrofuran), and esters (ethyl acetate) and halogenated hydrocarbons (dichloromethane) as well

Table 2.2 Ultraviolet/Visible Spectral Properties of Selected Carotenoids[a]

Carotenoid	Solvent	$\lambda_{max(nm)}$	ε	$E^{1\%}_{1cm}$
Canthaxanthin	Light petroleum	466	12400	2200
α-Carotene	Light petroleum	422, *444*, 474	150000	2800
β-Carotene	Light petroleum	425, *453*, 479	139000	2592
	Ethanol	*450*	135800	2529
β-Cryptoxanthin	Light petroleum	425, *452*, 479	132000	2386
Lutein	Hexane	*444*	147300	2589
	Ethanol	421, *445*, 475	145000	2550
Lycopene	Light petroleum	444, *472*, 502	185000	3450
Zeaxanthin	Light petroleum	426, *452*, 479	133400	2348
	Ethanol	*452*, 479	124990	2200

ε and $E^{1\%}$ values are given for the major wavelength, which is italicized when more than one peak is present. Data are for all-trans isomers. (For some compounds, molar extinction coefficients have been calculated from $E^{1\%}_{1cm}$ values in the original references or vice versa.)
[a]*More complete listings are given in Refs. 10–13.*

as acetonitrile. These compounds are not highly soluble in hexane and other hydrocarbon solvents, but are soluble in mixtures of hexane with the organic solvents given before. Acetone is a powerful solvent, but its own absorbance at 330 nm can interfere with spectrophotometric determination of these compounds; and acetone reacts with retinal. Chloroform should be avoided because it can form free radicals that attack these compounds. Dimethylsulfoxide is sometimes used to introduce retinoids and carotenoids into tissue cultures, but its low volatility makes it difficult to remove by evaporation if that is necessary. The glucuronides of retinol and retinoic acid are water soluble, however, as are some apocarotenoids such as bixin and norbixin.

Retinol bound to its plasma transport protein, retinol-binding protein (RBP), is surprisingly stable, presumably because the retinol ligand lies in a protective pocket in the protein. Retinol-RBP in dried blood spots (dried on filter paper) can be stable for at least three months at room temperature.[14]

Other chemical and physical properties of retinoids and carotenoids are determined by the presence or absence of polar groups, especially on the terminus of the retinoid chain. Alcohol groups (e.g., retinol) are readily esterified with fatty acids, and the esters are chemically more stable (less susceptible to oxidation) than the free alcohols. Examples of esters include retinyl acetate (not naturally occurring, but commercially available and completely bioavailable) and long-chain fatty acid esters such as retinyl palmitate (a major storage form of vitamin A in liver and other tissues). Xanthophylls (carotenoids with alcohol functional groups) can also be esterified; examples include lutein dipalmitate from marigold flowers. The esters are of course less polar than the free alcohol compounds but can be readily hydrolyzed to regenerate the parent alcohols. Aldehydes (e.g., retinal) can be readily formed by oxidation of primary alcohols in biological systems, or reduced to the parent alcohol; this facile biotransformation is important in the visual cycle. Aldehydes and ketones (e.g., canthaxanthin) are less polar than the corresponding alcohols. Carboxylic acids (e.g., retinoic acid or norbixin) are yet more polar. Carboxylic acids can of course be esterified with alcohols, although these esters are seldom found in biological systems. Reflecting these relative polarities, the general order of elution on reversed-phase high-performance liquid chromatography is carboxylic acids < alcohols < aldehydes and ketones < esters; for example, retinoic acid < retinol < retinal < retinyl palmitate.

Another characteristic of polyene systems is the facile ability to form positional isomers, that is, *cis/trans* forms (also referred to as *Zusammen/Entgegen*, Z/E). Interconversion of these isomers is catalyzed by light, so again it is important to protect these compounds from light to prevent accidental isomerization. There are differences in biological activity of retinoic acid isomers (all-*trans* retinoic acid vs 13-*cis* retinoic acid, isotretinoin) although they undergo racemization in biological systems. Isomerization of the retinal chromophore is a critical step in vision. Separation of these isomers can be an analytical challenge.

Handling and Storage

Retinoids in high concentrations are teratogenic (because of interference with appropriate gene expression) and should be handled with caution. Gloves should be worn when handling large quantities. Also, gloves provide some protection against other potential biohazards in biological samples. Carotenoids are not known to present biological hazards per se.

The polyene structure of retinoids and carotenoids also makes them very susceptible to attack by molecular oxygen and by radicals, especially in the presence of light. Crystalline compounds are reasonably stable during storage, but even they should be stored under an inert gas such as nitrogen or argon. Pure compounds and solutions should be stored below room temperature. Solutions can degrade overnight if not protected from oxygen and light. Integrity of stock solutions should be routinely monitored by examining their light absorption spectra before use. Peroxides readily attack retinoids and carotenoids. Including an antioxidant such as alpha-tocopherol or BHT (butylated hydroxytoluene) can help to protect the retinoid or carotenoid in solution but may interfere with some uses of the test compound (BHT may coelute with retinol under some HPLC conditions). Because these compounds decompose readily, concentrations of solutions should never be determined by mass (weight), but instead by spectrophotometry (making serial dilutions of stock solutions). This is very important in preparing solutions of retinoids and carotenoids as analytical standards or for use in tissue culture experiments.

Because retinoids and carotenoids are isomerized and degraded readily by light, bright lights should be avoided. In particular, sunlight should be avoided; windows should be covered well with aluminum foil or dark curtains. Laboratories should be provided with yellow (Gold F40) fluorescent lights or UV filters, which provide a comfortable level of lighting without damaging these compounds.[15] Failing that, labware should be amber-colored (low-actinic glassware) or wrapped with aluminum foil.

Retinoids and carotenoids are also isomerized and degraded by exposure to heat; stock compounds and samples should be kept frozen at or below −20°C until time of analysis, and kept on ice (4°C) during handling. In plasma and other biological samples, the presence of natural antioxidants protects these compounds, and they can be stored frozen for months or sometimes years.

Exposure to air or oxygen should be avoided at all times. Reference compounds should be stored under an atmosphere of argon or nitrogen (commercial nitrogen sometimes has enough residual oxygen to degrade these compounds), preferably in sealed ampules. If the intended use of the reference compound permits, incorporation of an antioxidant can be useful.

Exposure to acids should be avoided: Lewis acids (including sulfuric acid and phosphorus pentoxide and aluminum chloride) result in dehydration of retinol and retinyl esters, with rapid but transient formation of purple-blue complexes (the basis of the traditional Carr-Price assay for vitamin A). Retinol is readily dehydrated with acids such as hydrochloric acid to form anhydroretinol (a hydrocarbon with a retro-double bond structure). Glassware for use with retinoids and carotenoids should not be acid washed.

Biological Activities

The two major biological functions of retinoids are (1) in vision and (2) in control of gene expression. Interestingly, these functions are completely separate and rely on completely different biochemical mechanisms. (Some authors refer to vitamin A as an antioxidant, but the concentrations of retinoids in biological tissues are too low to be relevant as antioxidants.)

Vision

The role of vitamin A in vision is complex and will not be fully explored here. It should be noted, however, that human vitamin A deficiency is a major public health problem in parts of the world, sometimes to the extent of night blindness; assessment of vitamin A status is a major clinical concern in some countries. As noted earlier, simply determining plasma concentrations of retinol is not adequate for defining vitamin A status.

In brief, when a molecule of 11-*cis* retinal (attached to the protein opsin to form the rhodopsin complex) is struck by a photon of light, it rearranges to all-*trans* retinal. The resulting change in three-dimensional conformation of rhodopsin triggers a nerve impulse which is interpreted by the brain as a visual event. The visual cycle is completed as all-*trans* retinal is reisomerized to 11-*cis* retinal in a complex pathway which involves enzymatic reduction to retinol, esterification, isomerization, and ester hydrolysis and reoxidation to retinal, taking place in two different cell types.[16]

To support visual function, vitamin A must be provided to the animal at the oxidation level of alcohol (or ester) or aldehyde; retinoic acid cannot be biologically reduced to either of these forms and so cannot support vision.

Control of Gene Expression

Retinoids mediate their role in gene expression by binding to specific receptor proteins in the nucleus of the target cell; the retinoid-receptor complex then binds to specific regions of the nuclear DNA, either turning on or turning off gene expression (protein synthesis) from that gene (Fig 2.4). Specificity of binding of retinoid with the receptor protein (of which there are multiple types) and interactions among receptor proteins (including vitamin D receptor, thyroid receptor as examples) provide for exquisite control of gene expression.

Of the natural retinoid ligands controlling gene expression, all-*trans* retinoic acid and 9-*cis* retinoic acid are the most common. Alcohol, ester, and aldehyde forms of vitamin A do not have biological activity in gene expression except as precursors of the carboxylic acid forms. However, a multitude of synthetic retinoid analogs have been prepared and shown to have biological functions.[17]

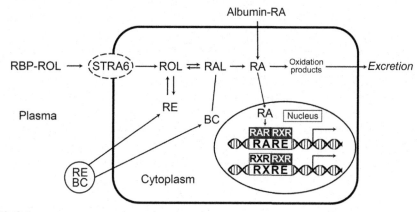

FIG. 2.4

Role of retinoids in regulation of gene expression. Retinol (ROL) is delivered to a receptive cell by retinol-binding protein (RBP) via a membrane receptor protein (STRA6). Alternatively, retinyl esters (RE) from plasma are hydrolyzed to retinol within the cell, or beta-carotene is cleaved to retinal (RAL) within the cell. Retinol is oxidized to retinal (RAL) which is in turn oxidized to retinoic acid (RA). Retinoic acid is transferred via cellular retinoic acid-binding proteins (not shown) to nuclear receptor proteins such as RAR (retinoic acid receptor protein) and RXR (retinoid X receptor protein). These nuclear receptor proteins, either as homodimers (RXR-RXR) or heterodimers (RAR-RXR) bind to response elements (RARE, RXRE) in specific genes, turning on or off gene transcription.

From Furr HC, Retinol: physiology. In: Caballero B, Finglas P, Toldra P, editors. Encyclopedia of food and health. Amsterdam: Elsevier/Academic Press; 2016. p. 597–603.

Retinoids have been applied in control of acne (especially 13-*cis* retinoic acid, Isotretinoin or Accutane) and for treatment of skin cancers. Acute promyelocytic leukemia responds to retinoic acid (tretinoin). Such results have encouraged synthesis and testing of a variety of synthetic retinoids, for example, Fenretinide (*N*-4-hydroxyphenylretinamide, 4-HPR) which has been used for treatment of some forms of breast cancer. However, prolonged high-dosage medical usage of retinoids can lead to symptoms of vitamin A toxicity.

Carotenoids are important as biological antioxidants and as precursors of vitamin A in foodstuffs. The xanthophylls lutein and zeaxanthin have been implicated in human macular degeneration because of their high concentrations in the macula of the eye, but a mechanistic connection has not yet been established.

Clinical Application of Measurement
Human Vitamin A Deficiency
Eye signs
Night blindness: Because of the role of vitamin A in visual function, impaired visual acuity in reduced illumination because of lack of adequate visual pigment (night blindness) is an indicator of impaired vitamin A status. This has proven difficult to quantitate, but can be useful as a qualitative indicator in assessing community health, especially for children.

Xerophthalmia: The other eye sign of vitamin A deficiency is xerophthalmia, "dry eyes," which has completely different etiology. The condition is manifest as dryness of the conjunctiva and cornea, preceding thickening and wrinkling of the conjunctiva with eventual ulceration and bacterial infection as epithelial integrity is destroyed. This condition develops, not as a consequence of the role of vitamin A in the visual cycle in the eye, but rather because of interference in the role of retinoids in gene expression: deficiency of retinoic acid results in enhanced production of collagenase in the cornea of the eye, leading to loss of integrity of the surface of the eye and subsequent bacterial infection. "Conjunctival Impression Cytology" (CIC) has been used as a method to detect vitamin A deficiency, but its application in the field is complicated by factors such as dusty environment.[18] Night blindness can be reversed by prompt prophylactic administration of vitamin A; xerophthalmia is irreversible and results in permanent blindness. Because vitamin A is also required for immune function, xerophthalmia is frequently accompanied by impaired immune response, tragically often resulting in death. In 2015 it was estimated that vitamin A deficiency resulted in 100,000 childhood deaths, mostly in South Asia and sub-Saharan Africa (around 2% of child deaths in these regions).[19]

Other Complications of Vitamin A Deficiency

It has been known since the 1920s, even before its chemical structure was elucidated, that vitamin A plays some role in immune response. A well-known example is the relationship between measles infection and vitamin A deficiency: vitamin A deficiency makes the subject more susceptible to measles infection and exacerbates the symptoms of the disease; and measles infection seems to worsen vitamin A status.[20] The mechanism is not yet clear, but most likely involves a role of retinoids in control of gene expression. Children are born with low liver reserves of vitamin A and accumulate vitamin A stores only slowly, so they are at greater risk of these problems.[21]

Assessing Vitamin A Status

Human vitamin A deficiency is a major public health problem in many parts of the world, including parts of Central America, sub-Saharan Africa, and South Asia and Southeast Asia.[22] Body stores of vitamin A usually increase across the lifespan, so deficiency is most likely in children (because of inadequate time to build up body stores) and pregnant and lactating women (because of the increased demand by the fetus and nursing infant). Vitamin A deficiency is often associated with poverty, high incidence of infectious diseases, limited infrastructure, and food insecurity.[23] Identification of areas of vitamin A deficiency is the important first step in planning nutrition intervention programs.

Worldwide, the most common cause of vitamin A deficiency is low intake of preformed vitamin A and of provitamin A carotenoids; dietary assessment can be an important first step in recognizing an individual or a region at risk of vitamin A inadequacy. In affluent societies, vitamin A inadequacy may be due to inadequate absorption (impaired bile secretion leading to general lipid malabsorption; reduced absorption due to bariatric surgery or short bowel syndrome; or chronic pancreatitis—cystic fibrosis results in malabsorption).[24] Liver failure impairs storage of vitamin A resulting in deficiency.

Hypervitaminosis A is rare, but is seen occasionally. Symptoms can include headache, nausea and vomiting, double vision, muscle and joint pain, hair loss. As noted, serum retinol concentrations are not a responsive indicator of total vitamin A stores, but high concentrations of retinyl esters in serum are indicative of vitamin A toxicity. Use of synthetic retinoids (e.g., for dermatologic treatment) can result in vitamin A toxicity,[25, 26] including teratogenesis.

Vitamin A in excess of immediate requirements is stored as retinyl esters in liver and also in adipose tissue (although the concentration of vitamin A in adipose tissue is low, the total mass of adipose tissue in the typical human assures that the total mass of vitamin A can be appreciable).[27] Kinetic studies using tracer-labeled vitamin A have shown that there is constant turnover of stored vitamin A and exchange with plasma retinol. Although retinol can be bound by serum albumin, most is transported by plasma retinol-binding protein (RBP) which is synthesized in liver and released to maintain nearly constant concentrations of the holo-RBP complex over a wide range of liver vitamin A concentrations (vitamin A status); simply measuring plasma or serum concentrations of retinol (or RBP) does not give a sensitive measure of vitamin A status until liver reserves of vitamin A are nearly depleted. Nevertheless, serum retinol concentration is the most commonly used biochemical indicator of vitamin A status because samples are easy to obtain and analyze.

Plasma retinol and RBP concentrations are lower in childhood than in adults, increasing with maturity.[28]

Retinol-binding protein is a negative acute-phase protein, that is, its plasma concentration decreases during inflammation. Thus measurement of serum retinol or RBP concentrations to assess vitamin A status in nutrition surveys should be accompanied by measurement of another acute-phase protein such as C-reactive protein (CRP) or alpha-1-acid glycoprotein (AGP, orosomucoid) to minimize false positives.[29]

Because as much as 90% or more of total body vitamin A resides in liver in the well-nourished human, liver concentration is considered the best index of vitamin A status (Table 2.3. Criteria of vitamin A status). Analysis of autopsy liver samples can give estimates of vitamin A status in a population.[32–34] Only in unusual circumstances can this tissue be sampled, however. Therefore creative approaches have been developed to estimate total body stores (with liver vitamin A stores as proxy). Isotope dilution assays in theory provide good estimates: a known amount of a tracer dose of vitamin A (with a radioactive or stable-isotope label) is administered to the subject; after a suitable period for mixing of the tracer with body stores, a sample of body vitamin A (i.e., a sample of plasma retinol) is analyzed for its tracer content (tracer/tracee ratio). In practice this approach is complicated by metabolism and loss of tracer during the mixing period. Recent advances in mathematical modeling of vitamin A kinetics are providing methods for facilitating this calculation.[35, 36]

Dose-response assays (Relative Dose Response, RDR) are also useful in estimating vitamin A status. The underlying principle depends on accumulation of apo-retinol-binding protein in liver when there is insufficient retinol available for formation and release of holo-RBP. An oral dose of retinol or analog is absorbed and transported to the liver, where it is promptly incorporated into apo-RBP and released as holo-RBP into the bloodstream. A transient increase in plasma retinol-RBP is seen, with maximum at ~5 h; comparison of retinol-RBP concentration at 5 h with that 0 time (time of dosing) is an index of vitamin A sufficiency, with elevated ratios indicating vitamin A deficiency. A similar approach (Modified Relative Dose Response, MRDR) uses vitamin A_2 (3,4,-didehydroretinol) as marker, in which case the ratio of plasma dehydroretinol/retinol at 5 hours is measured.[37]

Breastmilk concentrations of retinyl esters can be a good indicator of vitamin A status for population assessment.[38] Breastmilk vitamin A tends to reflect intake more than total body stores and is not susceptible to infection and inflammation so much as plasma retinol concentrations.[39]

A very good overall review of methods of assessment of vitamin A status has been provided by Tanumihardjo et al.[40]

Analytical Methods
Sample Preparation
Plasma/serum collection and storage

Blood should be protected from light immediately after it is drawn and centrifuged to remove blood cells. Hemolysis should be avoided; alternatively, the blood can be allowed to clot in a cool, dark place. Plasma or serum (either can be used) should

Table 2.3 Criteria of Vitamin A Status[30, 31]

Status	Plasma Retinol Concentration	Fasting Plasma Retinyl Esters Concentration	Liver Stores	Breast Milk Vitamin A Concentration	Clinical Signs	Vulnerable Groups
Deficient	< 0.35 µmol/L	< 5 µg/mL (0.02 µmol/L)	Severely depleted (<0.017 µmol/g)	< 0.35 µmol/L	Night blindness; other ocular manifestations; skin dryness	Preschool-age children; pregnant and lactating women with low vitamin A intakes; patients with chronic liver diseases or alcoholism
Marginal	0.35–0.7 µmol/L	Low (5–20 µg/mL; 0.02–0.07 µM)	Depleted (0.017–0.07 µmol/g)	0.35–1.05 µmol/L	None (positive dose response test)	Children, pregnant women in vulnerable populations
Adequate Excessive	>1.05–3 µmol/L > 3 µmol/L	2%–20% of retinol	0.07–0.7 µmol/g) High (0.7–1.05 µmol/g)	1.05–3.5 µmol/L	None Elevated liver enzymes in plasma	Long-term supplement use; frequent use of foods high in preformed vitamin A
Toxic	May be normal or elevated	May be higher than retinol concentrations	Very high (>1.05 µmol/g)		Headache; bone/ joint pain; elevated liver enzymes and clinical signs of liver disease	Food faddists and users of high-dose vitamin A supplements; patients treated with retinoids

be removed carefully to avoid hemolysis and transferred to a separate tube. If the sample is to be stored instead of being analyzed promptly, there should be minimal headspace to reduce oxidation. Sealed tubes can be stored at or below −20°C.

Sample extraction

Because retinoids are usually present in biological tissues in quite low concentrations, extraction is usually necessary. Depending on the analyte and the sample matrix, sometimes extraction with organic solvent is sufficient. Thus a known mass of liver tissue can be ground with anhydrous sodium sulfate (to macerate the tissue and to dehydrate the sample) under dichloromethane or dichloroethane or hexane/2-propanol[41]; the retinyl esters are released from the tissue and dissolved in the solvent. An aliquot of the solvent extract can be analyzed directly, or concentrated if necessary.

The major blood transport form of vitamin A in mammals is as retinol, bound to retinol-binding protein (see earlier). In almost all procedures, an aliquot of serum or plasma is treated with a chaotropic substance such as an alcohol to denature RBP; the released retinol is extracted with an immiscible organic solvent such hexane. Multiple extractions with organic solvent are usually required. An internal standard (see later) can be incorporated in the denaturing solvent.

Saponification

Release of retinoids and carotenoids from biological tissues may be simplified by general hydrolysis of the sample matrix. Vitamin A is stored in a variety of tissues, but especially in liver, in the form of retinyl esters. There is usually a mixture of long-chain fatty acid esters, the specific composition depending on the tissue and the species; in human and rat liver, retinyl palmitate is the predominant form but other esters (oleate, linoleate, linolenate, myristate, laurate) are also identifiable. In some cases it is useful to identify and quantitate the specific fatty acid esters. However, in most analyses it is more important to simply quantitate the total vitamin A, by saponifying and measuring the retinol component. Saponification also simplifies the sample matrix and sometimes is necessary to facilitate extraction. Similarly, xanthophylls are present in some plant tissues as esters with long-chain fatty acids; again, saponification markedly simplifies sample extraction. Acid-catalyzed hydrolysis should be avoided because of the likelihood of formation of anhydroretinol from retinol. If the sample does not contain sufficient amounts of natural antioxidants, it is recommended that an antioxidant be added before saponification.

Techniques for Qualitative and Quantitative Analysis

Quantitation of standards

As noted before, retinoids and carotenoids should not be quantitated by mass (weight) because of their instability. Instead, standard solutions are conveniently quantitated by spectrophotometry because of their high molar extinction coefficients (Tables 2.1 and 2.2). The full absorption spectrum should be determined instead of only the absorbance at the λ_{max}, to disclose any degradation of the standard. Both quantitative analysis (concentration) and qualitative evaluation (purity) are readily determined.

Biological Assay

In early studies of vitamin A, administration of test sources of vitamin A to deficient animals was very helpful in monitoring purification procedures and determining vitamin A content of foodstuffs. This is no longer necessary for vitamin A itself, but can still be useful for determining bioefficacy of provitamin A carotenoids and foodstuffs containing them. Studies of effects of retinoid analogs in cell culture systems may be considered a contemporary form of bioassay.

Carr-Price Assay

Retinoids and carotenoids, when exposed to a Lewis acid such as antimony trichloride in solution, produce an intense but transient blue color which can be quantitated in a colorimeter or spectrophotometer. This assay is very sensitive but requires considerable skill and the use of toxic chemicals.[42]

Fluorometry

As noted before, retinol and retinyl esters absorb light at 325 nm and emit fluorescent radiation at ~ 480 nm. This can be a rapid means of quantitating vitamin A in certain applications, such as small blood samples (the so-called Futterman assay)[43] and in fortified food products if interferences are not present. Unfortunately, esters of ferulic acid (oryzanol) in plant oils fluoresce identically, complicating analysis of fortified grain products (unpublished observations).

Thin-Layer Chromatography

Thin-layer chromatography (TLC) (on silica plates and on reversed-phase plates) can be used with great success in monitoring synthetic procedures for retinoids and carotenoids. The fluorescence of retinol and retinyl esters facilitates their detection (using commercial lamps emitting at 366 nm). The old technique of immersing the TLC plate in an iodine chamber will disclose all organic compounds on the plate, including fingerprints. However, TLC does not lend itself to quantitative analysis.

Gas Chromatography

Gas chromatography is very seldom used for the analysis of vitamin A compounds since they are very susceptible to heat and decompose immediately in typical GC injection ports. Cold on-column injection of retinoids (both underivatized and derivatized) has been used with success, however on capillary GC columns.[44] This method has been coupled with mass spectrometry for measurement of labeled retinol in isotope-dilution assessment of vitamin A status. Trimethylsilyl derivatives of retinol have been used for isotope dilution assessment of human vitamin A status by gas chromatography-mass spectrometry. Carotenoids are insufficiently volatile to be analyzed by GC.

High Performance Liquid Chromatography

Column chromatography was first developed by Mikhail Tswett as a means of separating carotenoid photosynthetic pigments.[45] The refinement of HPLC in the 1970s revolutionized the separation and quantitative analysis of retinoids and carotenoids. Mild temperatures and mild organic solvents ensure that samples are not degraded during analysis. The high molar absorbances of these compounds provide quantitation of minute quantities by light absorbance detectors (i.e., small sample requirements); photodiode array detectors or mass spectroscopic detectors assist qualitative analysis and confirmation of chromatograph peak identity. Relatively short analysis times allow ready analysis of multiple samples. Automated injectors and computer-controlled data systems facilitate repetitive analyses. Ultra-high-performance liquid chromatography (or ultra-high pressure liquid chromatography, UPLC) has smaller sample requirements and provides faster analysis times.

Because of the great popularity of HPLC for analysis of retinoids and carotenoids and the variety of samples analyzed, it is difficult to summarize suitable procedures. Some reviews of the methodology are given in Table 2.4.

Table 2.4 Useful Reviews of HPLC Analysis of Retinoids and Carotenoids

McCormick, A. M., Napoli, J. L., & DeLuca, H. F. (1980) High-pressure liquid chromatography of vitamin A metabolites and analogs. Methods Enzymol. 67: 220–233

Taylor, R. F. (1983) Chromatography of carotenoids and retinoids. Adv. Chromatogr. 22: 157–213

Lambert, W. E., Nelis, H. J., De Ruyter, M. G. M., & De Leenheer, A. P. (1985) Vitamin A: retinol, carotenoids, and related compounds. In: *Modern Chromatographic Analysis of the Vitamins* (De Leenheer, A. P., Lambert, W. E., & De Ruyter, M. G. M., eds.), pp. 1–72. Marcel Dekker, Inc., New York

Bhat, P. V. & Sundaresan, P. R. (1988) High-performance liquid chromatography of vitamin A compounds. CRC Crit. Rev. Anal. Chem. 20: 197–219

Khachik, F., Beecher, G. R., & Goli, M. G. (1991) Separation, identification, and quantification of carotenoids in fruits, vegetables and human plasma by high performance liquid chromatography. Pure Appl. Chem. 63: 71–80

Craft, N. E. (1992) Carotenoid reversed-phase high-performance liquid chromatography methods: reference compendium. Methods Enzymol. 213: 185–205

De Leenheer, A. P. & Nelis, H. J. (1992) Profiling and quantitation of carotenoids by high-performance liquid chromatography and photodiode array detection. Methods Enzymol. 213: 251–265

Furr, H. C., Barua, A. B., & Olson, J. A. (1992) Retinoids and carotenoids. In: *Modern chromatographic analysis of the vitamins* (Nelis, H. J., Lambert, W. E., & DeLeenheer, A. P., eds.), pp. 1-71. Marcel Dekker, New York

De Leenheer, A. P., Lambert, W. E., & Meyer, E. (1993) Chromatography of retinoids. In: *Retinoids: progress in research and clinical applications* (Livrea, M. A. & Packer, L., eds.), pp. 551–568. Marcel Dekker, New York

Lesellier, E., Tchapla, A., Marty, C., & Lebert, A. (1993) Analysis of carotenoids by high-performance liquid chromatography and supercritical fluid chromatography. Journal of Chromatography 633: 9–23

Furr, H. C., Barua, A. B., & Olson, J. A. (1994) Analytical methods. In: *The Retinoids* (Sporn, M. B., Roberts, A. B., & Goodman, D. S., eds.), pp. 179–209. Raven Press, New York

Table 2.4 Useful Reviews of HPLC Analysis of Retinoids and Carotenoids—cont'd

Pfander, H., Riesen, P., & Niggli, U. (1994) HPLC and SFC of carotenoids - scope and limitations. Pure Appl. Chem. 66: 947–954

Britton,G.; Liaaen-Jensen,S.; Pfander, H (eds) (1995) *Carotenoids. 1A. Isolation and Analysis*, pp. 1–317. Birkhauser, Basel

Wyss, R. (1995) Chromatographic and electrophoretic analysis of biomedically important retinoids. Journal of Chromatography B: Biomedical Applications 671: 381–425

Barua, A. B. & Furr, H. C. (1998) Properties of retinoids: structure, handling and preparation. Molecular Biotechnology 10: 167–182

Eitenmiller, R. R. & Landen, W. O. (1998) *Vitamin Analysis for the Health and Food Sciences*, pp. 1–501. CRC Press, Boca Raton, FL

Napoli, J. L. & Horst, R. L. (1998) Quantitative analyses of naturally occurring retinoids. In: *Retinoid Protocols* (Redfern, C. P. F., ed.), pp. 29–40. Humana Press, Totowa, NJ

Kimura, M. & Rodriguez-Amaya, D. B. (1999) Sources of errors in the quantitative analysis of food carotenoids by HPLC. Arch. Latinoam. Nutr. 49: 58S–66S

Mercadante, A. Z. (1999) Chromatographic separation of carotenoids. Arch. Latinoam. Nutr. 49: 52S–57S

Arnhold, T., Nau, H., & Ruehl, R. (2000) Vitamin A. In: *Modern Analytical Methodologies in Fat- and Water-Soluble Vitamins* (Song, W. O., Beecher, G. R., & Eitenmiller, R. R., eds.), pp. 3–49. John Wiley & Sons,Inc., New York

Barua, A. B., Olson, J. A., Furr, H. C., & van Breemen, R. B. (2000) Vitamin A and carotenoids. In: *Modern Chromatographic Analysis of Vitamins* (De Leenheer, A. P., Lambert, W. E., & van Bocxlaer, J. F., eds.), pp. 1–74. Marcel Dekker, Inc., New York

Eitenmiller, R. R. & Landen, W. O. (2000) Simultaneous determination of fat-soluble vitamins in food, feed, and serum. In: *Modern Analytical Methodologies in Fat- and Water-Soluble Vitamins* (Song, W. O., Beecher, G. R., & Eitenmiller, R. R., eds.), pp. 171-221. John Wiley & Sons Inc., New York

Garwin, G. G. & Saari, J. C. (2000) High-performance liquid chromatography analysis of visual cycle retinoids. Methods Enzymol. 316: 313–324

Oliver, J. & Palou, A. (2000) Chromatographic determination of carotenoids in foods. J.Chromatogr.A. 881: 543–555

Craft, N. E. (2001) Chromatographic techniques for carotenoid separation. In: *Current Protocols in Food Analytical Chemistry* (Wrolstad, R. E., Acree, T. E., An, H., Decker, E. A., Penner, M. H., Reid, D. S., Schwartz, S. J., Shoemaker, C. F., & Sporns, P., eds.), pp. F2.3.1–F2.3.15. John Wiley & Sons, New York

Gundersen, T. E. & Blomhoff, R. (2001) Qualitative and quantitative liquid chromatographic determination of natural retinoids in biological samples. J.Chromatogr.A. 935: 13–43

Minguez-Mosquera, M. I., Hornero-Mendez, D., & Perez-Galvez, A. (2002) Carotenoids and provitamin A in functional foods. In: *Methods of Analysis for Functional Foods and Nutraceuticals* (Hurst, W. J., ed.), pp. 102-157. CRC Press, Boca Raton, FL

Su, Q., Rowley, K. G., & Balazs, N. D. (2002) Carotenoids: separation methods applicable to biological samples. J. Chromatogr. B Analyt. Technol. Biomed. Life Sci. 781: 393–418

Furr, H. C. (2004) Analysis of retinoids and carotenoids: problems resolved and unsolved. J.Nutr. 134: 281S–285S

Gundersen, T. E. (2006) Methods for detecting and identifying retinoids in tissue. J.Neurobiol. 66: 631–644

Rivera, S. M. & Canela-Garayoa, R. (2012) Analytical tools for the analysis of carotenoids in diverse materials. J. Chromatogr. A. 1224: 1–10

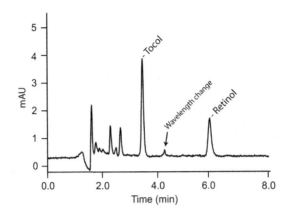

FIG. 2.5

Normal-phase HPLC of retinol from a dried blood spot. HPLC conditions: Column Chromegabond Diol 100 × 3.0 mm, 3 μm particle size; mobile phase 93 Hexane/5.4 Ethyl acetate/1.6 isopropanol at 1.0 mL/min. Absorbance detection at 300 nm until a change to 325 nm at 4.2 min. Tocol was used as internal standard.

From N. Craft (unpublished data).

Normal-phase (straight-phase, or adsorption chromatography) refers to use of polar stationary phases (e.g., silica or alumina particles) with nonpolar mobile phases (such as hexane with small concentrations of modifiers such as ethyl acetate, methanol, acetic acid). This modality is particularly powerful for separation of geometric isomers (e.g., 11-*cis* retinal from all-*trans* retinal in visual tissues). Fig. 2.5 illustrates normal-phase HPLC analysis of retinol from a dried blood spot.

Reversed-phase (partition) chromatography uses nonpolar stationary phases (e.g., octadecylsilane—ODS or C18) with polar mobile phases (usually based on methanol or acetonitrile, with a variety of modifiers such as water and low concentrations of other organic solvents). This modality is especially useful for dealing with sample components with a wide range of polarities, such as biological extracts and has become the most used form of HPLC for retinoids and carotenoids. Reversed-phase HPLC is readily amenable to solvent gradient elution to extend the range of components that can be analyzed in a single run. Current technology provides facile analysis of retinol, retinyl esters, and common carotenoids (as well as tocopherols) from human plasma in a single chromatographic run (Fig. 2.6).[46] Simple analyses, such as retinol from plasma extracts, can be easily accomplished on conventional columns (5-μm particle size, columns of 15–25 cm length); more demanding separations use smaller-diameter stationary phase particles, in shorter columns to reduce backpressure. Triacontyl (C30) polymeric stationary phases are available for difficult separations of carotenoids.[47]

Detectors for high performance liquid chromatography

Absorbance detectors are the most generally useful, because of the high molar extinction coefficients of retinoids and carotenoids, much higher than most other

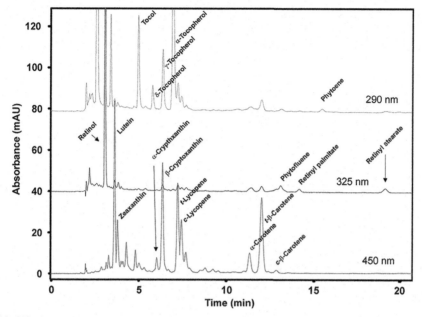

FIG. 2.6

Simultaneous analysis of retinol, tocopherols, and carotenoids in human serum by reversed-phase HPLC. Column 25-cm x 4.0-mm Spherisorb, 3-μm particle size with Ti frits; mobile phase acetonitrile (792 mL), isopropanol (60 mL), p-dioxane (148 mL, stabilized with 200 ppm BHT), and triethylamine (1 mL), with ammonium acetate (100 mM). Absorbance monitored at 290 nm (upper trace), 325 nm (middle trace) and 450 nm (lower trace). Tocol was used as internal standard.[46]

components in biological samples. They require minimal maintenance and provide excellent stability. Photodiode-array absorbance detectors allow qualitative analysis as well. Electrochemical detectors have been used successfully for analysis of retinoids and carotenoids but are not in common use. Fluorescence detectors are useful for analysis of retinol and retinyl esters, but most other retinoids and carotenoids are not fluorescent which can be an advantage or a disadvantage. Refractive index detectors are "universal" (which is often a disadvantage) but much less sensitive; they can be useful for preparative chromatography.

Mass spectrometry has become the "universal detector," providing both qualitative and quantitative analysis. This method is essential for analyzing isotopically labeled vitamin A, for example in isotope-dilution assessment of vitamin A status or for estimation of vitamin A equivalence of beta-carotene in specific foodstuffs. Older style "electron-impact" ionizers in mass spectrometers invariably resulted in dehydration of the retinoid, such that the "anhydroretinol" fragment (m/z 268) appeared to be the parent ion rather than the true retinol fragment (m/z 286). More recent "soft ionization" techniques produce more recognizable parent ions in the mass spectrum.

Standards and internal standards for HPLC

All-*trans* retinol, retinyl acetate and retinyl palmitate, all-*trans* retinal, and all-*trans* retinoic acid are readily available from commercial sources for use as standards; beta-carotene and some other carotenoids are available also. However, many retinoids and carotenoids of interest are not available commercially and must be obtained by chemical synthesis, isolation from biological sources, or gift.

The National Institute of Standards and Technology (NIST) in the United States sells a reference standard, Standard Reference Material (SRM) 968f: Fat-Soluble Vitamins in Frozen Human Serum (NIST SP 260-188), which is particularly useful for methods evaluation of plasma analysis.[48]

Use of an appropriate internal standard can greatly improve quantitative precision, to correct for losses during extraction and analysis. The ideal internal standard has chemical structure and physical properties nearly identical to those of the test compound, but can be readily separated and differentiated. Of course the compound used as internal standard should not be naturally present in the sample. A known amount is added to the sample before extraction. For analysis of retinol, retinyl acetate (or retinyl propionate) is useful if the sample is not saponified; it has identical spectrophotometric and fluorescence properties and can be easily separated by HPLC, either "normal-phase" or "reversed-phase." So-called 15-methylretinol and 15,15-dimethylretinol (and butyl analogs) have absorbance and fluorescence spectra identical to retinol and can be readily synthesized.[49] Anhydroretinol[50] and the naturally occurring 3,4-didehydroretinol (vitamin A_2) can be used as internal standards also.[51] Longer chain retinyl esters (e.g., retinyl pentadecanoate or retinyl heptadecanoate) can be readily synthesized and used in analysis of naturally occurring vitamin A esters (e.g., from liver samples).[52] Methyl retinoate can be used as internal standard for analysis of retinol but it has slightly different spectral properties and does not fluoresce. For mass spectrometric analysis of retinoids (e.g., in isotopedilution assessment of vitamin A status), a different isotopomer of the target compound may be chosen as internal standard. Canthaxanthin is commercially available and is a suitable internal standard for carotenoids.

Analyses for Retinol-Binding Protein

As noted before, retinol in blood is transported by retinol-binding protein (RBP); the molar ratio is usually about 0.9 moles retinol/mole RBP, so determination of RBP concentrations in blood is a very useful proxy for retinol concentration. The fluorescence of the retinol ligand is enhanced when it is bound by this protein, enhancing sensitivity.[43, 53–55] A small, portable fluorometer has been adapted for analysis of retinol-RBP in small blood samples.[56] However, HPLC techniques are so versatile and readily available that these approaches are seldom used.

For laboratories which do not have HPLC capability, enzyme-linked immunosorbent assays (ELISA) can be useful to quantitate RBP (and indirectly, serum retinol).[57, 58] These analyses have low limits of detection.[59] This procedure has been applied to analysis of blood spots dried on filter paper, providing much greater flexibility in sample storage and transport for vitamin A assessment surveys since refrigeration is not required.[60, 61]

Size-exclusion high-performance chromatography can be used for analysis of the intact plasma retinol-RBP complex without denaturation and extraction, exploiting the fluorescent properties of the retinol-RBP complex for specificity and sensitivity.[62, 63] In a somewhat similar manner, capillary electrophoresis with laser-induced fluorescence detection can be used for analysis of minuscule samples of retinol-RBP,[64] but neither technique has achieved widespread application because conventional extraction and HPLC analysis has proven so useful and the equipment is widely available.

Anticipated Directions

Technical improvements in ultra-high-pressure-liquid-chromatography (ultra-HPLC) will lead to increased sensitivity and decreased sample volume requirements as well as decreased analysis time per sample. Improved sensitivity in mass spectrometric detectors will be applicable both to ultra-HPLC and also to capillary electrophoresis separations of retinoids. Accelerator mass spectrometry is exquisitely sensitive; it may not be practical for routine analysis but will continue to be useful for studies of retinoid and carotenoid metabolism and kinetics.

It has already been shown that expression of a great number of genes is dependent on retinoids and carotenoids; it seems likely that analysis of relative expression of certain indicator proteins should give a sensitive functional measure of vitamin A status in humans.

References

1. Wolf G. A history of vitamin A and retinoids. *FASEB J* 1996;**10**:1102–7.
2. Semba RD. Vitamin A as "anti-infective". therapy, 1920-1940. *J Nutr* 1999;**129**:783–91.
3. Green HN, Mellanby E. Vitamin A as an anti-infective agent. *Br Med J* 1928;**2**:691–6.
4. Wolf G. The discovery of the visual function of vitamin A. *J Nutr* 2001;**131**:1647–50.
5. Arens JF, van Dorp DA. Synthesis of some compounds possessing vitamin A activity. *Nature* 1946;**157**:190–1.
6. Institute of Medicine (US) Panel on Micronutrients. *Dietary reference intakes for vitamin A, vitamin K, arsenic, boron, chromium, copper, iodine, iron, manganese, molybdenum, nickel, silicon, vanadium, and zinc. A report of the Panel on Micronutrients, Subcommittees on Upper Reference Levels of Nutrients and of Interpretation and Uses of Dietary Reference Intakes, and the Standing Committee on the Scientific Evaluation of Dietary Reference Intakes.* Washington, DC: National Academy Press; 2002. p. 1–728.
7. Furr HC, Barua AB, Olson JA. Analytical methods. In: Sporn MB, Roberts AB, Goodman DS, editors. *The retinoids.* New York: Raven Press; 1994. p. 179–209.
8. Barua AB, Furr HC. Properties of retinoids: structure, handling and preparation. *Mol Biotechnol* 1998;**10**:167–82.

9. Barua AB, Olson JA, Furr HC, van Breemen RB. Vitamin A and carotenoids. In: De Leenheer AP, Lambert WE, van Bocxlaer JF, editors. *Modern chromatographic analysis of vitamins*. New York: Marcel Dekker; 2000. p. 1–74.

10. De Ritter E, Purcell AE. Carotenoid analytical methods. In: Bauernfeind JC, editor. *Carotenoids as colorants and vitamin A precursors*. New York: Academic Press; 1981. p. 815–923.

11. Furr HC, Barua AB, Olson JA. Retinoids and carotenoids. In: Nelis HJ, Lambert WE, DeLeenheer AP, editors. *Modern chromatographic analysis of the vitamins*. New York: Marcel Dekker; 1992. p. 1–71.

12. Craft NE, Soares JH. Relative solubility, stability, and absorptivity of lutein and beta-carotene in organic solvents. *J Agric Food Chem* 1992;**40**:431–4.

13. Zscheile FP, White JW, Beadle BW, Roach JR. The preparation and absorption spectra of five pure carotenoid pigments. *Plant Physiol* 1942;**17**:331–46.

14. Huang Y, Clements PR, Gibson RA. Robust measurement of vitamin A status in plasma and blood dried on paper. *Prostaglandins Leukot Essent Fat Acids* 2015;**102–103**:31–6.

15. Landers GM, Olson JA. Absence of isomerization of retinyl palmitate, retinol, and retinal in chlorinated and nonchlorinated solvents under gold light. *J Assoc Off Anal Chem* 1986;**69**:50–5.

16. Kiser PD, Golczak M, Palczewski K. Chemistry of the retinoid (visual) cycle. *Chem Rev* 2014;**114**:194–232.

17. Uray IP, Dmitrovsky E, Brown PH. Retinoids and rexinoids in cancer prevention: from laboratory to clinic. *Semin Oncol* 2016;**43**:49–64.

18. Underwood BA. Biochemical and histological methodologies for assessing vitamin A status in human populations. *Methods Enzymol* 1990;**190**:242–51.

19. Stevens GA, Bennett JE, Hennocq Q, Lu Y, De Regil LM, Rogers L, Danaei G, Li G, White RA, Flaxman SR, Oehrle SP, Finucane MM, Guerrero R, Bhutta ZA, Then-Paulino A, Fawzi W, Black RE, Ezzati M. Trends and mortality effects of vitamin A deficiency in children in 138 low-income and middle-income countries between 1991 and 2013: a pooled analysis of population-based surveys. *Lancet Glob Health* 2015;**3**:e528–36.

20. Sesay FF, Hodges MH, Kamara HI, Turay M, Wolfe A, Samba TT, Koroma AS, Kamara W, Fall A, Mitula P, Conteh I, Maksha N, Jambai A. High coverage of vitamin A supplementation and measles vaccination during an integrated Maternal and Child Health Week in Sierra Leone. *Int Health* 2015;**7**:26–31.

21. Hilger J, Goerig T, Weber P, Hoeft B, Eggersdorfer M, Carvalho NC, Goldberger U, Hoffmann K. Micronutrient intake in healthy toddlers: a multinational perspective. *Nutrients* 2015;**7**:6938–55.

22. WHO. *Global prevalence of vitamin A deficiency in populations at risk 1995–2005*. Geneva: World Health. Organization; 2009. p. 1–68.

23. Stephensen CB. Vitamin A, infection, and immune function. *Annu Rev Nutr* 2001;**21**:167–92.

24. Faustino JF, Ribeiro-Silva A, Dalto RF, Souza MM, Furtado JM, Rocha GM, Alves M, Rocha EM. Vitamin A and the eye: an old tale for modern times. *Arq Bras Oftalmol* 2016;**79**:56–61.

25. Lam J, Polifka JE, Dohil MA. Safety of dermatologic drugs used in pregnant patients with psoriasis and other inflammatory skin diseases. *J Am Acad Dermatol* 2008;**59**:295–315.

26. Thielitz A, Gollnick H. Topical retinoids in acne vulgaris: update on efficacy and safety. *Am J Clin Dermatol* 2008;**9**:369–81.

27. Wongsiriroj N, Jiang H, Piantedosi R, Yang KJ, Kluwe J, Schwabe RF, Ginsberg H, Goldberg IJ, Blaner WS. Genetic dissection of retinoid esterification and accumulation in the liver and adipose tissue. *J Lipid Res* 2014;**55**:104–14.

28. Goodman DS. Plasma retinol-binding protein. In: Sporn MB, Roberts AB, Goodman DS, editors. *The Retinoids*. vol. 2. Orlando, FL: Academic Press; 1984. p. 41–88.

29. Thurnham DI. Inflammation and vitamin A. *Food Nutr Bull* 2015;**36**:290–8.

30. Ross AC. Vitamin A and carotenoids. In: Shils ME, Shike M, Ross AC, Caballero B, Cousins RJ, editors. *Modern nutrition in health and disease*. Philadelphia: Lippincott Williams & Wilkins; 2006. p. 351–75.

31. Tanumihardjo SA. Vitamin A: biomarkers of nutrition for development. *Am J Clin Nutr* 2011;**94**:658S–65S.

32. Underwood BA, Siegel H, Weisell RC, Dolinski M. Liver stores of vitamin A in a normal population dying suddenly or rapidly from unnatural causes in New York City. *Am J Clin Nutr* 1970;**23**:1037–42.

33. Dahro M, Gunning D, Olson JA. Variations in liver concentrations of iron and vitamin A as a function of age in young American children dying of the sudden infant death syndrome as well as of other causes. *Int J Vitam Nutr Res* 1983;**53**:13–8.

34. Schindler R, Friedrich DH, Kramer M, Wacker HH, Feldheim W. Size and composition of liver vitamin A reserves of human beings who died of various causes. *Int J Vitam Nutr Res* 1988;**58**:146–54.

35. Green MH. Evaluation of the "Olson equation", an isotope dilution method for estimating vitamin A stores. *Int J Vitam Nutr Res* 2014;**84**(Suppl. 1):9–15.

36. Green MH, Ford JL, Green JB, Berry P, Boddy AV, Oxley A, Lietz G. A retinol isotope dilution equation predicts both group and individual total body vitamin A stores in adults based on data from an early postdosing blood sample. *J Nutr* 2016;**146**:2137–42.

37. Tanumihardjo SA, Furr HC, Erdman Jr. JW, Olson JA. Use of the modified relative dose response (MRDR) assay in rats and its application to humans for the measurement of vitamin A status. *Eur J Clin Nutr* 1990;**44**:219–24.

38. Engle-Stone R, Haskell MJ, Nankap M, Ndjebayi AO, Brown KH. Breast milk retinol and plasma retinol-binding protein concentrations provide similar estimates of vitamin A deficiency prevalence and identify similar risk groups among women in Cameroon but breast milk retinol underestimates the prevalence of deficiency among young children. *J Nutr* 2014;**144**:209–17.

39. Tanumihardjo SA. Assessing vitamin A status: past, present and future. *J Nutr* 2004;**134**:290S–3S.

40. Tanumihardjo SA, Russell RM, Stephensen CB, Gannon BM, Craft NE, Haskell MJ, Lietz G, Schulze K, Raiten DJ. Biomarkers of nutrition for development (BOND)—vitamin A review. *J Nutr* 2016;**146**:1816S–48S.

41. Radin NS. Extraction of tissue lipids with a solvent of low toxicity. *Methods Enzymol* 1981;**72**:5–7.

42. Eitenmiller RR, Landen WO. *Vitamin analysis for the health and food sciences*. Boca Raton, FL: CRC Press; 19981–501.

43. Futterman S, Swanson D, Kalina RE. A new, rapid fluorometric determination of retinol in serum. *Investig Ophthalmol* 1975;**14**:125–30.

44. Furr HC, Clifford AJ, Jones AD. Analysis of apocarotenoids and retinoids by capillary gas chromatography-mass spectrometry. *Methods Enzymol* 1992;**213**:281–90.

45. Engelhardt H. One century of liquid chromatography. From Tswett's columns to modern high speed and high performance separations. *J Chromatogr B Analyt Technol Biomed Life Sci* 2004;**800**:3–6.

46. Craft NE, Furr HC. Improved HPLC analysis of retinol and retinyl esters, tocopherols, and carotenoids in human serum samples for the NHANES. *FASEB J* 2004;**18**:A534. [abs.].

47. Sander LC, Sharpless KE, Craft NE, Wise SA. Development of engineered stationary phases for the separation of carotenoid isomers. *Anal Chem* 1994;**66**:1667–74.

48. Thomas JB, Duewer DL, Mugenya IO, Phinney KW, Sander LC, Sharpless KE, Sniegoski LT, Tai SS, Welch MJ, Yen JH. Preparation and value assignment of standard reference material 968e fat-soluble vitamins, carotenoids, and cholesterol in human serum. *Anal Bioanal Chem* 2012;**402**:749–62.

49. Tosukhowong P, Olson JA. The synthesis, biological activity and metabolism of 15-[6, 7-14C2]- and 15-[21-3H] methyl retinone, 15-methyl retinol and 15- dimethyl retinol in rats. *Biochim Biophys Acta* 1978;**529**:438–53.

50. Wallingford JC, Underwood BA. Rapid preparation of anhydroretinol and its use as an internal standard in determination of liver total vitamin A by high-performance liquid chromatography. *J Chromatogr* 1986;**381**:158–63.

51. Tanumihardjo SA, Penniston KL. Simplified methodology to determine breast milk retinol concentrations. *J Lipid Res* 2002;**43**:350–5.

52. Ross AC. Separation and quantitation of retinyl esters and retinol by high-performance liquid chromatography. *Methods Enzymol* 1986;**123**:68–74.

53. Peterson PA, Rask L. Studies on the fluorescence of the human vitamin A-transporting plasma protein complex and its individual components. *J Biol Chem* 1971;**246**:7544–50.

54. Goodman DS, Leslie RB. Fluorescence studies of human plasma retinol-binding protein and of the retinol-binding protein-prealbumin complex. *Biochim Biophys Acta* 1972;**260**:670–8.

55. Glover J, Moxley L, Muhilal H, Weston S. Micro-method for fluorimetric assay of retinol-binding protein in blood plasma. *Clin Chim Acta* 1974;**50**:371–80.

56. Chaimongkol L, Pinkaew S, Furr HC, Estes J, Craft NE, Wasantwisut E, Winichagoon P. Performance of the CRAFTi portable fluorometer comparing with the HPLC method for determining serum retinol. *Clin Biochem* 2011;**44**:1030–2.

57. Erhardt JG, Estes JE, Pfeiffer CM, Biesalski HK, Craft NE. Combined measurement of ferritin, soluble transferrin receptor, retinol binding protein, and C-reactive protein by an inexpensive, sensitive, and simple sandwich enzyme-linked immunosorbent assay technique. *J Nutr* 2004;**134**:3127–32.

58. Gorstein JL, Dary O, Pongtorn, Shell-Duncan B, Quick T, Wasanwisut E. Feasibility of using retinol-binding protein from capillary blood specimens to estimate serum retinol concentrations and the prevalence of vitamin A deficiency in low-resource settings. *Public Health Nutr* 2008;**11**:513–20.

59. Hix J, Martinez C, Buchanan I, Morgan J, Tam M, Shankar A. Development of a rapid enzyme immunoassay for the detection of retinol-binding protein. *Am J Clin Nutr* 2004;**79**:93–8.

60. Fujita M, Brindle E, Shofer J, Ndemwa P, Kombe Y, Shell-Duncan B, O'Connor KA. Retinol-binding protein stability in dried blood spots. *Clin Chem* 2007;**53**:1972–5.

61. Baingana RK, Matovu DK, Garrett D. Application of retinol-binding protein enzyme immunoassay to dried blood spots to assess vitamin A deficiency in a population-based survey: the Uganda Demographic and Health Survey 2006. *Food Nutr Bull* 2008;**29**:297–305.

62. Furr HC, Olson JA. A direct microassay for serum retinol (vitamin A alcohol) by using size-exclusion high-pressure liquid chromatography with fluorescence detection. *Anal Biochem* 1988;**171**:360–5.

63. Burri BJ, Kutnink MA. Liquid-chromatographic assay for free and transthyretin-bound retinol-binding protein in serum from normal humans. *Clin Chem* 1989;**35**:582–6.
64. Shi H, Ma Y, Humphrey JH, Craft NE. Determination of vitamin A in dried human blood spots by high- performance capillary electrophoresis with laser-excited fluorescence detection. *J Chromatogr B Biomed Appl* 1995;**665**:89–96.

Further Reading

65. Furr H, Retinol C. Physiology. In: Caballero B, Finglas P, Toldra P, editors. *Encyclopedia of food and health*. Amsterdam: Elsevier/Academic Press; 2016. p. 597–603.

Methods for assessment of Vitamin D

3

Graham Carter*, David J. Card[†]

North West London Pathology, Imperial College Healthcare NHS Trust, Charing Cross Hospital, London, United Kingdom, [†]Nutristasis Unit, Viapath, St. Thomas' Hospital, London, United Kingdom

Chapter Outline

Structure and Function

Vitamin D is the generic name for a group of antirachitic compounds of which the most biologically important are cholecalciferol (vitamin D_3) and ergocalciferol (vitamin D_2). Both molecules belong to a class of compounds known as seco-steroids which have a similar structure to steroids except that carbons 9 and 10 of the B ring are not joined.[1] The resulting *cis*-triene structure gives rise to the characteristic absorption spectrum of vitamin D and its metabolites, and the propensity for UV light-induced conformational changes.[2]

Laboratory Assessment of Vitamin Status. https://doi.org/10.1016/B978-0-12-813050-6.00003-6

In humans the naturally occurring form of vitamin D is cholecalciferol, formed by the action of UV light on the skin which converts 7-dehydrocholesterol (pro-vitamin D) into previtamin D. Previtamin D undergoes thermal isomerization to vitamin D_3. Ergocalciferol, found in invertebrates, plants, and fungi, is synthesized by the action of UV light on ergosterol and undergoes similar transformations. Chemically, these two forms of vitamin D differ in the side chain; vitamin D_2 has an additional double bond between carbons 22 and 23 and a methyl group on carbon 24. Cutaneously synthesized cholecalciferol is transported to the liver bound to plasma proteins, approximately 95% to the specific vitamin D-binding protein (DBP) and 4% to plasma albumin. Less than 1% is unbound. The relative contribution of dietary sources of vitamin D are low, the majority of vitamin D in the body being derived from cutaneous photosynthesis. The foods richest in vitamin D are oily fish such as herrings and sardines with lower concentrations found in liver, eggs, and dairy produce. Because vitamin D is a lipid, absorption of dietary vitamin D via diffusion through the jejunum is a highly bile salt-dependent process.[3] Orally ingested vitamins D_3 and D_2 are primarily transported with chylomicrons and lipoproteins[4] and are thus presented to the liver differently. In the liver, both forms undergo hydroxylation at carbon 25 to give 25-hydroxyvitamin D_3 (25-OHD$_3$) and 25-hydroxyvitamin D_2 (25-OHD$_2$). A further hydroxylation occurs mainly, but not exclusively, in the kidney to form the bioactive molecules 1,25-dihydroxyvitamin D_3 (1,25(OH)$_2$D$_3$) and 1,25-dihydroxyvitamin D_2 (1,25(OH)$_2$D$_2$). Both forms of the native vitamin are used to supplement vitamin D-deficient subjects, although vitamin D_2 has been more commonly used in the United States. However, vitamin D_3 is now widely available as a food supplement and there is some evidence that it is more effective than vitamin D_2, possibly because it is more tightly bound and cleared less rapidly from the circulation than vitamin D_2,[5,6,7] although this view is not universally accepted.[8]

The formation of 25-OHD in the liver is not specifically regulated to control vitamin D status and the hydroxylation of ingested vitamin D is virtually quantitative until the amount exceeds about 2000 international units (IU) per day.[9] This, coupled with its relatively long half-life in the circulation of approximately three weeks[10] explains why the serum concentration of total 25-OHD (25-OHD$_3$ plus 25-OHD$_2$) is universally used as an index of vitamin D nutrition. Circulating concentrations of 1,25(OH)$_2$D are tightly regulated and concentrations in vitamin D deficiency can be misleading as concentrations can be low, normal, or even high[10]; its half-life is only about four to six hours so that serum concentrations will only reflect recent exposure to vitamin D.[11] Thus measurement of 1,25(OH)$_2$D is of little clinical value except in conditions such as vitamin D-dependent rickets and the hypercalcemia associated with certain granulomatous conditions such as sarcoidosis, psoriasis, and tuberculosis.[12] In the laboratory assessment of vitamin D status, there are two other hydroxylated metabolites that need to be considered, either because of their clinical importance or because of their impact on the accurate measurement of 25-OHD; these are (1) serum 24,25-dihydroxyvitamin D_3 [24,25(OH)$_2$D$_3$] which is present in concentrations of up to about 10% of total 25-OHD and (2) serum 3-epi-25-OHD$_3$ which, in adults, is usually less than 5% of total 25-OHD but can be very high in neonates.[13] External Quality Assessment Schemes can be a valuable source of information on assay specificity and accuracy.

Nomenclature

It is worth reiterating that the D vitamins are a group of compounds and the term "vitamin D" should not be used on laboratory request forms or in scientific articles as shorthand for specific metabolites. This can lead to confusion and delays in diagnosis, as well as the extra expense if the request is misinterpreted. The use of two different units of measurement adds a further complication; SI units are mainly used in Europe (e.g., nmol/L for 25-OHD and pmol/L for 1,25(OH)$_2$D) and mass units in the United States (ng/mL and pg/mL). Supplement manufacturers add to the confusion by often stating values in international units (IU); for vitamin D$_3$, one IU = 0.025 µg.

Dietary Intake

Sunlight is the major contributor to vitamin D status. Foods highest in vitamin D are oily fish with other foods containing insignificant quantities.[14] Ultraviolet light within the wavelength ranges of 290–315 nm can be used to synthesize vitamin D in the skin; sunlight that has passed through glass cannot be used to make vitamin D as these wavelengths are filtered out.[15]

Clinical Application of Measurement

In the past, vitamin D deficiency was associated only with rickets, the most extreme clinical manifestation of the disease, which in the developed world is now rare. Rickets was first referred to in the 1600s and in infants was later found to be curable by cod-liver oil or sunlight. Rickets was eventually linked to vitamin D in the early 20th century shortly after its identification in cod-liver oil.[16] Recent understanding of long-term suboptimal status has exposed a number of health issues to which it is now thought vitamin D status makes a significant contribution. Many diseases, some of which are common in old age such as cardiovascular disease, osteomalacia, type II diabetes, poor immunity, and poor mental health have been connected with suboptimal vitamin D status.[15] There is also evidence that maternal vitamin D status affects fetal development leading to diseases later in life such as schizophrenia.[15]

Despite strong evidence that vitamin D status is important in health, proof that supplementation with vitamin D improves health outcomes is lacking. For example, in their systematic review and meta-analysis, Reid et al found that there is little evidence that vitamin D supplementation improves bone mineral density.[17] Bone mineral density decreases with age, and the rate of loss is likely to be modulated by vitamin D status rather than reversed, therefore the results of such reviews are not entirely unexpected. It is likely therefore that vitamin D supplementation is only effective as a preventative measure and until the results of longitudinal clinical trials or cohort studies are available this part of the puzzle will be unresolved. With vitamin D deficiency thought to be such a common problem it is possible supplementation or fortification of foods will become widely used. In this scenario we may see a trail of evidence similar to other general supplementation scenarios such as addition of folic acid to flour or fluoride to water supplies.

Optimal serum concentrations of 25-OHD have been estimated at 75 nmol/L,[18] for which there is now good consensual agreement based on the evidence currently available. Reference ranges from "healthy populations" are not considered reliable due to the endemic nature of suboptimal status. It therefore follows that supplementation with vitamin D leading to increased serum concentrations would prevent diseases associated with suboptimal status. However, supplementation with vitamin D has not yet been shown to be effective in reversing diseases associated with vitamin D deficiency,[19] which may be due to irreversible effects as a result of long-term suboptimal status. It is likely that osteomalacia in old age is to some extent a result of lifelong subclinical vitamin D insufficiency. Supplementation therefore may be more effective if administered prophylactically in high-risk groups, for example, those living in countries with low UV radiation, darker skin pigmentation, skin covered by clothing, and so on. Prescription and self-administration of vitamin D is becoming more common and there are suggestions that blanket supplementation of the general population may be beneficial.[20]

A possible side effect of increased vitamin D supplementation in the general population is increased rates of vitamin D overdose and toxicity. Although vitamin D supplementation has been shown to be safe,[21] toxicity is possible when high doses of vitamin D are taken.[22] Due to homeostatic regulation of cutaneous photosynthesis, vitamin D toxicity cannot be induced by sunlight. Toxicity occurs through altered calcium homeostasis with hypercalcemia a common feature. Parathyroid hormone (PTH) should also be measured in order to rule out primary hyperparathyroidism (PTH is frequently suppressed in vitamin D toxicity). The US Institute of Medicine published guidelines in 2011, identifying risks for some outcomes at 25-OHD concentrations above 125 nmol/L.[23] However, this concentration is highly conservative and may be impractical to implement in routine clinical practice because of the high number of cases requiring further investigation for toxicity. Vitamin D toxicity case studies where 25-OHD concentrations have been measured suggest concentrations in the 100s to 1000s nmol/L are common, therefore a cutoff for further investigation of toxicity of 500 nmol/L is suggested here.

Analytical Methods

It is now generally accepted that measurement of 25-OHD is the best available marker of vitamin D status and this will form the main focus of the following discussion on analytical methods.

Total 25-hydroxyvitamin D

By common consent, 25-OHD is a difficult analyte to quantify accurately; it is likely this arises from the molecule's hydrophobic nature, its existence in several molecular forms, and its high affinity for DBP. Historically, it has been difficult to assess the accuracy of results due to the absence of an accepted definitive reference method. This problem has been mitigated by the introduction by the National Institute of Standards and Technology (NIST) of Standard Reference Materials (SRMs) in 2009 and the acceptance of the NIST, University of Ghent and the Centers for Disease Control and Prevention (CDC) LC-MS/MS methods as Reference Measurement Procedures (RMPs) by the Joint Committee for Traceability in Laboratory Medicine (JCTLM).[24] The formation of the Vitamin D Standardization Program (VDSP) and its offshoot, the Centres for Disease Control (CDC) Certification Program in 2010 has played an important role in encouraging kit manufacturers and some larger laboratories to standardize their methods to the RMPs.[25] Smaller laboratories using LC-MS/MS methods can purchase NIST SRMs[26] to ensure traceability to the RMPs. The introduction of the RMPs spawned accuracy-based external quality assessment (EQA) schemes (e.g., CAP ABDV and DEQAS) which are a relatively inexpensive way of monitoring the validity of results produced by clinical and research laboratories. In future, overall assay standardization should become less of an issue and attention will likely be diverted to achieving the accuracy of individual results, a far more intractable problem. To some extent, the use of summary statistics in method comparisons, particularly the correlation coefficient and regression equations given in most kit inserts, has disguised marked method-related differences in the results of individual samples (these are better revealed by Bland-Altman plots).[27] These apparently random errors may arise from matrix effects (interference from other unidentified components of the sample) or the presence of other metabolites which cross-react in the assay, depending on the specificity of the binding agent (usually an antibody). Modern fully automated assays have necessarily abandoned sample extraction and seem particularly prone to matrix effects.[28] DEQAS has identified triglycerides as a particular problem in some assays, even when present in fairly modest amounts.[29]

Issues relating to the measurement of 25-OHD can usefully be divided into three main categories: preanalytical, analytical, and postanalytical.

Preanalytical
Time of collection
Given the relatively long half-life of 25-OHD it is unsurprising that results appear to be unaffected by the time of collection or whether the subject is fasting. However, there is a well-known seasonal variation in blood concentrations (higher in summer than in winter)[10] and this should be borne in mind when interpreting results. Published guidelines on vitamin D intake do not take this into account presumably because vitamin D requirements remain the same throughout the year.

Blood sampling-dried blood spot/capillary blood
Venous blood is normally used for analysis but methods have been developed for measuring 25-OHD on dried blood spots (DBS) or capillary blood. These usually involve the extraction of the blood spot with an organic solvent and measurement of

25-OHD by LC-MS/MS.[30,31] Confounding factors include the need to correct for the hematocrit using an average value, the quality of the paper, the size of punch, and the location of the punch in the blood spot.[32] Long-term stability of 25-OHD in previously prepared dried blood spots is difficult to assess but one study showed good agreement between plasma and dried blood spots that had been stored for 19 years.[33] While some studies have shown a good overall correlation between venous blood and dried blood spot values, differences can be large in individual subjects,[33] which is likely to be influenced by the drying process. Some commercial laboratories have produced DBS kits (which include automatic lancets) for subjects to collect their own samples onto absorbent paper or in some kits into a microtube; these can then be posted back to the laboratory for analysis. One disadvantage to this approach is that there is currently no quality assessment scheme that can check the overall accuracy of these methods, the blood collection stage of which is usually in the hands of unskilled operators. Despite the problems mentioned before, the use of DBS analysis is potentially useful in epidemiological studies.

A novel approach to vitamin D screening uses a gold nanoparticle-based colorimetric immunoassay on a test strip. The colorimetric change is captured by a smartphone camera and compared with the color of a reference area on the strip; the 25-OHD concentration is measured by a specially developed smartphone app.[34]

Saliva

Measurement of 25-OHD in saliva has some potential advantages, in particular the ease of sample collection and the possibility that salivary concentrations might reflect those of free or unbound metabolite (see later). However, like earlier interest in salivary steroids, this approach has not proved entirely satisfactory, not least because saliva contains variable amounts of albumin and therefore bound 25-OHD. An investigation in 1989[35] showed that salivary concentrations corresponded to about 1.2% of total 25-OHD in serum but showed no relationship with directly measured serum-free 25-OHD. Salivary concentrations varied throughout the day and were possibly related to recent vitamin D intake. There is currently little serious interest in salivary measurements.

Blood sampling-sample tubes

Most commercial suppliers of ligand-binding assays claim that their method can be used on serum or plasma, often without providing supporting evidence. For logistical reasons tubes containing separating gels are popular in routine clinical laboratories but their suitability for certain steroid methods has been questioned.[36] The effect of separator gels on 25-OHD methods was investigated by DEQAS in 2009; the ligand-binding assays in use at the time were largely unaffected but some users of HPLC methods reported spuriously high results.[37] More surprising was the finding that the sample tube itself could influence the results of LC-MS/MS assays. A paper published in 2009 reported that serum from Sarstedt Monovette tubes gave spuriously high 25-OHD values in a routine LC-MS/MS assay due to ion suppression of the internal standard. This was apparently due to the leaching of oleamide, a so-called slipping agent used to release the plastic tubes from heated metal molds.[38]

Blood for the measurement of vitamin D metabolites is probably best collected in plain glass tubes without anticoagulants or separating gel, which have been screened for their suitability.

Blood sampling-sample stability

Although pure vitamin D and its metabolites are rapidly degraded in direct light, 25-OHD in serum is remarkably stable, probably because it is tightly bound to DBP and plastic tubes are relatively opaque to UV radiation.[2] Indeed samples of whole blood or serum can be left for days at ambient temperature without any significant change in 25-OHD concentration.[39,40] Samples stored frozen for over 40 years retained racial and seasonal differences which provided indirect evidence of stability over long time periods.[41] Concentration of 25-OHD also seem unaffected by repeated freeze-thaw cycles.[42,43] Fewer studies have been done for other metabolites but the evidence suggests that $1,25(OH)_2D$ is similarly stable.[44] However, it should be remembered that results produced by modern nonextraction assays can be affected by other constituents of the sample. Changes in the sample matrix (e.g., pH) which inevitably occur during storage might conceivably affect some assays and not others. It is virtually impossible for every method to be tested in a single study and kit manufacturers and users have a responsibility to check sample stability using their own method.

Analytical

Factors affecting the accuracy of 25-OHD measurements

Accurate measurement of 25-OHD has been complicated by several major problems including the lipophilic nature of the molecule which makes assays prone to nonspecific interference from other lipids, including lipoproteins. Early competitive protein-binding assays, which incorporated solvent extraction and chromatography steps,[45] were relatively unaffected by these so-called matrix effects but attempts at abandoning the preliminary purification steps were unsuccessful.[46] By their very nature, matrix effects are unpredictable and will vary from sample to sample. This problem is evident from many of the method comparisons, where good correlations cannot disguise large method-related differences in individual samples (Figs. 3.1 and 3.2). Undeterred by the failure of early attempts at simplifying assays, manufacturers have been under pressure to produce easy-to-use automated methods with a high sample throughput, arguably at the expense of accuracy and precision. A fundamental issue with all 25-OHD assays is the preparation of appropriate standards. In traditional steroid assays, aliquots of a stock standard, usually prepared in ethanol, would be added to analyte-free human serum. Unfortunately preparation of analyte-free serum, traditionally by stripping with activated charcoal, has proved virtually impossible due to the very high affinity of vitamin D metabolites to DBP.[2] The use of an alternative standard matrix such as bovine serum albumin (BSA) may result in the standards behaving differently to the samples. BSA may also contain DBP and bound 25-OHD as a contaminant. Furthermore, in a problem apparently unique to

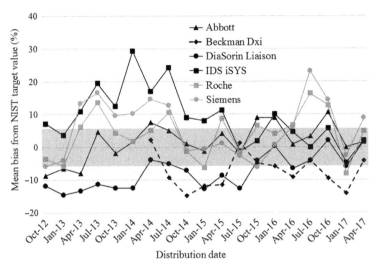

FIG. 3.1

Method bias of total 25-OHD results submitted to DEQAS. Mean % bias from NIST assigned values for the major automated ligand-binding assays (October 2012 to April 2017).

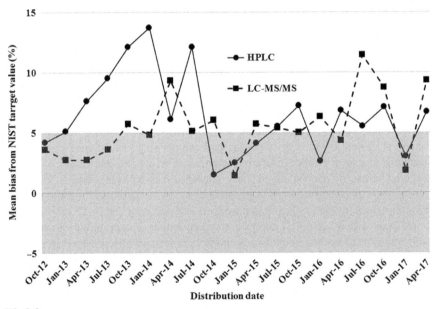

FIG. 3.2

Mean % bias from NIST assigned values for HPLC and LC-MS/MS methods (October 2012 to April 2017).

vitamin D, the addition of ethanolic solutions of 25-OHD$_3$ or 25-OHD$_2$ to human serum results in underrecovery of the added metabolite.[47,48] The reason for this is obscure, but may be related to the lipophilic nature of the molecules and the immediate uptake of exogenous material by serum lipids. For whatever reason, the added 25-OHD does not appear to enter and equilibrate with endogenous 25-OHD; this has profound implications for the preparation of standards and the validity of in vitro recovery experiments. Given these problems, it is essential that all immunoassay standards are ultimately calibrated with a reference measurement procedure or the NIST SRMs. Underrecovery of vitamin D metabolites from spiked sera is also a potential problem for those EQA schemes that, unlike DEQAS, distribute samples containing exogenous vitamin D metabolites.

A common feature of automated (nonextraction) 25-OHD assays is the need to irreversibly displace the analyte from its binding protein which, in the case of tightly bound 25-OHD, will involve disrupting the DBP, usually by a shift in pH. Manufacturers rarely divulge full details of the technique but the assumption, generally unsupported by published evidence, is that 100% of the 25-OHD is released from the DBP or, if not, that the percentage released is consistent from sample to sample. An investigation by Heijbower et al in 2012 suggested that this may not be true of certain automated methods the results of which were influenced by the DBP concentration of the sample.[49] Another potential problem arises from the presence in some human sera of anti-animal (heterophylic) antibodies which might cause interference in nonextraction assays.[50]

Historical Methods
Ligand-binding assays

The laboratory assessment of vitamin D status did not become widely available until the early 1970s following the commercial availability of high specific activity tritiated 25-OHD (*formerly* Amersham, New England Nuclear). One of the first published methods was that of Haddad and Chyu,[45] which used rachitic rat kidney cytosol as the source of binding protein, although the authors suggested that rachitic rat serum could also be used. This, and other competitive protein binding-based methods, shared common characteristics; procedural losses were monitored by the addition of ^{3}H-25-OHD to the sample before extraction with an organic solvent and chromatography on silicic acid or Sephadex. The Haddad and Chyu method used an ether extraction; this was convenient as the ether formed a clean upper layer which could be decanted after freezing the lower aqueous layer. The disadvantage of using ether was the low solubility of 25-OHD and, in the original method, the need to do three successive extractions. In a method developed by one of the authors (Carter, unpublished) the number of extractions was reduced by increasing the pH of the sample prior to extraction. At pH 10.0, a single extraction could remove 95% of the 25-OHD. Nevertheless, the need for chromatography and liquid scintillation counting restricted the use of this and similar methods to specialist laboratories. Attempts at abandoning the chromatography step[51] were unsuccessful due to nonspecific interference from serum lipids.[46]

Immunoassays

Probably the most significant advances in assay methodology in recent times were the development of an antibody to 25-OHD and, later, an iodinated tracer which were used in a radioimmunoassay described by Hollis and Napoli in 1985.[52,53] The antibody was raised against a synthetic analog (23,24,25,26,27-pentanor vitamin D-C(22)-carboxylic acid) coupled to bovine serum albumin. Because the structures of 25-OHD$_3$ and 25-OHD$_2$ differ only in the side chain, the antibody should theoretically be cospecific for both forms of the molecule. This is not the case with 3-epi-25-OHD where the structure is modified in the A-ring; immunoassays generally do not detect this form of the molecule. In practice it is the specificity of the assay rather than the antibody that is important and this could be affected by other factors. An indication of 25-OHD methods currently available can be seen from the results returned by participants in DEQAS, the largest specialist EQA scheme (Fig. 3.3 and Table 3.1). This list (not exhaustive) indicates that the majority of laboratories are now using fully automated commercial assays. Many major manufacturers of clinical chemistry analyzers have developed fully automated methods for 25-OHD, but as with other analytes, most factors affecting performance such as standardization and specificity are beyond the control of the user. Users have a responsibility to properly evaluate the published information and select a method that is fit for its intended use. External quality assurance schemes can be a valuable source of information on how the method performs in the 'real world' rather than that of the manufacturer's laboratory or a small group of specialist institutions.

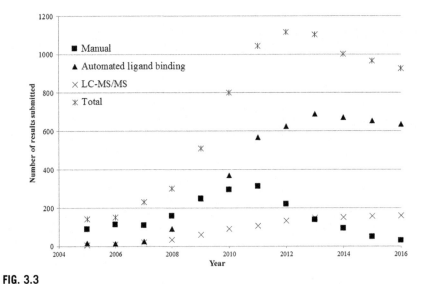

FIG. 3.3

Number of DEQAS participants submitting 25-OHD results (2005–2016).

Table 3.1 Method Timeline for 25-OHD: Year Methods First Appeared in DEQAS and Number of Results Submitted in July 2017

Date of Initial Participation	Method	Number of Returns (July 2017)
Oct 1989	Chromatographic competitive protein-binding assay	0
April 1991	HPLC	18
April 1993	IncStar RIA (until January 1999)	0
July 1999	DiaSorin RIA (formerly IncStar)	0
	IDS RIA	4
July 2001	Nichols Advantage (discontinued in April 2006)	0
Oct 2002	IDS EIA (OCTEIA)	12
April 2004	DiaSorin Liaison	0
Oct 2005	LC-MS/MS	150
Jan 2006	IDS Automated EIA	2
April 2007	DiaSorin Liaison Total	214
Oct 2007	Roche 25-OHD$_3$	0
July 2008	DIASource 25-OHD$_3$ RIA (formerly BioSource)	0
Jan 2009	IDS iSYS automated chemiluminescence immunoassay	63
Jan 2011	Abbott Architect	9
April 2011	Siemens ADVIA Centaur	60
	Roche Total 25-OHD	165
Jan 2012	Diazyme 25-OHD EIA	0
	DiaSource Total 25-OHD RIA	1
	DiaSource Total 25-OHD ELISA	1
July 2012	Euroimmun ELISA	10
Oct 2012	DRG ELISA	3
	Tosoh AIA	3
Jan 2013	Ortho Total 25-OHD	4
	Immunodiagnostik ELISA	0
April 2013	Quidel Microvue 25-OHD	0
July 2013	SNIBE Maglumi 25-OHD	2
Jan 2014	Beckman Access 2 Total 25-OHD	4
	Beckman Unicel Dxi Total 25-OHD	32
April 2014	Diazyme 25-OHD Chemistry Analyzers	3
	Fujirebio Lumipulse G 25-OHD	3
April 2015	bioMerieux 25-OHD	4
	Bio-Rad BioPlex 2200	1
	Qualigen Fstpak IP 25-OHD	1
Oct 2015	DRG Hybrid XL	1
Jan 16	Abbott Architect New Kit	71
July 2016	Human 25-OHD ELISA	1
	Organtec Alegria 25-OHD	2
Jan 17	Roche 25-OHD total II (cobas e801)	8
April 17	IDS iSYS New	6

Chromatographic assays

Gas chromatography-mass spectrometry (GC-MS) enjoyed limited use in the 1970s and 1980s.[54,55] The long columns (typically 15–30 m) offered excellent resolution and the ability to measure several metabolites in a single run. However, because of their low volatility, vitamin D metabolites required derivatization prior to chromatography making the sample preparation more complex; this, combined with lengthy run and column equilibration times, makes them unsuitable for routine clinical use. Nevertheless, these methods were considered the "gold standards" of their time.

High performance liquid chromatography methods for 25-OHD were introduced in the late 1970s.[56,57] Early methods used a UV detection system utilizing the unique cis-triene chromophore of vitamin D metabolites. A perceived advantage of liquid chromatography methods is the ability to separate and quantitate 25-OHD$_3$ and 25-OHD$_2$. Procedural losses were monitored by the addition of an internal standard, usually tritiated 25-OHD$_3$; this assumed that 25-OHD$_3$ and 25-OHD$_2$ behave identically in the extraction and chromatography stages, although this has been questioned.[57] Due to their relative complexity and need for specialist equipment and staff, these methods have never been widely used in clinical laboratories.

High performance liquid chromatography coupled to UV has been largely superseded by liquid chromatography with mass spectrometric detection, usually multiple mass analyzers in tandem (LC-MS/MS). Methods for measuring 25-OHD first appeared in the early 1990s.[58,59] At its simplest, LC-MS/MS involves the extraction of the analyte from the sample matrix and chromatographic retention, typically using reversed phase chromatography columns. Once eluted from the column the vitamin D metabolites are ionized before being selected and isolated in the first mass analyzer. Selected "precursor" ions are bombarded with argon gas in a collision chamber and product ions are further selected and isolated in the final mass analyzer. For 25-OHD$_3$ and 25-OHD$_2$, the selected precursor and product ions can be the simple dehydration products (m/z 401→383 and 413→395 respectively) or other smaller fragment ions, depending on the type of ionization source used and whichever can give the better signal:noise ratio with minimal potential interference. Mass spectrometry offers the advantage of the selection and separate detection of 25-OHD$_3$ and 25-OHD$_2$ although isobars such as 3-epi-25-OHD$_3$ must be chromatographically resolved prior to MS analysis for their presence to be recognized. Failure to chromatographically resolve the 3-epimer is believed to contribute to the positive bias of 25-OHD$_3$ results seen in many routine LC-MS/MS assays. Those laboratories that wish to resolve the 3-epimer typically need to extend their chromatographic run time although there are recently published methods that have reduced this to as little as five minutes.[60]

The sensitivity of LC-MS/MS assays for vitamin D metabolites can be enhanced by derivatization of the molecules using so-called dienophiles which react with the cis-triene structure found in all vitamins D. This technique which can increase sensitivity by 10–100 fold is particularly useful in research assays designed to detect multiple vitamin D metabolites some of which are present in low concentrations.[61]

Liquid chromatography-mass spectrometry is now regarded as the gold standard and three methods[62–64] have been accepted as reference method procedures by the

Joint Committee for Traceability in Laboratory Medicine (JCTLM)[24]. The popularity of LC-MS/MS methods has increased in recent years (Fig 3.3) and is likely to receive a boost when fully automated methods bring them within reach of the nonspecialist laboratory. For the routine clinical laboratory, barriers to adoption of the technique have been compelling; for example, until recently instruments were supplied for research use only, offering no regulatory approval of the system or the assays performed upon it. At least some of these problems should be addressed with the launch of "closed" LC-MS/MS methods such as the recently introduced Cascadion SM Clinical Analyser.[65] The Cascadion system was designed to provide the features expected of a conventional clinical chemistry analytical platform, working from primary sample tubes, minimal or no off-instrument sample preparation, random access for the analysis of urgent samples, and traceability of results to recognized international standards. With assays such as the Cascadion 25-OH vitamin D becoming available and the recent changes in the in vitro Diagnostic guidelines which state that laboratory-developed tests need to be "as good as" cleared assays, the quality of laboratory-developed tests will also inevitably be challenged.

Free 25-OHD

Like other steroid hormones, most 25-OHD is bound to serum proteins and the free fraction (less than 0.1% of total 25-OHD) is thought to be the fraction that enters the cells, although some tissues, particularly the kidney, are able to internalize DBP-bound 25-OHD due to the presence of the endocytic receptors megalin and cubilin.[66,67] Despite the widespread acceptance of total 25-OHD concentrations as the biomarker for vitamin D status, some studies have suggested that measurement of 25-OHD alone may not adequately reflect the relationship between vitamin D exposure and certain health outcomes. In particular, it is difficult to reconcile the observation that, in general, black Americans despite having lower total 25-OHD have good or better bone health than that of the white population.[68] This conundrum was addressed by Powe et al who in a paper published in 2013 provided evidence that DBP was low in African Americans but that the "bioavailable" 25-OHD calculated from the total 25-OHD and DBP concentrations was similar to their white counterparts.[69] However, these results have been challenged on the basis that the monoclonal antibody-based ELISA assay used in the Powe study underestimated the DBP due to the underrecovery of the GcF1 genotype, the most prevalent DBP variant found in African Americans. Other methods using polyclonal antibodies have proved more reliable as they bind a larger number of genetic variants than monoclonal antibodies.[70]

Monoclonal antibodies are used in a novel direct assay for Free 25-OHD (Future Diagnostics); this is a two-step immunoassay done in a microtiter plate coated with a monoclonal antibody to "vitamin D." Firstly, after addition of the sample, free 25-OHD$_3$ and 25-OHD$_2$ is bound to the cell wall without significantly disturbing the in vivo equilibrium between free and bound 25-OHD. The second step is to wash away excess serum and allow a known amount of biotinylated 25-OHD to react with the unoccupied binding sites of the monoclonal antibody. The unbound biotinylated 25-OHD is removed by a second washing and a streptavidin peroxidase conjugate added. After incubation with a chromogenic substrate the reaction is stopped and the absorbance (at 450 nm) is measured using a plate spectrophotometer. The concentration of free 25-OHD is inversely proportional to the absorbance. Hemolyzed and lipemic samples should not be used but the kit insert makes no mention of possible interference from biotin. The Future Diagnostics assay is only suitable for serum and its accuracy and reliability have yet to be independently assessed. The assay has not had FDA approval for use in diagnostic procedures and is stated to be for research purposes only.

The role of free 25-OHD in assessing vitamin D status is likely to remain controversial until assays are properly validated and standardized.

24,25-dihydroxyvitamin D

Until recently hydroxylation of 25-OHD to $24,25(OH)_2D$ was regarded as simply the first step in a five-step degradation process to produce water-soluble excretion products of vitamin D. With the advent of LC-MS/MS, particularly of methods that can simultaneously measure several metabolites,[60,61] the importance of $24,25(OH)_2D$ has been reassessed. Unsurprisingly, there is a strong association between $24,25(OH)_2D_3$ and $25-OHD_3$ and it has been suggested that the ratio of the 2 metabolites might provide a more sensitive index of vitamin D status than 25-OHD alone.[60,61] Further impetus to measuring $24,25(OH)_2D$ came with the realization that mutations of the CYP24A1 gene (24-hydroxylase) leading to reduced catabolism of 25-OHD were an important cause of the hypercalcemia found in Idiopathic Infantile Hypercalcaemia.[60] Research into $24,25(OH)_2D$ has been confined to laboratories using LC-MS/MS, as the metabolite cannot be measured by nonchromatographic immunoassays. However, most antibodies currently used in measuring 25-OHD cross-react with $24,25(OH)_2D$ by at least 100% and there is evidence that the strong correlation between the two metabolites might be the cause of the positive bias of some immunoassays, at least at high concentrations.[71]

3-epi-25-hydroxyvitamin D₃

The clinical relevance of 3-epi-25-hydoxyvitamin D (3-epi-25-OHD) is currently unknown and its importance lies in its potential interference in LC-MS/MS assays for 25-OHD. Because 3-epi-25-OHD has the same mass as 25-OHD, the two metabolites have to be resolved at the chromatographic stage. This can require longer columns which may reduce sample throughput, an important consideration in many clinical laboratories. A study carried out by DEQAS several years ago suggested that many routine LC-MS/MS methods do not separate the two metabolites and that the presence of significant amounts of the 3-epimer may contribute to the persistent positive mean bias of LC-MS/MS methods for 25-OHD.[71] Immunoassay methods for 25-OHD do not detect the 3-epimer, probably because antibodies are raised against an epitope in the side chain which is identical in both molecules.

1,25-dihydroxyvitamin D

As mentioned previously, measurement of 1,25-dihydroxyvitamin D ($1,25(OH)_2D$ has a very limited role in the assessment of vitamin D status. Accurate measurement of 1,25(OH)$_2$D is challenging due to the low concentrations in the circulation (picomolar amounts compared to the nanomolar amounts of 25-OHD). The first methods introduced in the 1970s were laborious radioreceptor assays (RRA) that involved the harvesting of the vitamin D receptor from the intestines of rachitic chicks and the extraction of large volumes of serum.[72] Approximately 160 laboratories currently participate in the DEQAS EQA scheme. Numbers of results submitted have increased following the introduction by DiaSorin of their fully automated method for use on their Liaison instrument.

Functional Markers

At this time there are no functional markers of vitamin D status that show equivalent diagnostic clinical utility to comparable markers such as methylmalonic acid for vitamin B$_{12}$ (Chapter 12) or PIVKA-II for vitamin K (Chapter 5) status assessment. Calcium concentrations become altered in both vitamin D deficiency and overload but are a late marker of these conditions which should trigger confirmatory analysis by measurement of 25-OHD and other markers involved in calcium homeostasis. Calcium concentrations can become altered in many conditions such as primary hyperparathyroidism, malignancy, diuretic use, renal impairment and so cannot be considered a specific marker of vitamin D status. Similarly, bone mineral density measurements and parathyroid hormone concentrations should not be used as frontline tests to investigate vitamin D status; rather they should be used within the mix of calcium and bone markers to fully establish the clinical condition where a range of issues may occur.[73] These markers have been employed more successfully in order to establish reference limits for 25-OHD. Parathyroid hormone has found particular utility in children[74] where there is less data for markers of vitamin D status and clinical endpoints available.

Postanalytical

Most clinical chemistry laboratories will provide reference ranges to help with the interpretation of results. Traditionally, reference ranges were derived from results obtained on a sample of the local population and are specific to the analytical method and the catchment area it serves. Where the data are normally distributed, the reference range is commonly defined by the mean ± 2 standard deviations and will include approximately 95% of the results. This approach is unsatisfactory for the interpretation of 25-OHD results as vitamin D deficiency is widespread; reference ranges so constructed would lead to the underdiagnosis of vitamin D deficiency. Interpretation of results relies on published guidelines such as those produced in 2011 by the Institute of Medicine (IOM)[75] and the Endocrine Society.[76] The guidelines differ in the recommended values for 25-OHD concentrations that should be aimed for after treatment (IOM 50 nmol/L, ES 75 nmol/L), largely because the IOM focused on bone health whereas the Endocrine Society considered the wider implications of vitamin D deficiency. The guidelines were produced before any serious attempt at method standardization. There have been suggestions that where samples still exist, the data from the original studies should be retrospectively corrected using algorithms published by the VDSP. In brief, this involves remeasuring 25-OHD on a representative selection of samples from the original study using a method traceable to the RMPs. Regression equations can be constructed and the constants used to convert all the original values to those that would have been given by a standardized method.[77] Only by retrospective standardization of past results or the future use of standardized assays can reliable guidelines be established.[77] However, this is of little interest to the average clinician who simply wishes to know if his patient needs a vitamin D supplement and what 25-OHD level he should be aiming for. There is a general consensus that a 25-OHD concentration below 50 nmol/L indicates vitamin D insufficiency or deficiency. Given the variability of 25-OHD assays and the relatively low toxicity of vitamin D, it has been suggested that clinicians should aim for a 25-OHD level of 75–100 nmol/L in supplemented patients.[78]

For children there is a lot less evidence available for optimal 25-OHD concentrations; however, the general consensus (with much debate) is that the limits established for adults are applicable to children. There is evidence that children may benefit from doses titrated to higher serum 25-OHD concentrations but this has yet to be firmly established.[74]

Best Practice

Despite the analytical challenges referred to earlier in this chapter, measurement of total 25-OHD (the sum of 25-OHD$_3$ and 25-OHD$_2$) is likely to remain the first line of approach to assessing vitamin D status for the foreseeable future. Tandem mass spectrometry remains the gold standard for measuring 25-OHD in human serum due to their superior accuracy and precision. Tandem mass spectrometry also has the added advantages that 25-OHD$_2$ and 25-OHD$_3$ can be quantified separately where necessary.

The advantage of using an automated method over LC-MS/MS is the capacity for high sample throughput, although progress has been made recently with the introduction of front-end automated sample preparation LC-MS/MS methods.[79] Issues concerning assay standardization are being addressed and the availability of total 25-OHD on most clinical chemistry analyzers makes these assays very attractive to a busy routine laboratory. Measurement of free 25-OHD and 24,25(OH)$_2$D might prove useful in certain clinical situations but are likely to remain the province of specialist laboratories.

Choice of Method

Selection of a suitable method will be influenced by several factors including workload, origin of samples, and skills of available staff. Laboratories receiving a large number of samples from patients supplemented with vitamin D$_2$ will require a method cospecific for 25-OHD$_3$ and 25-OHD$_2$. A laboratory receiving a high proportion of pediatric samples should ensure that the method does not include 3-epi-25-OHD$_3$ in total 25-OHD measurements. All methods should be traceable to the NIST SRMs or the RMPs and should provide long-term consistency of results, independent of kit lot number. Some EQA schemes (e.g., DEQAS) ask participants to provide the lot number of the kit used to analyze its samples and this can provide valuable information on the stability of results over time.

It should be self-evident that instruments and methods need to be chosen on scientific grounds. With restricted budgets, laboratory managers are keen to reduce costs by putting as many assays as possible on existing instruments even when the 25-OHD method may not be fit for purpose. This should be resisted.

Standardization and External Quality Assessment

The introduction of External Quality Assessment schemes undoubtedly played a major role in improving assay performance in clinical laboratories. There are several national and commercial surveys for 25-OHD; DEQAS is the largest specialist international scheme having approximately 800 members in over 50 countries. DEQAS is organized in the UK and was started after international and national surveys demonstrated serious shortcomings in 25-OHD assay performance.[80–82] DEQAS was extended in 1992 to include a scheme for $1,25(OH)_2D$. A pilot scheme for $24,25(OH)_2D$ was started in 2015 and approximately 10 participants are returning results (Table 3.2). Details of how DEQAS operates can be found on its website (www.deqas.org). In brief, five samples of liquid human sera are distributed quarterly and participants are given approximately five weeks to return results. Data are statistically trimmed to produce an All-Laboratory Trimmed Mean (ALTM), SD, and CV.[83] On a small number of results analyzed in 1994, the 25-OHD ALTM was shown to be a good surrogate for the "true" value given by a GC-MS reference method. In 2013 DEQAS in collaboration with NIST and the US Office of Dietary Supplements became an Accuracy-based scheme for 25-OHD, replacing the ALTM performance target with the "true" value assigned by the NIST RMP. For the first time participants and manufacturers could assess the accuracy of results by comparison with those given by an internationally recognized reference method. Currently, there is no RMP for $1,25(OH)_2D$ and EQA results are judged against the Method Mean. An RMP for $24,25(OH)_2D_3$ has been developed[84] and values have been assigned to NIST SRMs.

In 2016 a supplementary 25-OHD scheme was started for participants in the UK and Republic of Ireland; three samples are sent out quarterly in between those of the main distributions but are not assessed. The idea of the supplementary scheme is to reveal assay problems which can be addressed promptly and before analysis of samples sent in the main distribution.

While the main purpose of EQA schemes is to monitor the validity of the results produced by their participants, they are in a unique position to check the performance characteristics of individual methods. For example, DEQAS has investigated the effects of anticoagulants and serum separating gels[37] and the cross-reactivity of other vitamin D metabolites in 25-OHD assays.

The changing pattern of method usage is illustrated in Fig 3.3 and Table 3.1. A marked increase in fully automated assays in 2008 was followed by a sharp decline in

Table 3.2 24,25-Dihydroxyvitamin D Results for 25-OHD Samples 516–520 Distributed in July 2017 (Trimmed Data)

DEQAS Lab No.	Method	Sample 516 24,25OH-D3 nmol/L	Sample 517 24,25OH-D3 nmol/L	Sample 518 24,25OH-D3 nmol/L	Sample 519 24,25OH-D3 nmol/L	Sample 520 24,25OH-D3 nmol/L
52	LC-MS/MS	2.2	4.2	5.7	1.2	5.7
106	LC-MS/MS	3.9	6.7	9.2	2.1	9.8
528	LC-MS/MS	3.4	6.5	9.4	1.7	9.4
1455	LC-MS/MS	4.7	7.6	11.5	2.6	10.7
1479	LC-MS/MS	3.0	6.8	7.6	1.6	9.1
1751	LC-MS/MS	4.5	6.8	7.6	1.6	9.1
1864	LC-MS/MS	2.2	5.5	7.0	1.2	7.4
2123	LC-MS/MS	4.0	6.9	9.3	2.3	9.4
2211	LC-MS/MS	3.4	5.6	8.2	<2.5	7.8
2258	LC-MS/MS	6.7	14.3	18.4	5.3	17.0
Median		3.7	6.7	9.3	1.9	9.4
Mean		3.6	6.5	9.0	1.9	9.5
SD		0.8	0.7	1.4	0.6	1.5
CV%		22.4	10.5	15.7	29.2	15.6
n		8	8	8	8	8

manual methods. There has also been a steady increase in LC-MS/MS methods since 2008. Overall, the total number of participants in the 25-OHD scheme has declined in recent years. Reasons for this are uncertain but some will be due to laboratory closures/amalgamations or rationalizing of specialist assays. To more easily fulfill accreditation requirements, some laboratories may choose schemes where performance is judged against peer-group means rather than the "true" values of an accuracy-based scheme such as DEQAS.

Accuracy and Precision of 25-OHD Assays

Long-term assay performance of 25-OHD$_2$ methods is given in Figs. 3.1 and 3.2 which shows trends in mean distribution bias and imprecision (Figs. 3.4 and 3.5) since NIST target values became available in October 2012. Individual values (Figs. 3.6 and 3.7) represent the mean % bias of the samples in each distribution. Bias of individual samples is calculated from the submitted result (X) and the NIST assigned target value for the sample (TV): % Bias = {(X−TV)/TV)*100}. The shaded areas represent limits of acceptable performance suggested by Stöckl et al.[85] and adopted by the VDSP. For comparison purposes, samples containing 25-OHD$_2$ have been omitted as the Abbott Architect is known to recover only about 80% of 25-OHD$_2$. In general the results are erratic but have shown a recent improvement. In April 2017 five of the six fully automated methods had a bias within the VDSP limits. At 6.7% and 9.3%, respectively, the bias of the HPLC and LC-MS/MS assays were outside the VDSP limits in April 2017.

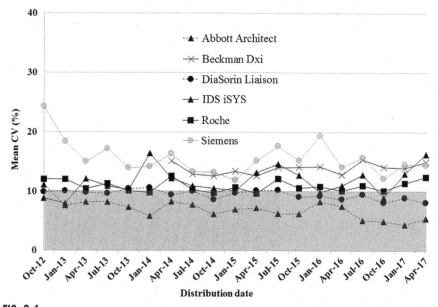

FIG. 3.4

Mean CV% for the major automated ligand-binding assays (October 2012 to April 2017).

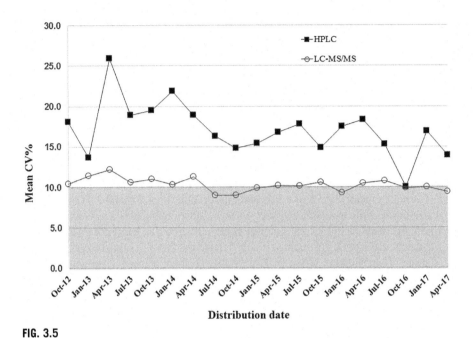

FIG. 3.5

Mean CV% for the chromatographic assays (October 2012 to April 2017).

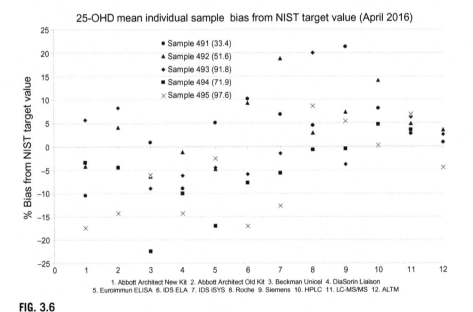

FIG. 3.6

April 2016. Sample 495 contained endogenous 25OHD$_2$, 35.3% of the total 25OHD (NIST target value 97.6 nmol/L).

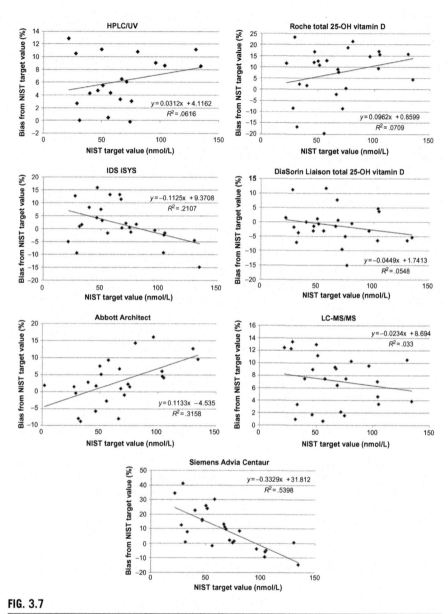

FIG. 3.7

% Bias vs NIST target values (25-OHD samples 496–520).

The problem of using mean bias to assess accuracy is that, by definition, the results give no indication of the sample-to-sample variation in bias which may be concentration dependent, affected by other unknown constituents of the sample (matrix effects) or the presence of other vitamin D metabolites. This is illustrated in Fig 3.6 which gives the method-related bias for each of the samples distributed in April 2016. Sample 495 (symbol x) contained endogenous 25-OHD$_2$ amounting to 35.3% of the total 25-OHD (97.6 nmol/L). Of the ligand-binding assays, both Abbott methods, the DiaSorin Liaison, the IDS EIA and IDS iSYS showed a greater negative bias for this sample than for the other samples in the April 2016 distribution. This was also apparent in the HPLC-UV methods. This suggests (but does not prove) that the methods in use in April 2016 underrecovered 25-OHD$_2$ despite most manufacturers claiming cospecificity for 25-OHD$_3$ and 25-OHD$_2$.

Potential interference of 3-epi-25-OHD and 24,25(OH)$_2$D in 25-OHD assays

Most, if not all, ligand-binding assays show a high cross-reactivity with 24,25(OH)$_2$D (100% or more in some assays) and some routine chromatographic assays fail to resolve the 3-epimer. Since both metabolites are strongly correlated with 25-OHD and concentrations of 24,25(OH)$_2$D can be as high as 10% of that of 25-OHD, bias of ligand-binding assays might be expected to be higher in samples with high concentrations of 25-OHD. The 3-epimer does not cross-react in immunoassays[86] but, since the mass is the same as 25-OHD, LC-MS/MS methods which do not resolve the 3-epimer from 25-OHD are also likely to give spuriously high results in samples with high concentrations of 25-OHD.

DEQAS investigated concentration-dependent bias in the most commonly used methods by plotting data gathered from five distributions (April 2016–July 2017) (Fig. 3.7). Only the Abbott Architect showed evidence of a concentration-dependent increase in bias perhaps due to cross-reactivity of 24,25(OH)$_2$D, known to be high in this method. More surprising is the Siemens Advia Centaur which shows a marked decrease in bias as 25-OHD concentrations increase. The reason for this is obscure but there are likely to be other factors which contribute to a concentration-dependent bias (negative or positive) in immunoassays.

Summary

1. Measurement of total 25-OHD is the best parameter for assessing vitamin D nutrition.
2. Measurement of 1,25(OH)$_2$D should not be used to assess vitamin D status.
3. 25-OHD methods should produce results traceable to a Reference Measurement Procedure and should be cospecific for 25-OHD$_3$ and 25-OHD$_2$. 3-epi-25-OHD should not be included in measurements of total 25-OHD.
4. Fully automated 25-OHD methods are susceptible to interference from other constituents of the sample.
5. Laboratories measuring 25-OHD should participate in an accuracy-based External Quality Assessment scheme.

Appendix
Commonly Used Commercial Automated Assays for Total 25-OHD

The following information on some of the more widely used assays comes largely from the manufacturers' own data.

Abbott

Abbott produces a chemiluminescent immunoassay kit for use on its ArchitectTM instrument. The method uses a monoclonal antibody coated onto paramagnetic particles and an acridinium label. The assay is said to be suitable for serum and most types of plasma, including that from blood collected in separator gel tubes. Calibrators are prepared in a matrix containing human serum and PBS buffer and the assay is standardized against the NIST SRM 2972. Method comparison to assigned reference values (LC-MS/MS) revealed a correlation coefficient of 0.99, a slope of 1.02, and an intercept of -0.99 (manufacturer's data). Cross-reactivity of 25-OHD_2 was assessed using endogenous (nonspiked) samples analyzed by LC-MS/MS. The stated cross-reactivity of 25-OHD_2 is 80.5% at a concentration of 62.4 nmol/L

The cross-reactivity of $24,25(OH)_2D_3$ was 101.9%–189.2% at a 25-OHD concentration of 50 nmol/L and that of 3-epi-OHD_3 1.3% at 250 nmol/L. According to the kit package insert, the Architect assay is susceptible to interference from triglyceride at concentrations >500 mg/dl; a triglyceride concentration of 800 mg/dL resulted in a bias of -10.2% at a 25-OHD concentration of approximately 75 nmol/L.

DiaSorin

The Hollis and Napoli assay was the basis of the first commercial 25-OHD RIA kit, marketed originally by Incstar, and then by the DiaSorin Corporation until 2016. DiaSorin introduced a chemiluminescence assay for use on their LIAISON analyzer in 2004. This competitive binding assay was designed with the same antibody as the radioimmunoassay but no longer included a sample extraction step. The current version of the LIAISON 25-OH Vitamin D Total assay was introduced in 2011. It is said to have good sensitivity and precision, and no human antianimal antibody interference. The manufacturers claim cospecificity for 25-OHD_3 and 25- OHD_2. The "3-epimer" of 25-OHD is stated as having 1.3% cross-reactivity. Standards provided in the kits are prepared from human serum from which endogenous 25-OHD has been removed by a proprietary charcoal stripping method and to which purified 25-OHD_3 has been added. The stock 25-OHD solution is calibrated by UV spectrophotometry and the concentrations of working standards are said to be verified by LC-MS/MS. The lot-to-lot reproducibility of the LIAISON Total methods is monitored by use of a common QC patient panel. In 2015 and subsequent years, the LIAISON Total assay has met the VDSP certification requirements without any need for changes to the assay or its calibration.

Immunodiagnostic Systems (IDS)

A radioimmunoassay and an enzyme immunoassay (EIA) are marketed by IDS but like all manual assays, the number of results submitted to the DEQAS EQA has declined in recent years (Fig. 3.3 and Table 3.1). In the RIA kit, calibrators are prepared by adding 25-OHD$_3$ to charcoal-stripped, lipid-free human serum, although full details are not provided. Values are assigned from a set of primary reference standards prepared from a stock 25-OHD solution calibrated by UV spectrophotometry. In the RIA, 25-OHD$_2$ cross-reacts by only 75%. A fully automated chemiluminescence method was introduced in 2009 for use on the iSYS automated platform. Working calibrators for both EIA and automated kits are prepared in a "human serum buffer matrix." According to the manufacturer the methods are cospecific for 25-OHD$_3$ and 25-OHD$_2$ and are traceable to the NIST Reference Measurement Procedure. Both methods are said to be unaffected by serum-separating gel or the anticoagulants heparin and EDTA. There was no significant interference from triglyceride up to a level of 500 mg/dL.

Roche

Roche is one of the few manufacturers to market a CPB method for 25-OHD and launched a new version in 2017 for use on their Cobas e analyzers. The new assay uses a ruthenium-labeled DBP as the capture component and a biotinylated 25-OHD derivative to form the signal complex; it is directly standardized against the ID-LC-MS/MS RMP of the University of Ghent. Whereas the older version (Elecsys Vitamin D total) had a high cross-reactivity with 24,25(OH)$_2$D, the package insert for the new assay states a cross-reactivity of only 13.7% at 24,25(OH)$_2$D concentrations of 240 nmol/L. This is achieved by the use of a specific monoclonal blocking antibody.

Siemens Healthineers

Siemens Healthineers offer 25-OHD reagent kits for use on their ADVIA Centaur and Dimension EXL (LOCI Module) instruments. The ADVIA Centaur method uses an antifluorescein monoclonal antibody covalently bound to paramagnetic particles, an anti-25-OHD monoclonal antibody labeled with an acridinium ester, and a vitamin D analog labeled with fluorescein. The method is standardized using internal standards traceable to the NIST standard reference materials (SRM 2972). The cross-reactivity of 25-OHD$_2$ is given as 104% and 3-epi-25-OHD$_3$ as 1.1% at a level of 250 nmol/L. No information is given for 24,25(OH)$_2$D. The method is stated to be suitable for serum, EDTA, and heparin plasma. There is less than a 10% change in results at a triglyceride concentration of 540 mg/dL.

The Dimension EXL LOCI method is a competitive chemiluminescent assay using a biotinylated monoclonal capture antibody and a streptavidin-based detection system. Internal standards are traceable to the NIST SRM2972. The cross-reactivity of 25-OHD$_2$ and 24,25(OH)$_2$D$_2$ is given as 89% and approximately 2%, respectively The cross-reactivity of 3-epi-25-OHD$_3$ is 2.5% at a concentration of 130 nmol/L. There is said to be less than a 10% change in results at a triglyceride concentration of 686 mg/dL.

References

1. International Union of Pure and Applied Chemistry and International Union of Biochemistry and Molecular Biology. *IUPAC-IUB joint commission on biochemical nomenclature: Nomenclature of vitamin D recommendations.* http://www.chem.qmul.ac.uk/iupac/misc/D.html; 1981. (Accessed August 5, 2003).
2. Hollis BW. Measuring 25-hydroxyvitamin D in a clinical environment: challenges and needs. *Am J Clin Nutr* 2008;**88**:507–10.
3. Hollander D, Muralidhara KS, Zimmerman A. Vitamin D-3 intestinal absorption in vivo: influence of fatty acids, bile salts, and perfusate pH on absorption. *Gut* 1978;**19**:267–72.
4. Vieth R. Vitamin D supplementation, 25-hydroxyvitamin D concentrations, and safety. *Am J Clin Nutr* 1999;**69**:842–56.
5. Trang NM, Cole DEC, Rubin LA, Pierratos A, Siu S, Vieth R. Evidence that vitamin D3 increases serum 25-hydroxyvitamin D more efficiently than does vitamin D2. *Am J Clin Nutr* 1998;**68**:854–8.
6. Armas LAG, Hollis BW, Heaney RP. Vitamin D2 is much less effective than vitamin D3 in humans. *J Clin Endocrinol Metab* 2004;**89**:5387–91.
7. Houghton LA, Vieth R. The case against ergocalciferol (vitamin D) as a vitamin supplement. *Am J Clin Nutr* 2006;**84**:694–7.
8. Holick MF, Biancuzzo RM, Chen TC, Klein EK, Young A, Bibuld D, et al. Vitamin D2 is as effective as Vitamin D3 in maintaining circulating concentrations of 25-hydroxyvitamin D. *J Clin Endocrinol Metab* 2008;**93**:677–81.
9. Heaney RP, Armas LAG, Shary JR, Bell NH, Binkley N, Hollis BW. 25-Hydroxylation of vitamin D: relation to circulating vitamin D under various input conditions. *Am J Clin Nutr* 2008;**87**:1738–42.
10. Cannell JJ, Hollis BW. Use of vitamin D in clinical practice. *Altern Med Rev* 2008;**13**:619.
11. Wooten AM. Improving the measurement of 25-hydroxyvitamin D. *Clin Biochem Rev* 2005;**26**:33–6.
12. Lips P. The relative value of 25(OH)D and 1,25(OH) D measurements. *Endocr Abstr* 2009;**20**:11.
13. Singh RJ, Taylor RL, Reddy S, Grebe SKG. C-3 epimers can account for a significant proportion of total circulating 25- hydroxyvitamin D in infants, complicating accurate measurement and interpretation of vitamin D status. *J Clin Endocrinol Metab* 2006;**91**:3055–61.
14. Calvo MS, Whiting SJ, Barton CN. Vitamin D intake: a global perspective of current status. *J Nutr* 2005;**135**:310–6.
15. Holick MF. Vitamin D deficiency. *N Engl J Med* 2007;**357**:266–81.
16. O'Riorden JLH, Bijvoet OLM. Rickets before the discovery of vitamin D. *BoneKEy Rep* 2014;**478**:1–6.
17. Reid IR, Bolland MJ, Grey A. Effects of vitamin D supplements on bone mineral density: a systematic review and meta-analysis. *Lancet* 2014;**383**:146–55.
18. Bischoff-Ferrari HA. Optimal serum 25-hydroxyvitamin D levels for multiple health outcomes. *Adv Exp Med Biol* 2014;**810**:500–25.
19. Reid IR, Bolland MJ, Grey A. Effects of vitamin D supplements on bone mineral density: a systematic review and meta-analysis. *Lancet* 2014;**383**:146–55.
20. Scientific Advisory Committee on Nutrition (SACN). *Vitamin D and heath.* https://www.gov.uk/government/groups/scientific-advisory-committee-on-nutrition.
21. Vieth R. Vitamin D supplementation, 25-hydroxyvitamin D concentrations, and safety. *Am J Clin Nutr* 1999;**69**:842–56.

22. Smollin C, Srisansanee WJ. Vitamin D toxicity in an infant: case files of the University of California, San Francisco medical toxicology fellowship. *Med Toxicol* 2014;**10**:190–3.

23. Ross AC, Manson JE, Abrams SA, Aloia JF, Brannon PM, Clinton SK, et al. The 2011 report on dietary reference intakes for calcium and vitamin D from the Institute of Medicine: what clinicians need to know. *J Clin Endocrinol Metab* 2011;**96**:53–8.

24. Jones RD, Jackson C. The Joint Committee for Traceability in Laboratory Medicine (JCTLM) – its history and operation. *Clin Chim Acta* 2016;**453**:86–94.

25. Binkley N, Sempos CT. Standardizing vitamin D assays: the way forward. *J Bone Miner Res* 2014;**29**:1709–14.

26. https://www-s.nist.gov/srmors/view_detail.cfm?srm=972A; (Accessed November 8, 2017).

27. Bland JM, Altman DG. Statistical method for assessing agreement between two methods of clinical measurement. *Lancet* 1986;**1**:307–10.

28. Carter GD. 25-hydroxyvitamin D: a difficult analyte. *Clin Chem* 2012;**58**:486–8.

29. Carter GD, Berry J, Durazo-Arvizu R, Gunter E, Jones G, Jones J, et al. Quality assessment of vitamin D metabolite assays used by clinical and research laboratories. *J Steroid Biochem Mol Biol* 2017;**173**:100–4.

30. Eyles D, Anderson C, Ko P, Jones A, Thomas A, Burne T, et al. A sensitive LC/MS/MS assay of 25OH vitamin D3 and 25OH vitamin D2 in dried blood spots. *Clin Chim Acta* 2009;**403**:145–51.

31. Makowski AJ, Rathmacher JA, Horst RL, Sempos CT. Simplified 25-hydroxyvitamin D standardization and optimization in dried blood spots by LC-MS/MS. *J AOAC Int* 2017;**100**:1328–36.

32. Kvaskoffa D, Koa P, Similaa HA, Eyles DW. Distribution of 25-hydroxyvitamin D in dried blood spots and implications for its quantitation by tandem mass spectrometry. *J Chromatogr B Anal Technol Biomed Life Sci* 2012;**901**:47–52.

33. Heath AK, Williamson EJ, Ebeling PR, Kvaskoff D, Eyles DW, English DR. Measurements of 25-hydroxyvitamin D concentrations in archived dried blood spots are reliable and accurately reflect those in plasma. *J Clin Endocrinol Metab* 2014;**99**:3319–24.

34. Lee S, Oncescu V, Mancuso M, Mehta S, Erickson D. A smartphone platform for the quantification of vitamin D levels. *Lab Chip* 2014;**14**:1437–42.

35. Fairney A, Saphier PW. Studies on the measurement of 25-hydroxy vitamin D in human saliva. *Br J Nutr* 1987;**57**:13–25.

36. Ferry JD, Collins S, Sykes E. Effect of serum volume and time of exposure to gel barrier tubes on results for progesterone by Roche Diagnostics Elecsys 2010. *Clin Chem* 1999;**45**:1574–5.

37. Vitamin D external quality assessment scheme (DEQAS). *25- OHD report, July 2009 distribution*. London, UK: Charing Cross Hospital; 2009.

38. Bowron B, Powers V, Berresford P, Knox S, Thomas P. Ion suppression in an LC-MS/MS assay for 25-OH vitamin D caused by sample collection tubes: a solution to the problem. *Ann Clin Biochem* 2009;**46**:44.

39. Lissner D, Mason RS, Posen S. Stability of vitamin D metabolites in human blood serum and plasma. *Clin Chem* 1981;**27**:773–4.

40. Wielders JPM, Wijnberg FA. Preanalytical stability of 25(OH)-Vitamin D3 in human blood or serum at room temperature: solid as a rock. *Clin Chem* 2009;**55**:1584–5.

41. Bodnar LM, Catov JM, Wisnerand KL, Klebanoff MA. Racial and seasonal differences in 25-hydroxyvitamin D detected in maternal sera frozen for over 40 years. *Br J Nutr* 2009;**101**:278–84.

42. Antoniucci DM, Black DM, Sellmeyer DE. Serum 25-hydroxyvitamin D is unaffected by multiple freeze-thaw cycles. *Clin Chem* 2004;**51**:258–60.
43. Hyppönen E, Turner S, Cumberland P, Power C, Gibb I. Serum 25-hydroxyvitamin D measurement in a large population survey with statistical harmonization of assay variation to an international standard. *J Clin Endocrinol Metab* 2007;**92**:4615–22.
44. Vitamin D external quality assessment scheme (DEQAS). *1,25- (OH)2D report, July 2017 distribution*. London, UK: Charing Cross Hospital; 2017.
45. Haddad JG, Chyu KJ. Competitive protein-binding radioassay for 25-hydroxycholecalciferol. *J Clin Endocrinol Metab* 1971;**33**:554–7.
46. Dorantes LM, Arnaud SB, Arnaud CD, Kilgust KA. Importance of the isolation of 25-hydroxyvitamin D before assay. *J Lab Clin Med* 1978;**91**:791–6.
47. Carter GD, Jones JC, Berry JL. The anomalous behaviour of exogenous 25-hydroxyvitamin D in competitive binding assays. *J Steroid Biochem Mol Biol* 2007;**103**:480–2.
48. Horst R. Exogenous versus endogenous recovery of 25-hydroxyvitamins D2 and D3 in human samples using high-performance liquid chromatography and the DiaSorin LIAISON Total-D assay. *J Steroid Biochem Mol Biol* 2010;**121**:180–2.
49. Heijbower AC, Blankenstein MA, Kema IP, Buijs MM. Accuracy of 6 routine 25-hydroxyvitamin D assays: influence of vitamin D binding protein concentration. *Clin Chem* 2012;**58**:543–8.
50. Cavalier E, Carlisi A, Bekaert AC, Rousselle O, Chapelle JP. Human anti-animal interference in DiaSorin Liaison total 25(OH)-vitamin D assay: towards the end of a strange story? *Clin Chim Acta* 2012;**413**:527–8.
51. Belsey RE, DeLuca HF, Potts JT. A rapid assay for 25-OH-vitamin D without preparative chromatography. *J Clin Endocrinol Metab* 1974;**38**:1046–51.
52. Hollis BW, Napoli JL. Improved radioimmunoassay for vitamin D and its use in assessing vitamin D status. *Clin Chem* 1985;**31**:1815–9.
53. Hollis BW, Kamerud JQ, Selvaag SR, Lorenz JD, Napoli JL. Determination of vitamin D status by radioimmunoassay with an ^{125}I-labelled tracer. *Clin Chem* 1993;**39**:529–53.
54. De Leenheer AP, Cruyl AA. Vitamin D in plasma: quantitation by mass fragmentography. *Anal Biochem* 1978;**91**:293–303.
55. Coldwell RD, Porteous CE, Trafford DJ, Makin HL. Gas chromatography-mass spectrometry and the measurement of vitamin D metabolites in human serum or plasma. *Steroids* 1987;**49**:155–96.
56. Gilbertson TJ, Stryd RP. High-performance liquid chromatographic assay for 25-hydroxyvitamin D3 in serum. *Clin Chem* 1977;**23**:1700–4.
57. Stryd RP, Gilbertson TJ. Some problems in development of a high-performance liquid chromatographic assay to measure 25-hydroxyvitamin D_2 and 25-hydroxyvitamin D_3 simultaneously in human serum. *Clin Chem* 1978;**24**:927–30.
58. Watson D, Setchell KDR, Ross R. Analysis of vitamin D and its metabolites using thermospray liquid chromatography/mass spectrometry. *Biomed Chromatogr* 1991;**5**:153–60.
59. Vreeken RJ, Honing M, van Baar BLM, Ghijsen RT, de Jong GJ, Brinkman UA. On-line post-column Diels-Alder derivatization for the determination of vitamin D_3 and its metabolites by liquid chromatography/thermospray mass spectrometry. *Biol Mass Spectrom* 1993;**22**:621–32.
60. Kaufmann M, Gallagher JC, Peacock M, Schlingmann KP, Konrad M, DeLuca HF, et al. Clinical utility of simultaneous quantitation of 25-hydroxyvitamin D and 24,25-dihydroxyvitamin D by LC-MS/MS involving derivatization with DMEQ-TAD. *J Clin Endocrinol Metab* 2014;**99**:2567–74.

61. Jones G, Kaufmann M. Vitamin D metabolite profiling using liquid chromatography–tandem mass spectrometry (LC–MS/MS). *J Steroid Biochem Mol Biol* 2016;**164**:110–4.

62. Tai SS, Bedner M, Phinney KW. Development of a candidate reference measurement procedure for the determination of 25-hydroxyvitamin D3 and 25-hydroxyvitamin D2 in human serum using isotope-dilution liquid chromatography-tandem mass spectrometry. *Anal Chem* 2010;**82**:1942–8.

63. Stepman HC, Vanderroost A, van Uytfanghe K, Thienpont LM. Candidate reference measurement procedures for serum 25-hydroxyvitamin D3 and 25-hydroxyvitamin D by using isotope-dilution liquid chromatography-tandem mass spectrometry. *Clin Chem* 2011;**57**:441–8.

64. Mineva EM, Schleicher RL, Chaudhary-Webb M, Maw KL, Botelho JC, Vesper HW, et al. A candidate reference measurement procedure for quantifying serum concentrations of 25-hydroxyvitamin D_3 and 25-hydroxyvitamin D_2 using isotope-dilution liquid chromatography-tandem mass spectrometry. *Anal Bioanal Chem* 2015;**407**:5615–24.

65. http://www.thermofisher.com/cascadion; (Accessed March 12, 2018).

66. Marzolo MP, Farfan P. New insights into the roles of megalin/LRP2 and the regulation of its functional expression. *Biol Res* 2011;**44**:89–105.

67. Christensen EI, Birn H. Megalin and cubilin: multifunctional endocytic receptors. *Nat Rev Mol Cell Biol* 2002;**3**:258–68.

68. Barrett JA, Baron JA, Karagas MR, Beach ML. Fracture risk in the U.S. medicare population. *J Clin Epidemiol* 1999;**52**:243–9.

69. Powe CE, Evans MK, Wenger J, Zonderman AB, Berg AH, Nalls M, et al. Vitamin D-binding protein and vitamin D status of black Americans and white Americans. *N Engl J Med* 2013;**369**:1991–2000.

70. Nielson CM, Jones KS, Chun RF, Jacobs J, Wang Y, Hewison M, et al. Free 25-hydroxyvitamin D: impact of vitamin D binding protein assays on racial-genotypic associations. *J Clin Endocrinol Metab* 2016;**101**:2226–34.

71. Cashman KD, Hayes A, Galvin K, Merkel J, Jones G, Kaufmann M, et al. Significance of serum 24,25-dihydroxyvitamin D in the assessment of vitamin D status: a double-edged sword? *Clin Chem* 2015;**61**:636–45.

72. Hollis BW. Detection of vitamin D and its major metabolites. In: Feldman D, Glorieux FH, Pike JW, editors. *Vitamin D*. New York, NY: Academic Press; 1997.

73. Seamans KM, Cashman KD. Existing and potentially novel functional markers of vitamin D status: a systematic review. *Am J Clin Nutr* 2009;**89**:1997–2008.

74. Maguire JL, Birken C, Thorpe KE, Sochett EB, Parkin PC. Parathyroid hormone as a functional indicator of vitamin D sufficiency in children. *JAMA Pediatr* 2014;**168**:383–5.

75. Institute of Medicine (US). *Committee to review dietary reference intakes for vitamin D and calcium. Dietary intakes for calcium and vitamin D*. Washington, DC: National Acadamies Press; 2011.

76. Holick MF, Binkley NC, Bischoff-Ferrari HA, Gordon CM, Hanley DA, Heaney RP, et al. Evaluation, treatment, prevention of vitamin D deficiency: an Endocrine Society clinical practice guideline. *J Clin Endocrinol Metab* 2011;**96**:1911–30.

77. Sempos CT, Durazo-Arvizu RA, Binkley N, Jones J, Merkel JM, Carter GD. Developing vitamin D dietary guidelines and the lack of 25-hydroxyvitamin D assay standardization: the ever-present past. *J Steroid Biochem Mol Biol* 2016;**164**:115–9.

78. Binkley N, Carter GD. Toward clarity in clinical vitamin D status assessment: 25(OH)D Assay Standardization. *Endocrinol Metab Clin N Am* 2017;**46**:885–99.

79. Geib T, Meier F, Schorr P, Lammert F, Stokes CS, Volmer DA. A simple micro-extraction plate assay for automated LC-MS/MS analysis of human serum 25-hydroxyvitamin D levels. *J Mass Spectrom* 2015;**50**:275–9.
80. Jongen MJM, Van Ginkel FC, Vander Vijgh WJF, Kuiper S, Netelenbos JC, Lips P. An international comparison of Vitamin D metabolite measurements. *Clin Chem* 1984;**30**:399–403.
81. Mayer E, Schmidt-Gayk H. Inter-laboratory comparison of 25-hydroxyvitamin D determination. *Clin Chem* 1984;**30**:1199–204.
82. Carter GD, Short F. 25- hydroxyvitamin D assays: results of a national quality assessment. *J Endocrinol* 1988;**117**(suppl). Abstract 25.
83. Healy MJ. Outliers in clinical chemistry quality-control scheme. *Clin Chem* 1979;**25**:675–7.
84. Tai SS-C, Nelson MA. Candidate reference measurement procedure for the determination of (24R), 25-dihydroxyvitamin D3 in human serum using isotope- dilution liquid chromatography-tandem mass spectrometry. *Anal Chem* 2015;**87**:7964–70.
85. Stöckl D, Sluss PM, Thienpont LM. Specifications for trueness and precision of a reference measurement system for serum/plasma 25- hydroxyvitamin D analysis. *Clin Chim Acta* 2009;**408**:8–13.
86. Carter GD, Jones JC, Shannon J, Williams EL, Jones G, Kaufmann M, et al. 25-Hydroxyvitamin D assays: potential interference from other circulating vitamin D metabolites. *J Steroid Biochem Mol Biol* 2016;**164**:134–8.

Further Reading

87. Cavalier E, Lukas P, Crine Y, Peeters S, Carlisi A, Le Goff C, Gadisseur R, Delanaye P, Souberbielle J-C. Evaluation of automated immunoassays for 25(OH)-vitamin D determination in different critical populations before and after standardization of the assays. *Clin Chim Acta* 2014;**431**:60–5.
88. Binkley N, Dawson-Hughes B, Durazo-Arvizu R, Thamm M, Tian L, Merkel JM, et al. Vitamin D measurement standardization: the way out of the chaos. *J Steroid Biochem Mol Biol* 2017;**173**:117–21.

Methods for assessment of Vitamin E

4

Scott W. Leonard*, **Maret G. Traber***,†

*Linus Pauling Institute, Oregon State University, Corvallis, OR, United States, †College of Public Health and Human Sciences, Oregon State University, Corvallis, OR, United States

Chapter Outline

Laboratory Assessment of Vitamin Status. https://doi.org/10.1016/B978-0-12-813050-6.00004-8

Introduction

The purpose of this chapter is to give an overview of the structure and function of vitamin E, then to focus on the best methods for its measurement for a variety of biological samples. Historically, vitamin E was considered difficult to measure because of its instability and its lipophilicity. These challenges have largely been overcome, and with modern technologies it is possible to detect femtomole (10^{-15} mol) quantities with accuracy and precision.

Structure and Function

Structure

Vitamin E is a collective name for molecules that exhibit the antioxidant activity of α-tocopherol. Vitamin E occurs naturally in eight different forms: four tocopherols and four tocotrienols, which have similar chromanol structures: trimethyl (α-), dimethyl (β- or γ-), and monomethyl (δ-) (Fig. 4.1). The phytyl side chain in the tocopherols is saturated and the side chain in the equivalent tocotrienols is unsaturated.

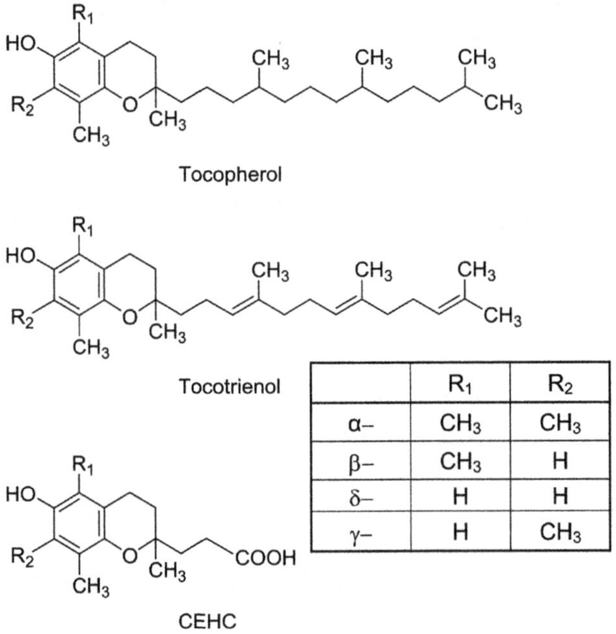

	R_1	R_2
α–	CH_3	CH_3
β–	CH_3	H
δ–	H	H
γ–	H	CH_3

FIG. 4.1

Structures of tocopherols, tocotrienols, and their respective carboxyethyl hydroxy chromanols.

Vitamin E supplements often contain chemically synthesized α-tocopherol or esters of α-tocopherol. Chemically synthesized α-tocopherol contains eight stereo-isomers, arising from the three chiral centers (2, 4′, and 8′) and is designated all racemic (*all-rac*)-α-tocopherol. The naturally occurring and most biologically active form, *RRR*-α-tocopherol represents only one of these eight stereoisomers. The other stereoisomers exhibit lower biological activity than *RRR*-α-tocopherol, with the 2*S*-forms generally having half the activity than the 2*R*-forms.[1] Ester forms such as α-tocopheryl acetate or succinate prevent oxidation and prolong shelf life, but are hydrolyzed in the gut and absorbed in the unesterified form.[2]

Function

Vitamin E functions in vivo as a chain-breaking antioxidant, preventing the propagation of free radical damage caused by peroxyl radicals.[3] Vitamin E especially protects polyunsaturated fatty acids (PUFA), such as docosahexaenoic acid,[4] within phospholipids of biological membranes and in plasma lipoproteins. Peroxyl radicals react 1000 times faster with vitamin E than with PUFA.[5] During the reaction, vitamin E is oxidized to a tocopheroxyl radical which if left unchecked can initiate tocopherol-mediated peroxidation, for example, of cholesterol esters in low density lipoprotein (LDL).[6] In vivo it is likely that the tocopheroxyl radical is reduced by hydrogen donors such as ascorbate (vitamin C). This phenomenon has led to the idea of vitamin E recycling, in which the antioxidant function of oxidized vitamin E is continuously restored by other antioxidants.[7]

Of the four tocopherols and four tocotrienols found in food, only α-tocopherol meets human vitamin E requirements.[8] Despite the fact that all of the vitamin E forms have similar antioxidant activities (within an order of magnitude), non-α-tocopherols are poorly recognized by the hepatic α-tocopherol transfer protein (α-TTP).[9] α-TTP is responsible for maintaining plasma α-tocopherol concentrations.[10] Hosomi and coworkers demonstrated that the relative affinities of various forms of vitamin E for α-TTP (calculated from the degree of competition with *RRR*-α-tocopherol) were as follows: *RRR*-α-tocopherol, 100%; β-tocopherol, 38%; γ-tocopherol, 9%; δ-tocopherol, 2%; and α-tocotrienol, 12%.[9] These investigators concluded that the affinity of α-TTP for the different vitamin E analogs is one of the critical determinants for plasma concentrations and, thus the biological activity of vitamin E.[11]

Catabolism of the various forms of vitamin E also appears to play a critical role in hepatic regulation of vitamer concentrations. The non-α-tocopherol forms of vitamin E are catabolized preferentially.[12, 13] The catabolites of vitamin E are the CEHC (2′-carboxyethyl-6-hydroxychroman) products of the respective forms of vitamin E, i.e., α-, β-, γ-, and δ-tocopherols (Fig. 4.1). Tocotrienols are catabolized to the same end products produced from tocopherol catabolism.[14] The various vitamin E forms are initially ω-oxidized by cytochrome P450 4F2 (CYP 4F2),[13] then following β-oxidation, CEHCs are sulfated or glucuronidated,[15–17] similar to other xenobiotics, and excreted in urine[3] or bile.[18]

Urinary vitamin E catabolites have been proposed as a biomarker of vitamin E status[19–21] but have not been widely studied as of yet.

Clinical Application of Measurements

Vitamin E has been said to be "a vitamin looking for a disease."[22] Fat-soluble vitamins, such as A, D, and K, have specific deficiency symptoms. Examples of their metabolic roles include vision, bone formation, and blood clotting, respectively, that are well established. Further, vitamins A and D are necessary ligands of nuclear receptors. Thus it was anticipated that vitamin E, as α-tocopherol, would have a nonantioxidant function, given its higher biologic activity compared with other vitamin E forms. However, in >90 years of investigation, this has not proven to be the case.

Adequate vitamin E intake is protective against (1) cardiovascular disease in certain patient subgroups,[23] such as those on hemodialysis[24] or in diabetic individuals; (2) persons with cognitive impairment[25] and neurological disease, such as Alzheimer's[26]; and (3) progression of nonalcoholic steatohepatitis.[27] Importantly, α-tocopherol has a significant role in the prevention of miscarriage in humans.[28] Vitamin E's role in maintaining health is not surprising given that the major function of vitamin E is that of an antioxidant and oxidative damage to lipids has been implicated in each of these disease conditions.

It is notoriously difficult to show adverse consequences of vitamin E deficiency in humans. Anemia and decreased erythrocyte survival[29] were widely accepted as signs of vitamin E deficiency until neurological abnormalities that responded to vitamin E supplementation were discovered.[30–32] Severe vitamin E deficiency in humans occurs as a result of genetic defects in the α-TTP,[33] causing the disorder ataxia with vitamin E deficiency (AVED).[34] The lack of functional α-TTP results in the rapid depletion of plasma α-tocopherol.[35] There are other genetic defects that can cause fat malabsorption such as those that impair microsomal triglyceride protein and apolipoprotein B function,[36] leading to vitamin E deficiency. Progressive severe deficiency in humans manifests as muscle deterioration, including the heart, ultimately resulting in death.[37]

Vitamin E adequacy during the first 1000 days of life is necessary for subsequent adult health and well-being.[38] Deficiency is seldom found in adults but is more frequently seen in children, likely because of limited α-tocopherol stores that are unable to support rapid growth, thereby allowing deficiency symptoms to be more readily apparent. A susceptible target tissue frequently affected by α-tocopherol deficiency is the developing nervous system. Neurologic abnormalities associated with vitamin E deficiency were detected and reversed with vitamin E supplementation in malnourished children in India.[39, 40] Additionally, low plasma α-tocopherol concentrations were associated with poorer cognitive function in children with cystic fibrosis, a group at high risk for malnutrition.[41] Conversely, improved cognitive function was shown in a group of children in China born to mothers with higher blood and cord serum vitamin E.[42]

Measuring Vitamin E Status in Humans

Circulating α-tocopherol concentrations can be difficult to interpret. In adults consuming a healthy diet plasma α-tocopherol concentrations average ~20 μmol/L, whereas individuals consuming supplements have higher concentrations of ~30 μmol/L or more.[19] The Institute Of Medicine defined <12 μmol/L to be the deficient/inadequate range for healthy adult.[8] Below this concentration there is an association with increased infection, anemia, stunting of growth, and poor outcomes during pregnancy for both the infant and the mother.[38] It is thus important to measure circulating vitamin E in at-risk populations and those showing signs of deficiency.

Assessment of plasma α-tocopherol concentrations can be complicated because α-tocopherol is transported in plasma lipoproteins and concentrations of cholesterol and lipoproteins, as well as α-tocopherol, increase with age.[43] As certain conditions such as malnutrition and infectious diseases can lower circulating cholesterol and its lipoprotein carriers, correction of plasma α-tocopherol for lipids is not always appropriate. A suggested best practice approach is to report both lipid standardized and uncorrected values.[44]

Historically, vitamin E status was assessed using hemolysis of the erythrocyte by hydrogen peroxide as a biological end point.[45] This methodology was developed because of the general consensus that vitamin E was important for erythrocyte membrane stability and the observation that anemia was present in animals fed a vitamin E-deficient diet. The challenge of measuring vitamin E in erythrocytes is caused by their high iron contents. Therefore another technique developed by Krishnamurthy and Bieri[46] involved measurements of ^{14}C-tocopherol distribution in rat erythrocytes following administration of dietary radioactive α-tocopherol. This indirect measurement was used because no satisfactory method for measuring tocopherol in erythrocytes was available.

More direct measurement techniques such as spectrophotometry[47,48] and spectrofluorometery[49] were developed to measure tocopherol, as was gas–liquid chromatography,[50] but it was not until the late 1970s that HPLC methods with various detection systems were developed for plasma and tissue tocopherol analysis.[51–55] Prior to 1980, there were only 20 papers published for measurements of vitamin E in biological samples using HPLC (PubMed). Over the next decade HPLC methodologies for measurement of molecules, such as fat-soluble vitamins, improved and showed increased sensitivity and selectivity. Publications for vitamin E analysis in biological samples using HPLC as cited in PubMed, reached approximately 200 papers in the 1980s, and since 1990, have reached well over 2000 papers. GC with flame ionization detection is still used for vitamin E measurements, but the GC equipment is now more likely to be coupled to a mass spectrometer detector.[56] The most popular methods for vitamin E analysis use HPLC for separation and then depending on sensitivity needed, various detection methods are available.

Sample Types and Sample Stability

Vitamin E is ubiquitous in human and animal tissues, and in the different blood components. Some representative reports are cited herein detailing concentrations in the different components of the blood[52,57] with plasma or serum being the most commonly analyzed, and in different tissues,[58–60] as well as in tissues from various experimental animals.[61–64] Vitamin E has also been measured in human feces.[65]

In the blood and tissues of normal healthy individuals, both α- and γ-tocopherol can readily be detected, as these are the major forms found in the body, with γ-tocopherol measured in subjects from the United States at roughly 10% of α-tocopherol.[52] Urine does not contain any vitamin E. As discussed under "Structure and Function," section the different forms of vitamin E are catabolized to the water-soluble CEHCs for excretion following excess intake.[3] Vitamin E catabolites have also been reported in the circulation,[16,66] the liver,[67] and feces[68–70] but urine is the best medium for analysis as the catabolites are concentrated in the urine for excretion, although there are reports of fecal vitamin E catabolites.[70]

Sufficient care should be taken during sample collection to prevent sample degradation and possible oxidation. α-Tocopherol is stable in whole blood stored overnight at 4°C,[71,72] and under different heat and light conditions during short-term storage,[73] with significant losses occurring in γ-tocopherol concentrations. Vitamin E catabolites are also stable following overnight storage at 4°C.[72] Tocopherols have been shown to be stable in plasma stored frozen at $\leq -70\,°C$ for 6 months to a year,[72] and summarized in Greaves et al.[74] Under ideal conditions, blood samples should be processed immediately following collection, and plasma and/or red cells stored at $\leq -70°C$ until analysis.

The information in the literature on tissue stability and vitamin E is largely available with regards to meat stability and improved quality during storage with increases in dietary vitamin E. Therefore, due to limited information, the best practice for tissue handling would be to follow procedures similar to the recommendations for blood.[72]

Preanalytical Sample Selection
Blood
Sample selection
Vitamin E can be measured accurately in plasma, serum, or red cells. Vitamin E catabolite concentrations have been reported in serum and plasma, but erythrocyte concentrations have not been reported.

Fasting
Both plasma α-tocopherol concentrations and α-tocopherol/lipid ratios increase after a meal and potentially with fasting, likely from α-tocopherol mobilization from tissue.[75] Since nutritional adequacy evaluations for plasma vitamin E should be reported both adjusted and unadjusted for cholesterol or total lipids,[44] there is no recommendation for or against collecting fasting blood samples for analysis.[74]

Anticoagulants
Plasma and erythrocytes are obtained from anticoagulated blood samples. Blood can be collected into tubes containing anticoagulant (EDTA, Na- or Li-heparin,

or citrate),[72,76] or plain tubes for serum collection.[76] Tubes can be either glass or plastic, but validation of tube types to guard against preanalytical sample contamination is strongly recommended.

Erythrocytes can be washed with EDTA-saline, then concentrations reported per packed cell volume, or cholesterol (measured separately in the same aliquot). Separation of blood components is performed using centrifugation as discussed.[52] Sample aliquots should be stored frozen in cryogenic vials at $\leq -70°C$ and processed within 6–12 months.

Urine

Samples can be collected as spot collections or 24 h collections as the vitamin E catabolite concentrations will need to be adjusted for creatinine.[77] No urine preservatives are necessary as the urine is slightly acidic and is high in the antioxidant, urate; moreover, preservatives may interfere with analysis. Samples should be kept cold (to discourage bacterial growth) and processed as soon as possible, generally within 24 h.

It is recommended that multiple aliquots of a 24 h collection sample are frozen as small aliquots are easier to thaw. Note that sample volumes are critical to measure (e.g., total sample, plus aliquot volumes). Samples should only be thawed and used once. Sample aliquots should be stored $\leq -70°C$ and processed within 6–12 months.

Tissue

Tissue samples collected for either vitamin E or catabolite analysis need to be frozen for storage. Small sections of the tissue should be placed into a cryogenic vial and flash frozen in liquid nitrogen. If it is deemed necessary, animal tissues can by washed using transcardial perfusion techniques prior to tissue excision. Tissue samples should be stored at $\leq -70°C$ and processed within 6 months.

Sample Preparation and Extraction

Frozen sample aliquots of plasma, serum or RBCs should be thawed at room temperature and mixed gently by hand inversion to ensure homogeneity prior to aliquoting and extraction. Tissue samples should be kept frozen and small pieces (~50–100 mg) cut with a blade and then weighed prior to analysis. Smaller tissue sections can be used for vitamin E analysis with adjustments to standard curves and injection volumes (details later). As catabolite concentrations are ~100–1000-fold lower than tocopherol concentrations, using less tissue for catabolite measurements will necessitate use of more sensitive detection methods, such as mass spectrometry[77] (see later for further details). Samples do not need to be protected from light during the extraction process, although caution is advised to avoid long delays during sample processing, especially if antioxidants are not used during extraction.

Blood and Tissue Vitamin E

Vitamin E can be successfully extracted from blood components (plasma, RBCs, etc.) and tissue using liquid–liquid (LL) partitioning techniques. Two different methods for preparing samples for extraction are most commonly used: (1) alcoholic KOH saponification or (2) detergents.

Saponification (method one)

Saponification with alcoholic potassium hydroxide is the first technique.[52] With this approach esterified fatty acids are released (from, e.g., triglycerides, phospholipids, and cholesteryl esters) and in the case of fortified food,[60] esterified vitamin E is cleaved to the free phenol form. The same protocol is followed for tissue samples of ~50 mg weight, but carefully weighed to allow expression per mg in the denominator. Using this method tissue samples do not need to be homogenized.

For white cell analysis a larger sample is required, such as a 10 mL blood sample.[52] Food can also be extracted for tocopherols using the saponification method by multiplying all of the assay reagents by a factor of five and starting with 1 g of food.[60]

Although the saponifying agent is usually potassium hydroxide,[60] sodium hydroxide can also be used.[78] Samples (0.1 mL thawed plasma, 0.5 mL thawed RBCs, or ~50 mg partially frozen tissue) are placed into 1 mL deionized water in 10 mL glass screwcap tubes and 2 mL ethanol (containing 1% ascorbic acid as an antioxidant) are added. Samples are vortexed briefly prior to the addition of 0.3 mL saturated (22 M) potassium hydroxide and then incubated in a water bath at 70°C for 30 min. Note: the sample must be completely dissolved, if not, a longer incubation time is required. Extreme cases, such as adipose tissue, may require increasing the amount of saponification reagents used, but be sure to maintain the indicated ratio of components.

Following incubation, samples are cooled on ice for 10 min prior to the addition of 1 mL water, containing 1% ascorbic acid (added to protect the vitamin E[79]), and 2 mL hexane (or heptane). Samples are mixed by hand inversion for 1 min and then the aqueous and organic layers are either allowed to separate on the benchtop for 10 min, or placed in a benchtop centrifuge at $200 \times g$ for 1 min at 4°C. (Caution is advised—at increased centrifuge speeds the weight of the screwcap lid will crush the glass tubes.) A known volume of the organic layer is drawn off using a glass pipette and transferred to a smaller glass tube, evaporated to dryness under a stream of N_2 gas and redissolved in 0.1–0.3 mL of methanol. Sample tubes should be vortexed for 10 s and then the methanol containing the dissolved sample residue transferred to low-volume HPLC injection vials. Glass vials are recommended but plastic can also be used if validated.

Detergents (method two)

The second method that can be used is very similar to the first, but with the addition of 0.1 M sodium dodecylsulfate (SDS) in place of the first water addition,[80] and with the deletion of the saponification step. The detergent solution causes the protein to precipitate and solubilizes the lipids. This protocol requires that tissue samples are homogenized prior to extraction. The volume of ethanol added should be equal to the combined aqueous volume. At this point the samples are mixed with hexane (or heptane) as described before and extracted similarly. An antioxidant such as pyrogallol or ascorbic acid can be added prior to extraction.[57,60,81]

Merits of methods one and two

The two sample preparation assays have several aspects in common, such as, protecting the samples during extraction with an antioxidant, protein precipitation with alcohol, and phase separation with hexane (or heptane). A lipophilic antioxidant such as butylated hydroxytoluene (BHT) can also be added prior to the hexane extraction step (such as 20 μL of a 5 M BHT solution) to protect the vitamin E in the organic phase.[62] BHT acts synergistically with tocopherol, working as a coantioxidant to regenerate vitamin E.[82] If samples contain fat, such as in food, saponification should be used. The SDS method is quicker for plasma/serum, but cannot be used for all sample types and tissue homogenization will need to be performed prior to extraction. SDS is not recommended for samples that will be analyzed by mass spectrometry.

Vitamin E Catabolites

The different isoforms of vitamin E, including the tocopherols and tocotrienols (α, β, δ, and γ for each), are metabolized to their respective CEHCs in the liver and transported through the circulation for excretion in the urine. For a more comprehensive discussion on the metabolism of vitamin E, refer to various reviews.[3,83,84] Vitamin E catabolism produces both long-chain catabolites (LCMs) and shorter chain end products (CEHCs and CMBHCs). Although recent reports[85–87] on the possible biological activities of LCMs are exciting, this section will focus on measurement of the end-product CEHCs as they are still believed to be the main catabolites.[56] The catabolites are excreted in the urine as conjugates and are most commonly measured following either acid or enzymatic deconjugation. The catabolites have been reported to be conjugated mainly to glucuronide or sulfate[3] and thus early studies utilized β-glucuronidase (with sulfatase activity) to study the catabolites.[12,88] More recently Li et al.[89] have reported on the benefits of acid hydrolysis for catabolite extraction, suggesting it to be a superior procedure as acid hydrolysis is more complete than enzyme hydrolysis. The author's laboratory has tested both protocols and still find enzyme hydrolysis as discussed in Lodge et al.[88] to be effective but care needs to be used in enzyme purchase and usage.[90] Acid hydrolysis also produces a lower background when using mass spectrometry (MS) (discussed later) and is a more cost-effective method. For tissue analysis, the only method tested to date is enzyme hydrolysis so that protocol will be discussed later. In this section, we will give detailed information on using HCl hydrolysis for catabolite extraction.

Acid hydrolysis of CEHC conjugates from urine

Urine aliquots are thawed at room temperature. To a 10 mL screwcap tube, add 1 mL urine and 1 mL 6 N HCl. Vortex briefly. Add 0.5 mL ascorbic acid solution (114 mM, 2% ascorbic acid) to protect the catabolites from oxidation during hydrolysis and then incubate for 60 min at 60°C in a hot water bath. Cool for 10 min on ice prior to extraction. At this point an internal standard should be added. Internal standard is not added earlier because it is subject to losses not incurred by the conjugated-CEHCs (unpublished observations, Traber and Leonard). Trolox (6-hydroxy-2,5,7,8-tetramethylchroman-2-carboxylic acid), a commercially available water-soluble derivative of vitamin E, is suitable for this application.[88] We suggest that the added amount be comparable to a mid-range CEHC amount that would be expected for the measured analytes. A suggested protocol is to add 0.02 mL 6 μM trolox (dissolved in ethanol, precision in the amount added is needed at this step). Samples are then extracted with 4 mL diethyl ether. After adding the ether, samples are mixed gently by hand inversion. Samples should not be shaken vigorously, as foaming will occur. Allow samples to separate on the benchtop for 10 min or place in a benchtop centrifuge at $200 \times g$ for 1 min at 4°C. Caution is advised—at increased centrifuge speeds the weight of the screwcap lid will crush the glass tubes. A known volume of the organic layer is drawn off using a pipet-aid and glass pipette and transferred to a smaller glass tube, evaporated to dryness under a stream of N_2 gas and redissolved in 0.1–0.3 mL of methanol:water (1:1). Sample tubes should be vortexed for a minimum of 10 s and then the solvent transferred to low-volume HPLC injection vials.

Standards for both α- and γ-CEHC catabolites are commercially available (Cayman Chemical, https://www.caymanchem.com/). External standard curves are prepared in the acceptable range for the detector chosen (see discussion later). Sample dilution during the assay can also be adjusted according to the detector chosen by substituting a portion of the urine sample volume with deionized water. For example, 0.2 mL of urine sample and 0.8 mL water.

Plasma, tissue, and fecal vitamin E catabolites

While consuming a nonsupplemented diet, plasma vitamin E catabolite concentrations are very low (~1000-fold lower than tocopherols),[77] with α-CEHC representing approximately 10% of the total CEHC.[16, 66] Supplementation has been shown to increase plasma catabolite concentrations,[16, 66, 91] and certain diseases and disease states, such as metabolic syndrome, may lower plasma CEHC concentrations.[21] While the plasma is not as concentrated as the urine with the vitamin E catabolites, it is still a valuable medium for measurement and important data can be attained. The methods for extraction are similar to methods used for urinary catabolite measurements, as detailed before, but with an increased need for sample concentration during extraction. Using more sensitive detectors, such as a mass spectrometer, is also helpful.

There are a few reports on the vitamin E catabolites in rat liver[15, 67, 92] and kidney,[92] but further work needs to be done looking at tissue distribution of the catabolites. Kiyose et al. developed a method which determines both unconjugated and conjugated CEHC in various tissues by making methyl esters of the CEHC (CEHC-Me).[92]

Extraction of tissue short chain catabolites

Here we will describe a method to extract the short chain catabolites following enzyme hydrolysis.[67] Tissue samples (50–100 mg) are placed into a 10 mL screwcap tube containing 2 mL of PBS with 0.5% EDTA and 0.1 mL ascorbic acid solution (57 mM, 1%). Samples are homogenized with a Polytron homogenizer (Brinkmann, Westbury, NY) using a 7 mm foam reducing generator. Once the samples are homogenized the trolox internal standard, as described before, and 0.1 mL β-glucuronidase (10 mg of β-glucuronidase (300,000 U/g β-glucuronidase activity and 10,000 U/g sulfatase activity) pH 6.8, in 0.1 mL of 10 mM acetate buffer) are added. Samples are incubated at 37°C for 1 h. Following incubation, samples are cooled on ice for 10 min and acidified by the addition of 0.01 mL of 12 N HCl (concentrated HCl). Samples are extracted with ether as detailed before under urine extraction. When extracting from tissue, following sample resuspension it is recommended that the samples be centrifuged and aliquots taken to remove any precipitates that may form. This method was developed for liver tissue but should work for any tissue.[77]

As mentioned in the previous section, there is interest in the measurement of LCMs. Jiang's group has performed a very detailed analysis of both conjugated and unconjugated forms of the LCMs in plasma and liver of rats following supplementation with γ-tocopherol.[15] The extraction protocol is similar to the method discussed before for tissue but with the deletion of the enzyme step and ethyl acetate used for extraction. LCM analysis has also been detailed by Galli's group using human plasma to look at the production of the first two catabolites produced during vitamin E catabolism following supplementation.[56, 91, 93] For these analyses, the enzyme method reported before for tissue extraction can be used with hexane/tert-butyl methyl ether (2/1, *v/v*) used for extraction. Short-chain[70] and long-chain[68] catabolites have also been measured in fecal collections from mice gavaged with either CEHC or γ- and δ-tocopherol, respectively. For these studies, fecal samples were homogenized in methanol, centrifuged at high speed to pellet the precipitate, and methanol extracts injected directly into a mass spectrometer. If investigative work is performed looking at LCMs a larger sample size should be used as the concentrations are very low, especially in the plasma.

Measures to be Used as Outcome Denominators

Plasma and erythrocyte lipid analysis

Plasma lipids should be measured for accurate vitamin E assessment.[44, 94, 95] As discussed under "Clinical Application of Measurement," section vitamin E is transported in plasma lipoproteins; therefore, concentrations of cholesterol and lipoproteins, as well as α-tocopherol increase with age.[43] Certain conditions, such as malnutrition and infectious diseases can also alter circulating lipid concentrations, thus it is recommended that both lipid adjusted and unadjusted vitamin E concentrations are reported.[44] Plasma cholesterol and triglycerides can be measured by standard clinic kits (Thermo Scientific).

Using saponification, the RBC or serum sample also contains unesterified cholesterol that can be measured using standard techniques and used as a denominator for reporting results.

Urinary creatinine

Urinary creatinine can be measured by spectrophotometry in spot-collected samples or 24 h collections. The authors of this chapter have found that a fairly consistent level of urinary CEHC is excreted during both day and night, so 24 h urine collections may not be necessary.[21] As the excretion rate of creatinine is known to be fairly constant, creatinine can be measured and used for standardization purposes. Standard clinical kits are available (Cayman Chemical and ThermoFisher).

Analytical (Vitamin E & Carboxyethyl-6-Hydroxychromanol) Chromatographic Instruments

Measurement of tocopherols by GC preceded that of HPLC but over the last 25 years the use of HPLC methods has become more common for vitamin E measurement. Silylating reagents such as N-methyl-N-trimethyl-silyltrifluoroacetamide (MSTFA) or N,O-(bis-trimethylsilyl) trifluoroacetamide (BSTFA)-trimethylchlorosilane (TMCS) (10:1 v/v) are used to derivatize tocopherols and CEHCs in biological sample extracts for GC separation, as described.[12, 58, 96, 97] Utilizing HPLC techniques, however, sample derivatization is not required.

Comparison of gas and liquid chromatography methods

Different detectors can be coupled to GC and will be discussed later. GC techniques may have slightly lower recoveries and limit of detection compared to LC[56] and a small amount of oxidation or sample loss may occur during the derivatization step.[98] Both LC and GC methods show nearly the same accuracy and precision.[98] Samples prepared for HPLC analysis are not derivatized, so are not susceptible to water hydrolysis and can be injected from the same vial multiple times, if necessary. Gas chromatography is still the main option for volatile compounds but for vitamin E and the CEHC catabolites, either LC or GC can be used. For a more in-depth discussion on use of GC for vitamin E determination, refer to several reviews.[99–101]

Gas chromatography methods

Tocopherols and tocotrienols are low molecular weight compounds with hydrophobic properties which allow them to be analyzed with GC techniques. Gas chromatography column technology has evolved from hand-packed glass columns to capillary columns. Capillary columns provide higher resolution increasing the separation and accuracy of the vitamin E isomers. Lipophilic compounds such as tocopherols and tocotrienols and their derivatives require the use of nonpolar chromatographic stationary phases. Most techniques employ fused silica with bonded dimethyl and phenyl groups. With these techniques satisfactory separation of the different isomers is achieved. As determined by their boiling points, the elution order is as follows: δ-tocopherol, β-tocopherol, γ-tocopherol, δ-tocotrienol, β-tocotrienol, γ-tocotrienol, α-tocopherol, and α-tocotrienol.[100] For a good review of the different columns that can be used for the determination of vitamin E by GC, refer to Bartosinska et al.[100]

High- and ultra-performance liquid chromatography

High performance liquid chromatography equipment can be equipped with multiple pumps and both isocratic and gradient solvent delivery methods can be used. An autosampler can be attached, with cooling ability if desired, allowing the user to prepare large batches of samples for overnight injection. If an autosampler is used, we recommend that the sample temperature be set between 5°C and 10°C. We also recommend that the column be kept above room temperature, such as 40°C in a column oven, to avoid temperature swings, as changes in temperature can have an effect on peak retention.

Newer LC technology has been developed allowing the user improved resolution, decreased run times, and increases in sensitivity.[102] The newer UPLC techniques with decreased run times allow for increased throughput and decreased solvent usage.

Columns and Mobile Phases

Normal phase vs reversed-phase separations

Both NP- and RP-HPLC are used for vitamin E analysis.[101, 103] NP methods are able to separate all eight isoforms of vitamin E but require the use of hexane and other more volatile solvents as mobile phases. In NP-HPLC the vitamin E forms are separated by adsorption according to the number of methyl groups on their chromanol rings. Generally, the more polar isoforms are retained longer on NP columns (retention: $\alpha < \beta \leq \gamma < \delta$).[103] These techniques are used when it is necessary to quantitate all vitamin E forms such as in the analysis of some foods and animal feeds. Since human and animal tissues and most foods do not contain all eight isoforms,[104] separation of all isoforms is not always necessary. For a comparison of different NP columns, see Kamal-Eldi et al.[103]

Reversed phase methods are most commonly used to measure the main forms of vitamin E (α- and γ-tocopherol) found in biological fluids and tissues. Reversed phase HPLC offers some practical advantages such as faster mobile phase equilibration, chromatographic reproducibility, and compatibility with electrochemical and highly sensitive MS detection techniques.[101]

Types of columns

Using RP-HPLC, baseline separation of tocotrienols and tocopherols (retention: $\delta < \gamma \leq \beta < \alpha$) (Fig. 4.2) can be achieved using traditional octadecylsilica (ODS; C18) columns and acetonitrile- or methanol-water-based mobile phases, except for β- and γ-isomers that are not resolved or are poorly resolved depending on mobile phase and column chosen.[101] A non-C18 stationary phase capable of resolving the γ−/β-tocopherol pair is pentafluorophenyl silica.[105] Although more expensive, chiral

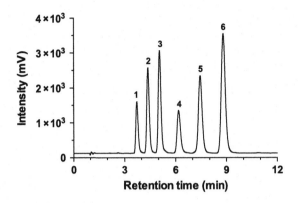

FIG. 4.2

Reversed-phase HPLC-ECD detection of tocopherols and tocotrienols in a standard mixture. Approximately 200–400 pmol injected of each: δ-tocotrienol (1), γ-tocotrienol (2), α-tocotrienol (3), δ-tocopherol (4), γ-tocopherol (5), and α-tocopherol (6).

chemistry columns can also be used to separate both β- and γ-isomers,[106] as well as the eight stereoisomers,[107] that are formed during the chemical preparation of synthetic vitamin E (all-*rac*-α-T).[36]

Typical column lengths used include 100, 150, and 250 mm. For analysis of samples which include both tocopherols and tocotrienols, longer columns should be used to achieve baseline separation for all compounds, but these will require longer run times. For more information on RP column and mobile phase choices, see Abidi.[101] Typical column core sizes range between 3 and 5 μm, with a smaller particle size of 2.1 μm used for MS detection, specifically electrospray ionization techniques. The smaller the core particle size the greater the system back pressure.

For more recently developed UPLC techniques, sub two micron core-shell columns have been developed producing good separation and resolution for both tocopherols[102] (Fig. 4.3) and CEHCs (Fig. 4.4) (Leonard and Traber unpublished results) with run times of half to one-third that of HPLC methodologies. When choosing an HPLC column, whether for NP or RP techniques, there are many acceptable columns for vitamin E and CEHCs to choose from.

Detectors

There are four main detectors used for vitamin E detection coupled to LC: UV absorbance, fluorescence (FL), electrochemical detection (ECD), and MS, with the level of sensitivity increasing in that order.[101] FL detection has been the most widely used for NP-HPLC, but for RP-HPLC, detector choice is more likely dependent on availability and study design. Because of their oxidative potential, tocol-derived antioxidants can easily be analyzed by HPLC-ECD which is more affordable than MS detection systems and more sensitive and selective than FL and UV. Techniques where UV is preferred is the simultaneous determination of vitamin E and cholesterol,[108] and

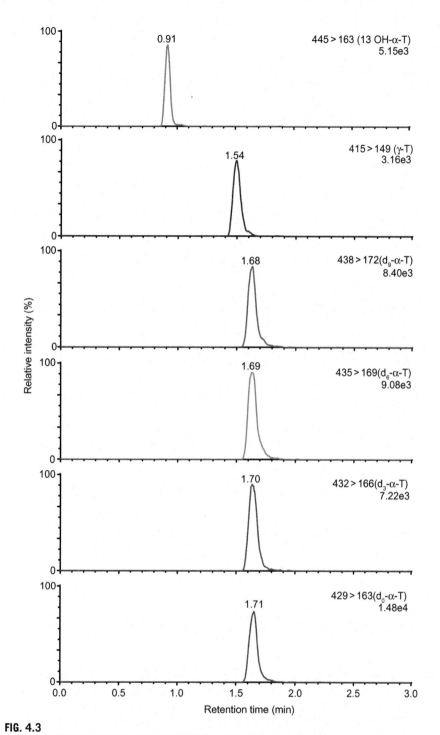

FIG. 4.3

Reversed-phase UPLC APCI(−) MS/MS detection of deuterium labeled (d) and unlabeled tocopherols in a standard mixture. Approximately 10–20 pmol injected of each: 13′-OH-α-tocopherol (13 OH-α-T), γ-tocopherol (γ-T), d_9-α-tocopherol (d_9-α-T), d_6-α-tocopherol (d_6-α-T), d_3-α-tocopherol (d_3-α-T), and d_0-α-tocopherol (d_0-α-T).

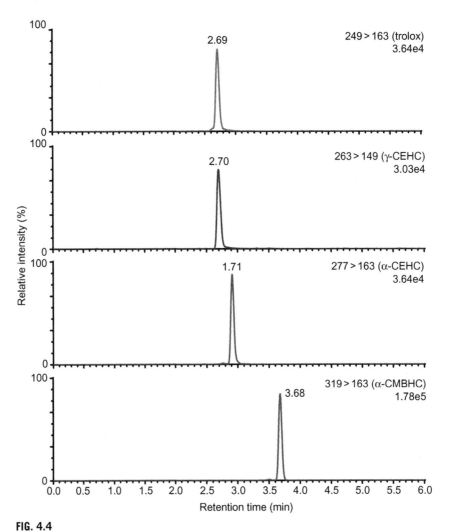

FIG. 4.4

Reversed-phase UPLC ESI(−) MS/MS detection of CEHCs in a standard mixture. Approximately 2–3 pmol injected for each. Detector response is greater for CMBHC.

when there is a need to determine esterified derivatives such as tocopheryl acetate[109] and succinate.[110] ECD can be used for vitamin E[60] and CEHC[88] detection, as well as related molecules found in biological samples such as nitrated γ-tocopherol,[111] and the oxidized quinone derivatives that are poorly fluorescent and present at concentrations too low to be assessed by UV.[56]

Any of the detectors mentioned before can be used for human plasma vitamin E analysis, unless metabolic studies using deuterium-labeled compounds are used. The measurement of deuterated tocopherols, and CEHCs, requires the use of MS techniques.

Synthesis of deuterium-labeled tocopherols[58] and CEHCs[112] has been very useful in obtaining important information regarding vitamin E status, requirements, and metabolism. Both GC[20, 97, 113] and LC coupled to MS[66, 114] and MSMS[12,115] techniques have been developed. Quadrupole instruments are most commonly used but a time-of-flight detector can also be used.[116] For a review of GC- and LCMS approaches for determining tocopherols and tocotrienols in biological samples, refer to Bartosinska et al.[100]

Ionization sources for MS include APCI, ESI, and APPI. APCI-MS is suitable for analysis of nearly neutral molecules of small to moderately large size.[117] While ESI is especially suited to charged or readily charged molecules. APCI is most frequently used in ionization techniques for vitamin E,[100] with negative mode shown to be superior to positive mode APCI for tocopherols[105] (see Fig. 4.3 for an example of APCI MS/MS). The addition of acidic modifiers to the mobile phase in positive or negative mode is not required,[100] and may have an ion suppressing effect (data not shown). ESI techniques are more suited for the vitamin E catabolites (CEHCs) due to the more polar nature of the molecules[21, 66, 68] (see Fig. 4.4 for a chromatographic example of ESI MSMS).

In terms of selectivity and sensitivity, MS detection is the most powerful method of detection for vitamin E and the vitamin E status biomarker, CEHC, in biological samples. Interest in MS techniques for vitamin E and other small molecule detection will continue to increase as MS detectors become more affordable and instrument footprints continue to shrink, requiring less laboratory space. But for now, as mentioned before, any of the detectors discussed can be used for detection of vitamin E in biological samples. Detector choice will depend mostly on instrumentation availability and the design of the study.

Interpretation of Results

In plasma samples collected from adults consuming a healthy diet, plasma α-tocopherol concentrations average ~20 μmol/L, whereas individuals consuming supplements may have double to triple this concentration.[19] The IOM has defined <12 μmol/L to be the deficient/inadequate range for healthy adults.[8] Although γ-tocopherol is the main form of vitamin E in the western diet, plasma concentrations are typically <10% of total tocopherols[52] and tocotrienols even lower.[118] A suggested best practice approach would be to report plasma vitamin E concentrations both corrected and uncorrected for lipids.[44] Lipid corrected normal values for plasma α-tocopherol are >0.8 mmol/mol total lipid (cholesterol plus triglycerides) or >2.8 mmol/mol cholesterol.[36]

Urine concentrations for the CEHC catabolites are reported per creatinine (Table 4.1).[77] Normal concentrations are 0.5–3.0 μmol/g creatinine. Following supplementation concentrations can increase to 10–30 μmol/g creatinine. Plasma α-CEHC concentrations are normally low, 10–30 nmol/L, and have been shown to increase up to 100–200 nmol/L with supplementation. Plasma γ-CEHC concentrations are normally 20–200 nmol/L and have been shown to increase up to 1000 nmol/L with supplementation.[77]

Table 4.1 Reference Value Range for Different Biological Samples

Biological Sample	Metabolite	Normal	Supplemented
Plasma (μmoL/L)	α-Tocopherol	12–30[a]	30–40[b]
Plasma (μmoL/L)	γ-Tocopherol	1–4[c]	4–15[d]
Urine (μmoL/g creatinine)[e]	α-CEHC	0.5–1	2–20
Urine (μmoL/g creatinine)[e]	γ-CEHC	0.5–3	10–30
Plasma (nmoL/L)[e]	α-CEHC	10–30	100–200
Plasma (nmoL/L)[e]	γ-CEHC	20–200	100–1000

[a]Ref. 120.
[b]Ref. 19.
[c]Ref. 121.
[d]Ref. 122.
[e]Ref. 77.

Antagonists and Interfering Compounds

A main antagonist for vitamin E analytical methodologies is the extraction of non-volatile or large molecules, such as high molecular weight proteins and lipid compounds, contaminating the chromatographic system and/or causing detection issues. One such concern is the extraction of plasma phospholipids.[98] Sample cleanup utilizing either LL or SPE techniques for isolation of the lipid fraction from the biological matrix is recommended. Adding a saponification step to the isolation process can help eliminate the macromolecule interferences of hydrophobic nature; however, it is important to prevent artifactual oxidative degradation by the addition of antioxidants such as ascorbic acid.

For CEHC analysis, it is important to ensure complete hydrolysis of the urinary conjugates. As discussed by Li et al.,[89] catabolite conjugate cleavage by hydrochloric acid is a very efficient and simple approach and sample preparations are compatible with different detection systems, including mass spectrometry. Urinary salts are known to cause ionization suppression when using MS techniques and care should be taken during sample cleanup to minimize salt in the injection vials by the addition of LL or SPE steps. LCMS ionization efficiency has also been shown to decrease depending on what source and what mobile phase modifiers are used.[105] For this reason, when performing vitamin E analysis using LCMS with an APCI source in negative mode, mobile phase modifiers such as acetic and formic acid should not be used.

Standardization

Method validation should be performed prior to establishing new laboratory vitamin E methodology. Methods should be tested for analyte recovery, linearity range, limit of quantitation, and accuracy and precision of known quality control material, such as NIST SRM 1950 Catabolites in Human Plasma, or NIST 968e Fat Soluble

Table 4.2 Physicochemical Data of Tocopherols[a], Tocotrienols[a], and CEHCs

Substance	MW	λ_{max} (nm)	E1%, 1 cm[b]	ε[b]
α-Tocopherol	430.7	292	75.8	3270
β-Tocopherol	416.7	296	89.4	3730
δ-Tocopherol	402.7	298	91.4	3810
γ-Tocopherol	416.7	298	87.3	3520
α-Tocotrienol	424.7	292	91.0	3870
β-Tocotrienol	410.7	295	87.5	3600
δ-Tocotrienol	396.7	292	103.0	4230
γ-Tocotrienol	410.7	298	83.0	3300
α-CEHC	278.7	292	120.2	3350
γ-CEHC	264.7	296	149.2	3950
Trolox	250.7	291	130.2	3260

[a]Modified from Podda et al.[60]
[b]Extinction coefficients are given for ethanol solutions.

Vitamins, Carotenoids, and Cholesterol in Human Serum. Laboratory standards should be prepared and determined spectrophotometrically (Table 4.2).[60] All forms of tocopherol and tocotrienol standards are readily available, as are some vitamin E catabolite standards (α-, γ-,δ-CEHC and α-CMBHC) (Cayman Chemical, Ann Arbor, MI). Large batches of commercially available quality control material, such as plasma from Valley Biomedical (Winchester, VA), are available for purchase, and can be analyzed routinely to create an in-house reference sample. And, laboratory groups should participate in External Quality Assurance Programs, such as the RCPAQAP (Royal College of Pathologists of Australia Quality Assurance Program or the DSQAP (Dietary Supplement Quality Assurance Program) in collaboration with the National Institutes of Health (NIH) Office of Dietary Supplements (ODS) or UK NEQAS Clinical Chemistry. Laboratory results should be reported according to the International System of Units.[119] For plasma vitamin E reporting, results are reported in μmol/L or mmol/mol lipid. There are no established reporting methods for CEHCs but for urine concentrations the most common reporting of results is in μmol/g creatinine.[77]

Conclusions

All forms of vitamin E can be measured in biological samples by GC or LC techniques; however, because of the need for sample derivatization, decreased sensitivity, and instrument availability, LC methodologies are far more commonly used. Several different LC detectors (UV, FL, ECD, and MS), and many different columns can be used for analyte selection, but the most sensitive and selective detection systems include ECD and MS detectors utilizing reversed phase chromatography. Similar sample extraction methods can be used for both LC and GC methodologies, with benefits obtained by incorporating a saponification step and artifactual oxidation prevention by the addition of an antioxidant such as ascorbic acid. Similarly, urinary biomarkers of vitamin E, the CEHC catabolites, are isolated from sample matrices using liquid-liquid extraction and can be chromatographed with either GC or LC instruments, with a preference for LC over GC, utilizing ECD or MS detectors. Plasma vitamin E concentrations should be reported both standardized to lipid concentrations as well as unadjusted, and urinary CHECs should be standardized to creatinine.

Laboratory assessment techniques for vitamin E have improved significantly in the last 40–50 years, with major improvements in sensitivity and selectivity. These, improvements, along with the addition of the newer vitamin E catabolite measurements discussed, will continue to improve our understanding of not only the status of vitamin E but also disease risk.

References

1. Weiser H, Vecchi M, Schlachter M. Stereoisomers of alpha-tocopheryl acetate. IV. USP units and alpha-tocopherol equivalents of all-rac-, 2-ambo- and RRR-alpha-tocopherol evaluated by simultaneous determination of resorption-gestation, myopathy and liver storage capacity in rats. *Int J Vitam Nutr Res* 1986;**56**:45–56.
2. Cheeseman KH, Holley AE, Kelly FJ, Wasil M, Hughes L, Burton G. Biokinetics in humans of RRR-alpha-tocopherol: the free phenol, acetate ester, and succinate ester forms of vitamin E. *Free Radic Biol Med* 1995;**19**:591–8.
3. Brigelius-Flohe R, Traber MG. Vitamin E: function and metabolism. *FASEB J* 1999;**13**:1145–55.
4. Choi J, Leonard SW, Kasper K, McDougall M, Stevens JF, Tanguay RL, et al. Novel function of vitamin E in regulation of zebrafish (Danio rerio) brain lysophospholipids discovered using lipidomics. *J Lipid Res* 2015;**56**:1182–90.
5. Kamal-Eldin A, Appelqvist LA. The chemistry and antioxidant properties of tocopherols and tocotrienols. *Lipids* 1996;**31**:671–701.
6. Upston JM, Neuzil J, Witting PK, Alleva R, Stocker R. Oxidation of free fatty acids in low density lipoprotein by 15-lipoxygenase stimulates nonenzymic, alpha-tocopherol-mediated peroxidation of cholesteryl esters. *J Biol Chem* 1997;**272**:30067–74.
7. Traber MG. Vitamin E. In: Shils ME, Shike M, Ross AC, Cabalerro B, Cousins RJ, editors. *Modern nutrition in health and disease.* 10th ed. Baltimore: Williams & Wilkins; 2006. p. 396–411.
8. Food and Nutrition Board and Institute of Medicine. *Dietary reference intakes for vitamin C, vitamin E, selenium, and carotenoids.* Washington, DC: National Academy Press; 2000529.
9. Hosomi A, Arita M, Sato Y, Kiyose C, Ueda T, Igarashi O, et al. Affinity for alpha-tocopherol transfer protein as a determinant of the biological activities of vitamin E analogs. *FEBS Lett* 1997;**409**:105–8.
10. Traber MG, Sokol RJ, Burton GW, Ingold KU, Papas AM, Huffaker JE, et al. Impaired ability of patients with familial isolated vitamin E deficiency to incorporate alpha-tocopherol into lipoproteins secreted by the liver. *J Clin Invest* 1990;**85**:397–407.
11. Traber MG, Manor D. Vitamin E. *Adv Nutr* 2012;**3**:330–1.
12. Swanson JE, Ben RN, Burton GW, Parker RS. Urinary excretion of 2,7, 8-trimethyl-2-(beta-carboxyethyl)-6-hydroxychroman is a major route of elimination of gamma-tocopherol in humans. *J Lipid Res* 1999;**40**:665–71.

13. Sontag TJ, Parker RS. Cytochrome P450 omega-hydroxylase pathway of tocopherol catabolism. Novel mechanism of regulation of vitamin E status. *J Biol Chem* 2002;**277**:25290–6.
14. Lodge JK, Ridlington J, Leonard S, Vaule H, Traber MG. Alpha- and gamma-tocotrienols are metabolized to carboxyethyl-hydroxychroman derivatives and excreted in human urine. *Lipids* 2001;**36**:43–8.
15. Jiang Q, Freiser H, Wood KV, Yin X. Identification and quantitation of novel vitamin E metabolites, sulfated long-chain carboxychromanols, in human A549 cells and in rats. *J Lipid Res* 2007;**48**:1221–30.
16. Stahl W, Graf P, Brigelius-Flohe R, Wechter W, Sies H. Quantification of the alpha- and gamma-tocopherol metabolites 2,5,7, 8-tetramethyl-2-(2′-carboxyethyl)-6-hydroxychroman and 2,7, 8-trimethyl-2-(2′-carboxyethyl)-6-hydroxychroman in human serum. *Anal Biochem* 1999;**275**:254–9.
17. Pope SA, Burtin GE, Clayton PT, Madge DJ, Muller DP. Synthesis and analysis of conjugates of the major vitamin E metabolite, alpha-CEHC. *Free Radic Biol Med* 2002;**33**:807–17.
18. Kiyose C, Saito H, Kaneko K, Hamamura K, Tomioka M, Ueda T, et al. Alpha-tocopherol affects the urinary and biliary excretion of 2,7,8-trimethyl-2 (2′-carboxyethyl)-6-hydroxychroman, gamma-tocopherol metabolite, in rats. *Lipids* 2001;**36**:467–72.
19. Lebold KM, Ang A, Traber MG, Arab L. Urinary alpha-carboxyethyl hydroxychroman can be used as a predictor of alpha-tocopherol adequacy, as demonstrated in the energetics study. *Am J Clin Nutr* 2012;**96**:801–9.
20. Schultz M, Leist M, Petrzika M, Gassmann B, Brigelius-Flohe R. Novel urinary metabolite of alpha-tocopherol, 2,5,7,8-tetramethyl-2(2′-carboxyethyl)-6-hydroxychroman, as an indicator of an adequate vitamin E supply? *Am J Clin Nutr* 1995;**62**:1527s–34s.
21. Traber MG, Mah E, Leonard SW, Bobe G, Bruno RS. Metabolic syndrome increases dietary alpha-tocopherol requirements as assessed using urinary and plasma vitamin E catabolites: a double-blind, crossover clinical trial. *Am J Clin Nutr* 2017;**105**:571–9.
22. Mason KE. The first two decades of vitamin E. *Fed Proc* 1977;**36**:1906–10.
23. Vardi M, Levy NS, Levy AP. Vitamin E in the prevention of cardiovascular disease: the importance of proper patient selection. *J Lipid Res* 2013;**54**:2307–14.
24. Boaz M, Smetana S, Weinstein T, Matas Z, Gafter U, Iaina A, et al. Secondary prevention with antioxidants of cardiovascular disease in endstage renal disease (SPACE): randomised placebo-controlled trial. *Lancet* 2000;**356**:1213–8.
25. McDougall M, Choi J, Magnusson K, Truong L, Tanguay R, Traber MG. Chronic vitamin E deficiency impairs cognitive function in adult zebrafish via dysregulation of brain lipids and energy metabolism. *Free Radic Biol Med* 2017;**112**:308–17.
26. Dysken MW, Sano M, Asthana S, Vertrees JE, Pallaki M, Llorente M, et al. Effect of vitamin E and memantine on functional decline in Alzheimer disease: the TEAM-AD VA cooperative randomized trial. *JAMA* 2014;**311**:33–44.
27. Sanyal AJ, Chalasani N, Kowdley KV, McCullough A, Diehl AM, Bass NM, et al. Pioglitazone, vitamin E, or placebo for nonalcoholic steatohepatitis. *N Engl J Med* 2010;**362**:1675–85.
28. Shamim AA, Schulze K, Merrill RD, Kabir A, Christian P, Shaikh S, et al. First-trimester plasma tocopherols are associated with risk of miscarriage in rural Bangladesh. *Am J Clin Nutr* 2015;**101**:294–301.
29. Farrell PM, Bieri JG, Fratantoni JF, Wood RE, di Sant'Agnese PA. The occurrence and effects of human vitamin E deficiency. A study in patients with cystic fibrosis. *J Clin Invest* 1977;**60**:233–41.

30. Cynamon HA, Milov DE, Valenstein E, Wagner M. Effect of vitamin E deficiency on neurologic function in patients with cystic fibrosis. *J Pediatr* 1988;**113**:637–40.

31. Elias E, Muller DP, Scott J. Association of spinocerebellar disorders with cystic fibrosis or chronic childhood cholestasis and very low serum vitamin E. *Lancet* 1981;**2**:1319–21.

32. Sitrin MD, Lieberman F, Jensen WE, Noronha A, Milburn C, Addington W. Vitamin E deficiency and neurologic disease in adults with cystic fibrosis. *Ann Intern Med* 1987;**107**:51–4.

33. Traber MG, Sokol RJ, Kohlschutter A, Yokota T, Muller DP, Dufour R, et al. Impaired discrimination between stereoisomers of alpha-tocopherol in patients with familial isolated vitamin E deficiency. *J Lipid Res* 1993;**34**:201–10.

34. Di Donato I, Bianchi S, Federico A. Ataxia with vitamin E deficiency: update of molecular diagnosis. *Neurol Sci* 2010;**31**:511–5.

35. Morley S, Cecchini M, Zhang W, Virgulti A, Noy N, Atkinson J, et al. Mechanisms of ligand transfer by the hepatic tocopherol transfer protein. *J Biol Chem* 2008;**283**:17797–804.

36. Traber MG. Vitamin E. In: Erdman J, Macdonald I, Zeisel S, editors. *Present knowledge in nutrition*. Singapore: International Life Sciences Institute: Wiley-Blackwell; 2012. p. 214–29.

37. Saito K, Matsumoto S, Yokoyama T, Okaniwa M, Kamoshita S. Pathology of chronic vitamin E deficiency in fatal familial intrahepatic cholestasis (Byler disease). *Virchows Arch A Pathol Anat Histol* 1982;**396**:319–30.

38. Traber MG. Vitamin E inadequacy in humans: causes and consequences. *Adv Nutr* 2014;**5**:503–14.

39. Kalra V, Grover J, Ahuja GK, Rathi S, Khurana DS. Vitamin E deficiency and associated neurological deficits in children with protein-energy malnutrition. *J Trop Pediatr* 1998;**44**:291–5.

40. Kalra V, Grover JK, Ahuja GK, Rathi S, Gulati S, Kalra N. Vitamin E administration and reversal of neurological deficits in protein-energy malnutrition. *J Trop Pediatr* 2001;**47**:39–45.

41. Koscik RL, Farrell PM, Kosorok MR, Zaremba KM, Laxova A, Lai HC, et al. Cognitive function of children with cystic fibrosis: deleterious effect of early malnutrition. *Pediatrics* 2004;**113**:1549–58.

42. Chen K, Zhang X, Wei XP, Qu P, Liu YX, Li TY. Antioxidant vitamin status during pregnancy in relation to cognitive development in the first two years of life. *Early Hum Dev* 2009;**85**:421–7.

43. Ford ES, Schleicher RL, Mokdad AH, Ajani UA, Liu S. Distribution of serum concentrations of alpha-tocopherol and gamma-tocopherol in the US population. *Am J Clin Nutr* 2006;**84**:375–83.

44. Traber MG, Jialal I. Measurement of lipid-soluble vitamins—further adjustment needed? *Lancet* 2000;**355**:2013–4.

45. Gordon HH, Demetry JP, Csapo G. Hemolysis in hydrogen peroxide of erythrocytes of premature infants: effect of alpha tocopherol. *AMA Am J Dis Child* 1952;**84**:472–4.

46. Krishnamurthy S, Bieri JG. The absorption, storage, and metabolism of alpha-tocopherol-C^{14} in the rat and chicken. *J Lipid Res* 1963;**4**:330–6.

47. Quaife ML, Scrimshaw NS, Lowry OH. A micromethod for assay of total tocopherols in blood serum. *J Biol Chem* 1949;**180**:1229–35.

48. Hashim SA, Schuttringer GR. Rapid determination of tocopherol in marco- and microquantities of plasma. Results obtained in various nutrition and metabolic studies. *Am J Clin Nutr* 1966;**19**:137–45.

49. Duggan DE. Spectrofluorometric determination of tocopherols. *Arch Biochem Biophys* 1959;**84**:116–22.

50. Bieri JG, Poukka RK. Red cell content of vitamin E and fatty acids in normal subjects and patients with abnormal lipid metabolism. *Int Z Vitaminforsch* 1970;**40**:344–50.

51. Bieri JG, Tolliver TJ, Catignani GL. Simultaneous determination of alpha-tocopherol and retinol in plasma or red cells by high pressure liquid chromatography. *Am J Clin Nutr* 1979;**32**:2143–9.

52. Hatam LJ, Kayden HJ. A high-performance liquid chromatographic method for the determination of tocopherol in plasma and cellular elements of the blood. *J Lipid Res* 1979;**20**:639–45.

53. Vatassery GT, Hagen DF. A liquid chromatographic method for quantitative determination of alpha-tocopherol in rat brain. *Anal Biochem* 1977;**79**:129–34.

54. De Leenheer AP, De Bevere VO, Cruyl AA, Claeys AE. Determination of serum alpha-tocopherol (vitamin E) by high-performance liquid chromatography. *Clin Chem* 1978;**24**:585–90.

55. Nilsson B, Johansson B, Jansson L, Holmberg L. Determination of plasma alpha-tocopherol by high-performance liquid chromatography. *J Chromatogr* 1978;**145**:169–72.

56. Torquato P, Ripa O, Giusepponi D, Galarini R, Bartolini D, Wallert M, et al. Analytical strategies to assess the functional metabolome of vitamin E. *J Pharm Biomed Anal* 2016;**124**:399–412.

57. Roxborough HE, Burton GW, Kelly FJ. Inter- and intra-individual variation in plasma and red blood cell vitamin E after supplementation. *Free Radic Res* 2000;**33**:437–45.

58. Burton GW, Traber MG, Acuff RV, Walters DN, Kayden H, Hughes L, et al. Human plasma and tissue alpha-tocopherol concentrations in response to supplementation with deuterated natural and synthetic vitamin E. *Am J Clin Nutr* 1998;**67**:669–84.

59. Traber MG, Leonard SW, Traber DL, Traber LD, Gallagher J, Bobe G, et al. Alpha-tocopherol adipose tissue stores are depleted after burn injury in pediatric patients. *Am J Clin Nutr* 2010;**92**:1378–84.

60. Podda M, Weber C, Traber MG, Packer L. Simultaneous determination of tissue tocopherols, tocotrienols, ubiquinols, and ubiquinones. *J Lipid Res* 1996;**37**:893–901.

61. Leonard SW, Terasawa Y, Farese Jr. RV, Traber MG. Incorporation of deuterated RRR- or all-rac-alpha-tocopherol in plasma and tissues of alpha-tocopherol transfer protein—null mice. *Am J Clin Nutr* 2002;**75**:555–60.

62. Christen S, Jiang Q, Shigenaga MK, Ames BN. Analysis of plasma tocopherols alpha, gamma, and 5-nitro-gamma in rats with inflammation by HPLC coulometric detection. *J Lipid Res* 2002;**43**:1978–85.

63. Williams CA, Kronfeldt DS, Hess TM, Saker KE, Waldron JN, Crandell KM, et al. Antioxidant supplementation and subsequent oxidative stress of horses during an 80-km endurance race. *J Anim Sci* 2004;**82**:588–94.

64. Cherian G, Traber MG, Goeger MP, Leonard SW. Conjugated linoleic acid and fish oil in laying hen diets: effects on egg fatty acids, thiobarbituric acid reactive substances, and tocopherols during storage. *Poult Sci* 2007;**86**:953–8.

65. Nierenberg DW, Lester DC, Colacchio TA. Determination of tocopherol and tocopherol acetate concentrations in human feces using high-performance liquid chromatography. *J Chromatogr* 1987;**413**:79–89.

66. Leonard SW, Paterson E, Atkinson JK, Ramakrishnan R, Cross CE, Traber MG. Studies in humans using deuterium-labeled alpha- and gamma-tocopherols demonstrate faster plasma gamma-tocopherol disappearance and greater gamma-metabolite production. *Free Radic Biol Med* 2005;**38**:857–66.

67. Leonard SW, Gumpricht E, Devereaux MW, Sokol RJ, Traber MG. Quantitation of rat liver vitamin E metabolites by LC-MS during high-dose vitamin E administration. *J Lipid Res* 2005;**46**:1068–75.

68. Jiang Q, Xu T, Huang J, Jannasch AS, Cooper B, Yang C. Analysis of vitamin E metabolites including carboxychromanols and sulfated derivatives using LC/MS/MS. *J Lipid Res* 2015;**56**:2217–25.

69. Zhao Y, Lee MJ, Cheung C, Ju JH, Chen YK, Liu B, et al. Analysis of multiple metabolites of tocopherols and tocotrienols in mice and humans. *J Agric Food Chem* 2010;**58**:4844–52.

70. Johnson CH, Slanar O, Krausz KW, Kang DW, Patterson AD, Kim JH, et al. Novel metabolites and roles for alpha-tocopherol in humans and mice discovered by mass spectrometry-based metabolomics. *Am J Clin Nutr* 2012;**96**:818–30.

71. Craft NE, Brown ED, Smith Jr. JC. Effects of storage and handling conditions on concentrations of individual carotenoids, retinol, and tocopherol in plasma. *Clin Chem* 1988;**34**:44–8.

72. Leonard SW, Bobe G, Traber MG. Stability of antioxidant vitamins in whole human blood during overnight storage at 4°C and frozen storage up to 6 months. *Int J Vitam Nutr Res* 2018; (In Press).

73. Clark S, Youngman LD, Chukwurah B, Palmer A, Parish S, Peto R, et al. Effect of temperature and light on the stability of fat-soluble vitamins in whole blood over several days: implications for epidemiological studies. *Int J Epidemiol* 2004;**33**:518–25.

74. Greaves RF, Woollard GA, Hoad KE, Walmsley TA, Johnson LA, Briscoe S, et al. Laboratory medicine best practice guideline: vitamins a, e and the carotenoids in blood. *Clin Biochem Rev* 2014;**35**:81–113.

75. Brouwer DA, Molin F, van Beusekom CM, van Doormaal JJ, Muskiet FA. Influence of fasting on circulating levels of alpha-tocopherol and beta-carotene. Effect of short-term supplementation. *Clin Chim Acta* 1998;**277**:127–39.

76. Key T, Oakes S, Davey G, Moore J, Edmond LM, McLoone UJ, et al. Stability of vitamins A, C, and E, carotenoids, lipids, and testosterone in whole blood stored at 4 degrees C for 6 and 24 hours before separation of serum and plasma. *Cancer Epidemiol Biomark Prev* 1996;**5**:811–4.

77. Leonard SW, Traber MG. Measurement of the vitamin E metabolites, carboxyethyl hydroxychromans (CEHCs), in biological samples. *Curr Protoc Toxicol* 2006;**29**:. 7.8.1–7.8.12, (Chapter 7: Unit 7.8).

78. Man EaG EF. Notes on the extraction and saponification of lipids from blood and blood serum. *J Biol Chem* 1937;**122**:77–88.

79. Buettner GR. The pecking order of free radicals and antioxidants: lipid peroxidation, alpha-tocopherol, and ascorbate. *Arch Biochem Biophys* 1993;**300**:535–43.

80. Burton GW, Webb A, Ingold KU. A mild, rapid, and efficient method of lipid extraction for use in determining vitamin E/lipid ratios. *Lipids* 1985;**20**:29–39.

81. Kayden HJ, Chow CK, Bjornson LK. Spectrophotometric method for determination of tocopherol in red blood cells. *J Lipid Res* 1973;**14**:533–40.

82. Marteau C, Favier D, Nardello-Rataj V, Aubry JM. Dramatic solvent effect on the synergy between alpha-tocopherol and BHT antioxidants. *Food Chem* 2014;**160**:190–5.

83. Jiang Q. Natural forms of vitamin E: metabolism, antioxidant, and anti-inflammatory activities and their role in disease prevention and therapy. *Free Radic Biol Med* 2014;**72**:76–90.

84. Galli F, Azzi A, Birringer M, Cook-Mills JM, Eggersdorfer M, Frank J, et al. Vitamin E: emerging aspects and new directions. *Free Radic Biol Med* 2017;**102**:16–36.

85. Birringer M, Lington D, Vertuani S, Manfredini S, Scharlau D, Glei M, et al. Proapoptotic effects of long-chain vitamin E metabolites in HepG2 cells are mediated by oxidative stress. *Free Radic Biol Med* 2010;**49**:1315–22.

86. Wallert M, Schmolz L, Koeberle A, Krauth V, Glei M, Galli F, et al. Alpha-tocopherol long-chain metabolite alpha-13'-COOH affects the inflammatory response of lipopolysaccharide-activated murine RAW264.7 macrophages. *Mol Nutr Food Res* 2015;**59**:1524–34.

87. Schmolz L, Wallert M, Rozzino N, Cignarella A, Galli F, Glei M, et al. Structure-function relationship studies in vitro reveal distinct and specific effects of long-chain metabolites of vitamin E. *Mol Nutr Food Res* 2017;**61**:https://doi.org/10.1002/mnfr.201700562. Epub 2017 Nov 3.

88. Lodge JK, Traber MG, Elsner A, Brigelius-Flohe R. A rapid method for the extraction and determination of vitamin E metabolites in human urine. *J Lipid Res* 2000;**41**:148–54.

89. Li YJ, Luo SC, Lee YJ, Lin FJ, Cheng CC, Wein YS, et al. Isolation and identification of alpha-CEHC sulfate in rat urine and an improved method for the determination of conjugated alpha-CEHC. *J Agric Food Chem* 2008;**56**:11105–13.

90. Freiser H, Jiang Q. Optimization of the enzymatic hydrolysis and analysis of plasma conjugated gamma-CEHC and sulfated long-chain carboxychromanols, metabolites of vitamin E. *Anal Biochem* 2009;**388**:260–5.

91. Giusepponi D, Torquato P, Bartolini D, Piroddi M, Birringer M, Lorkowski S, et al. Determination of tocopherols and their metabolites by liquid-chromatography coupled with tandem mass spectrometry in human plasma and serum. *Talanta* 2017;**170**:552–61.

92. Kiyose C, Saito K, Yachi R, Muto C, Igarashi O. Changes in the concentrations of vitamin E analogs and their metabolites in rat liver and kidney after oral administration. *J Clin Biochem Nutr* 2015;**56**:143–8.

93. Ciffolilli S, Wallert M, Bartolini D, Krauth V, Werz O, Piroddi M, et al. Human serum determination and in vitro anti-inflammatory activity of the vitamin E metabolite alpha-(13′-hydroxy)-6-hydroxychroman. *Free Radic Biol Med* 2015;**89**:952–62.

94. Ford L, Farr J, Morris P, Berg J. The value of measuring serum cholesterol-adjusted vitamin E in routine practice. *Ann Clin Biochem* 2006;**43**:130–4.

95. Winbauer AN, Pingree SS, Nuttall KL. Evaluating serum alpha-tocopherol (vitamin E) in terms of a lipid ratio. *Ann Clin Lab Sci* 1999;**29**:185–91.

96. Liebler DC, Burr JA, Philips L, Ham AJ. Gas chromatography-mass spectrometry analysis of vitamin E and its oxidation products. *Anal Biochem* 1996;**236**:27–34.

97. Galli F, Lee R, Dunster C, Kelly FJ. Gas chromatography mass spectrometry analysis of carboxyethyl-hydroxychroman metabolites of alpha- and gamma-tocopherol in human plasma. *Free Radic Biol Med* 2002;**32**:333–40.

98. Mottier P, Gremaud E, Guy PA, Turesky RJ. Comparison of gas chromatography-mass spectrometry and liquid chromatography-tandem mass spectrometry methods to quantify alpha-tocopherol and alpha-tocopherolquinone levels in human plasma. *Anal Biochem* 2002;**301**:128–35.

99. Poojary MM, Passamonti P. Improved conventional and microwave-assisted silylation protocols for simultaneous gas chromatographic determination of tocopherols and sterols: method development and multi-response optimization. *J Chromatogr A* 2016;**1476**:88–104.

100. Bartosinska E, Buszewska-Forajta M, Siluk D. GC-MS and LC-MS approaches for determination of tocopherols and tocotrienols in biological and food matrices. *J Pharm Biomed Anal* 2016;**127**:156–69.

101. Abidi SL. Chromatographic analysis of tocol-derived lipid antioxidants. *J Chromatogr A* 2000;**881**:197–216.
102. Chebrolu KK, Yousef GG, Park R, Tanimura Y, Brown AF. A high-throughput, simultaneous analysis of carotenoids, chlorophylls and tocopherol using sub two micron core shell technology columns. *J Chromatogr B Anal Technol Biomed Life Sci* 2015;**1001**:41–8.
103. Kamal-Eldi A, Gorgen S, Pettersson J, Lampi AM. Normal-phase high-performance liquid chromatography of tocopherols and tocotrienols. Comparison of different chromatographic columns. *J Chromatogr A* 2000;**881**:217–27.
104. Dial S, Eitenmiller RR. Tocopherols and tocotrienols in key foods in the US diet. In: Ong A, Niki E, Packer L, editors. *Nutrition, lipids, health and disease*. Champaign, IL: AOCS Press; 1995. p. 327–42.
105. Lanina SA, Toledo P, Sampels S, Kamal-Eldin A, Jastrebova JA. Comparison of reversed-phase liquid chromatography-mass spectrometry with electrospray and atmospheric pressure chemical ionization for analysis of dietary tocopherols. *J Chromatogr A* 2007;**1157**:159–70.
106. Fu JY, Htar TT, De Silva L, Tan DM, Chuah LH. Chromatographic separation of vitamin E enantiomers. *Molecules* 2017;**22**:. 10.3390/molecules22020233.
107. Kiyose C, Muramatsu R, Kameyama Y, Ueda T, Igarashi O. Biodiscrimination of alpha-tocopherol stereoisomers in humans after oral administration. *Am J Clin Nutr* 1997;**65**:785–9.
108. Lopez-Cervantes J, Sanchez-Machado DI, Rios-Vazquez NJ. High-performance liquid chromatography method for the simultaneous quantification of retinol, alpha-tocopherol, and cholesterol in shrimp waste hydrolysate. *J Chromatogr A* 2006;**1105**:135–9.
109. Thompson LB, Schimpf K, Baugh S. Determination of vitamins A and E in infant formula and adult/pediatric nutritional formula by HPLC with UV and fluorescence detection: first action 2012.09. *J AOAC Int* 2013;**96**:1407–13.
110. Good RL, Roupe KA, Fukuda C, Clifton GD, Fariss MW, Davies NM. Direct high-performance liquid chromatographic analysis of D-tocopheryl acid succinate and derivatives. *J Pharm Biomed Anal* 2005;**39**:33–8.
111. Leonard SW, Bruno RS, Paterson E, Schock BC, Atkinson J, Bray TM, et al. 5-Nitrogamma-tocopherol increases in human plasma exposed to cigarette smoke in vitro and in vivo. *Free Radic Biol Med* 2003;**35**:1560–7.
112. Mazzini F, Netscher T, Salvadori P. First synthesis of rac-(5-2H3)-alpha-CEHC, a labeled analogue of a major vitamin E metabolite. *J Org Chem* 2004;**69**:9303–6.
113. Ingold KU, Burton GW, Foster DO, Hughes L, Lindsay DA, Webb A. Biokinetics of and discrimination between dietary RRR- and SRR-alpha-tocopherols in the male rat. *Lipids* 1987;**22**:163–72.
114. Vaule H, Leonard SW, Traber MG. Vitamin E delivery to human skin: studies using deuterated alpha-tocopherol measured by APCI LC-MS. *Free Radic Biol Med* 2004;**36**:456–63.
115. Lauridsen C, Leonard SW, Griffin DA, Liebler DC, McClure TD, Traber MG. Quantitative analysis by liquid chromatography-tandem mass spectrometry of deuterium-labeled and unlabeled vitamin E in biological samples. *Anal Biochem* 2001;**289**:89–95.
116. Hall WL, Jeanes YM, Pugh J, Lodge JK. Development of a liquid chromatographic time-of-flight mass spectrometric method for the determination of unlabelled and deuterium-labelled alpha-tocopherol in blood components. *Rapid Commun Mass Spectrom* 2003;**17**:2797–803.
117. Byrdwell WC. Atmospheric pressure chemical ionization mass spectrometry for analysis of lipids. *Lipids* 2001;**36**:327–46.

118. O'Byrne D, Grundy S, Packer L, Devaraj S, Baldenius K, Hoppe PP, et al. Studies of LDL oxidation following alpha-, gamma-, or delta-tocotrienyl acetate supplementation of hypercholesterolemic humans. *Free Radic Biol Med* 2000;**29**:834–45.

119. Young DS. Implementation of SI units for clinical laboratory data. Style specifications and conversion tables. *Ann Intern Med* 1987;**106**:114–29.

120. Peter S, Friedel A, Roos FF, Wyss A, Eggersdorfer M, Hoffmann K, et al. A systematic review of global alpha-tocopherol status as assessed by nutritional intake levels and blood serum concentrations. *Int J Vitam Nutr Res* 2015;**85**:261–81.

121. Hensley K, Benaksas EJ, Bolli R, Comp P, Grammas P, Hamdheydari L, et al. New perspectives on vitamin E: gamma-tocopherol and carboxyelthylhydroxychroman metabolites in biology and medicine. *Free Radic Biol Med* 2004;**36**:1–15.

122. Yoshikawa S, Morinobu T, Hamamura K, Hirahara F, Iwamoto T, Tamai H. The effect of gamma-tocopherol administration on alpha-tocopherol levels and metabolism in humans. *Eur J Clin Nutr* 2005;**59**:900–5.

Methods for assessment of Vitamin K

Renata M. Górska,
Nutristasis Unit, Viapath, St. Thomas' Hospital, London, United Kingdom

Chapter Outline

Laboratory Assessment of Vitamin Status. https://doi.org/10.1016/B978-0-12-813050-6.00005-X

Structure and Function

Vitamin K is a term used to refer to a group of fat-soluble vitamins that are needed for the posttranslational modification of glutamate residues (Glu) to γ-carboxyglutamate (Gla) residues in 17 vitamin K-dependent proteins (VKDPs). γ-carboxyglutamate residues have unique calcium-binding properties that influence the folding of VKDPs and confer optimal biological activity.[1] Seven VKDPs play a role in blood coagulation, while osteocalcin (OC) and matrix Gla protein (MGP) support metabolic pathways critical for bone and cardiovascular health. The functions of the other Gla proteins have yet to be fully elucidated but are likely to include the regulation of vascular biology and the stimulation of cell proliferation.[2]

Vitamin K-dependent coagulation plasma proteins possess 10–12 GLA residues that are distributed over a c.45 amino acid peptide sequence that is referred to as the GLA domain, whereas the much smaller OC and MGP peptides contain three and five Gla residues, respectively.

Chemical Properties

Vitamin K is the generic term for homologous fat-soluble compounds derived from 2-methyl-1,4-naphthoquinone. The different forms of vitamin K have a distinct side chain attached at the C3-position of the naphthoquinone nucleus (Fig. 5.1).

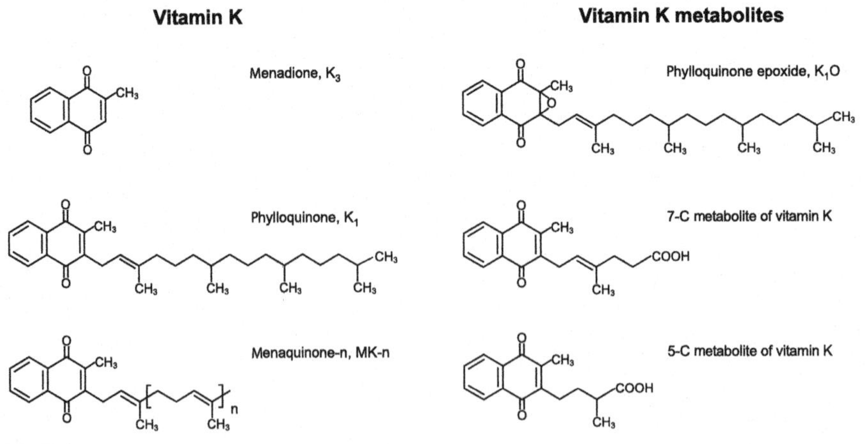

FIG. 5.1

Chemical structures of K vitamers and metabolites.

Vitamin K_1 (2-methyl-3-phytyl-1,4-naphthoquinone) is the trivial name given to phylloquinone and is synthesized by plants. It has a phytyl side chain consisting of four isoprene units, which contains 20 carbons with a single double bond. The phylloquinone found in natural products is present mainly as the trans isomer, while 10%–20% of the vitamin K used in food supplementation may be present as the cis isomer. Only the trans isomer of vitamin K is biologically active.[3] Phylloquinone is obtained from plants and is present in biological samples in the oxidized (naphthoquinone), reduced (naphthoquinol which is a highly fluorescent molecule), and epoxide (vitamin K_1 2,3 epoxide) forms.

The menaquinones, collectively known as vitamin K_2 (2-methyl-3-multiprenyl-1,4-naphthoquinone), are a series of compounds containing an unsaturated side chain consisting of a number (where $n = 4$–13) of repeating prenyl units. Each menaquinone is referred to as menaquinone-n, abbreviated MK-n. Menaquinones synthesized by bacteria may also contain one (or more) saturated double bonds in the side chain, e.g., MK-n (II-H2) and MK-n (VI-H4) have the second and sixth bond saturated, respectively. Menaquinones are formed by the intestinal microflora (e.g., *Escherichia coli*, *Staphylococcus aureus*, *Eubacterium lentum*), with the most common configuration being all-trans.[4]

2-Methyl-1,4-naphthoquinone (vitamin K_3, also referred to as menadione) does not occur in nature without a phytyl or multiprenyl side chain. Vitamin K activity is conferred on synthetic menadione in vertebrates through their ability to add a geranylgeranyl side chain at the C3-position (thus converting menadione into MK-4).[5] For this reason, menadione should be thought of as a provitamin. Interestingly, menadione has been shown to exhibit antitumor activity in rodent and human tumor cells.[6]

The functional 2-methyl-1,4-naphthoquinone group confers similar chemical properties for all K-vitamins. However, substantial differences may be expected with physiological processes such as intestinal absorption, cellular uptake, tissue distribution, and turnover. These differences reflect the different lipophilicity conferred on the vitamin by the side chain.

Vitamin K is insoluble in water; sparingly soluble in methanol; and readily soluble in ether, n-hexane, and chloroform. The molecules are sensitive to daylight and are sensitive to a strong alkaline medium, however are stable in a slightly acidic medium and under oxidizing conditions.

Nutritional Sources

The most abundant naturally occurring form of vitamin K in blood is phylloquinone. Dietary phylloquinone is obtained from plant sources. In general, green leafy vegetables including broccoli, spinach, and certain lettuces contain the highest phylloquinone concentrations of the vitamin (400–700 µg/100 g) and contribute to approximately 60% of total phylloquinone intake. The bioavailability of phylloquinone from vegetables is poor since the vitamin is tightly bound to chloroplast membranes in which it plays a key role in photosynthesis. Significant quantities of phylloquinone are also found in some nonleafy green vegetables, several vegetable oils, fruits, grain, and dairy products. Certain plant oils and margarine, spreads, and salad dressings derived from these plant oils are also important dietary sources (50–200 µg/100 g).[7] Booth et al. demonstrated that the daily restriction of phylloquinone intake to 35 µg/ day causes a rapid decrease in plasma phylloquinone and urinary excretion of Gla residues, and increases the abundance of undercarboxylated forms of the VKDPs without affecting classic measurements of blood coagulation.[8]

Hepatic stores of vitamin K are primarily in the form of menaquinones. Dietary intakes of menaquinones have been much less investigated. The best source of long-chain menaquinones are animal livers and fermented foods. These include some cheeses and natto, which are particularly rich in MK-7 (1000 µg/100 g). Chicken egg yolk (approx. 30 µg/100 g) and dairy products (e.g., butter approx. 15 µg/100 g) also provide relatively rich sources of menaquinones.[2]

Menadione, as water-soluble salts, is used as a feed supplement in animal farms and therefore may indirectly enter the human food chain and is converted to MK-4.[9] Formation of MK-4 from phylloquinone is a metabolic transformation that does not require bacterial-mediated conversion to menadione as an intermediate.[5]

The World Health Organization (WHO/FAO, 2004) derived a Recommended Nutrient Intake (RNI) of 1 µg/kg body weight per day of phylloquinone, corresponding to 55 µg/day for adult women and 65 µg/kg for adult men. This intake is considered as adequate to optimize the γ-carboxylation of the VKDPs involved in blood coagulation (but does not consider the γ-carboxylation status of extra-hepatic VKDPs not involved in coagulation).

Vitamin K Absorption and Metabolism

Vitamin K has an absorption pathway that is similar to the other fat-soluble vitamins. As with all nonpolar lipids, absorption takes place predominately in the proximal intestine and is dependent on bile and pancreatic secretion.[10] In healthy adults the efficiency of absorption of phylloquinone in its free form is approximately 80% and much less from plant tissues. Vitamin K is mainly transported by triglyceride-rich lipoproteins (TRL), with lesser amounts carried by very low-density lipoproteins (VLDL). Long-chain menaquinones are mainly transported by low-density lipoprotein particles (LDL)[6] (Fig. 5.2).

Intestinal Absorption and Entry Into Circulation

After absorption, vitamin K is emulsified by bile salts to form mixed micelles and incorporated into chylomicrons (CM) containing apoA and apoB-48. CM are subsequently secreted into the lymph and enter the blood circulation via the lacteals and thoracic duct. Circulating CM are cleared by lipoprotein lipase hydrolysis at the surface of capillary endothelial cells. After removal of the triglyceride core, smaller CM remnants (CR) are formed and reenter the circulation having lost much of the apoA and apoC but retaining vitamin K in the lipophilic core.[6]

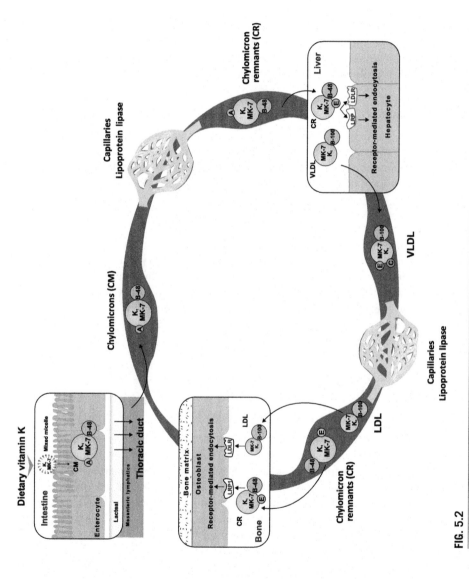

FIG. 5.2

Schematic illustration of absorption, transport, and cellular uptake of dietary vitamin K_1 (K_1) and menaquinone-7 (MK-7).

Uptake by Liver

In the liver CR follow the classical receptor-mediated pathway of endocytosis through LDL receptor (LDLR) and receptor-related protein 1 (LRP). The lipids are repackaged into VLDL containing apoB-100 and return to the circulation where they acquire apoC and apoE. Subsequent hydrolysis of triglycerides in the capillaries results in formation of VLDL remnants called intermediate-density lipoproteins (IDL) and liberation of free fatty acids. Further IDL modifications include the loss of apoC and apoE. This process provides transformation of the remnant VLDL particles into LDL. Vitamin K is presumed to still be located in the lipophilic core.[6]

Uptake by Bone

Dietary vitamin K is delivered to human bone through circulating lipoproteins such as CR and LDL. Osteoblasts express the LDLR and LRPI. Studies investigating vitamin K absorption showed that osteoblasts mainly obtain phylloquinone via the CR pathway and MK-7 via the LDL pathway.[6]

Excretion

Vitamin K is metabolized in the liver and excreted in the urine and bile predominately in the form of glucuronide conjugates of two major carboxylic aglycones, with 5–7C side chains, respectively. In studies using radiolabeled phylloquinone 20% of an injected dose of was recovered in the urine whereas about 40% was excreted in the feces via the bile; the proportion excreted was the same regardless of whether the injected dose was 1–45 mg.[10] The equivalent excretion data for menaquinones have not been determined.[6]

Clinical Application of Measurement
Classical Role of Vitamin K in Hemostasis

Seven of the known VKDPs have a crucial role in blood coagulation (procoagulant factors II, VII, IX, X and anticoagulant proteins C, S, and Z). They are all synthesized in the liver in a precursor form and require vitamin K-dependent posttranslational γ-carboxylation of multiple Glu residues to Gla, to gain full biological activity.[2] The Gla residues are located within a region of the protein referred to as the Gla domain, clustered in the N-terminal.[11] The role of these residues is to confer an efficient metal chelating site to bind calcium ions (Ca^{2+}). In the presence of calcium the coagulation factors undergo a conformational change that leads to the exposure of phospholipid-binding sites, essential for participation in coagulation.[2]

Unlike other fat-soluble vitamins, the body is unable to store vitamin K for long periods of time. Hepatic stores of phylloquinone are rapidly (<3 days) depleted without sufficient dietary intake.[12] However, the body is able to recycle vitamin K through a process known as the Vitamin K Cycle (Fig. 5.3).

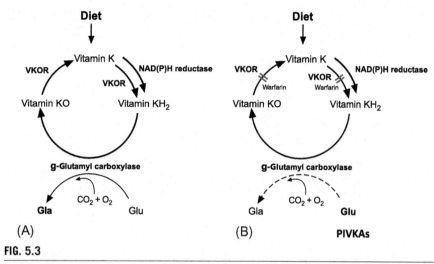

FIG. 5.3

The vitamin K epoxide cycle in (A) absence and (B) presence of warfarin.

The Vitamin K Cycle not only drives the conversion of Glu to Gla, but also recovers vitamin K for reuse in subsequent γ-carboxylation reactions.[6, 13] Vitamin K is converted to the active form vitamin K quinol—hydroquinone (KH$_2$) by the action of vitamin K epoxide reductase (VKOR) and an NAD(P)H-dependent quinone reductase. The metabolically active KH$_2$, molecular oxygen, and CO$_2$ are required for the carboxylation of Glu to Gla by vitamin K γ-glutamyl carboxylase (GGCX).[14] As a result of the carboxylation KH$_2$ is oxidized to vitamin K 2,3-epoxide (KO). Vitamin K 2,3-epoxide is subsequently recycled to vitamin K by VKOR. A clinically important property of the dithiol-dependent VKOR is its sensitivity to 4-hydroxycoumarin (e.g., warfarin) and 1,3-inandione structures. Drugs based on these structures have been developed as oral anticoagulants and are well known for their inhibition of VKOR activity. Alternatively, vitamin K may enter the cycle via an NAD(P)H-dependent vitamin K reductase activity, which is less sensitive to warfarin and this provides an additional pathway for the reduction of vitamin K to KH$_2$ in the presence of warfarin or other oral anticoagulant drugs.[2]

The vitamin K cycle conserves the pool of vitamin K that is available to function in the γ-carboxylation of VKDPs and may explain why the daily requirement for the vitamin is very low. The complete γ-carboxylation of native Glu residues cannot be achieved if the supply of vitamin K is insufficient and/or when the vitamin K cycle is disturbed. Deficiency results in the release of undercarboxylated species of the vitamin K-dependent coagulation proteins. The abundance of undercarboxylated species is broadly proportional to the degree of vitamin K deficiency or parallel the degree of any liver dysfunction.[2]

Undercarboxylated VKDPs are referred to as Protein Induced by Vitamin K Absence or Antagonism (PIVKAs). Proteins induced by vitamin K absence have impaired activity since they fail to bind calcium ions thus preventing the conformational change that allows binding to phospholipid membranes. The laboratory measurement of PIVKAs, e.g., PIVKA-II, is the most useful and sensitive homeostatic marker of hepatic subclinical vitamin K deficiency since PIVKA-II is detectable in plasma before any changes occur in conventional coagulation tests.[15]

Antagonists of Vitamin K Action

The history of the discovery of the anticoagulant action of coumarins has been well described.[16, 17] Warfarin is currently the most common oral anticoagulant employed in thromboprophylaxis and acts by limiting the hepatic bioavailability of the active form of vitamin K to GGCX for use in the posttranslational modification of Gla proteins. The site of warfarin action is VKOR, which recycles inactive KO back to the active hydroquinone form. Anticoagulation results because species of undercarboxylated vitamin K-dependent coagulation factors (II, VII, IX, and X) with impaired biological function are released into the circulation at the expense of native protein with full activity. The 1,3-indandiones, e.g., phenindione, are another group of vitamin K antagonists that despite their hepatic toxicity have also been used clinically. The aim of warfarin therapy is to induce anticoagulation to a degree that prevents thromboembolic events, but without increasing the risk of hemorrhagic complications.[18, 19]

Vitamin E is a known vitamin K antagonist although the mechanism of the interactions is unclear. Vitamin K and α-tocopherol share a common metabolic pathway and up-regulation of these pathways after administration of high doses of vitamin E may increase the rate of vitamin K catabolism.[20] It has also been shown that the addition of vitamin E to the diet of patients on coumarin anticoagulant therapy increases the incidence of hemorrhagic events.[21] Booth et al. investigated patients who were not on anticoagulant therapy but supplemented with 1000 IU of vitamin E, interestingly, a high-dose of vitamin E increased PIVKA-II concentrations in adults.[22] These findings indicate that the interactions between vitamin K and E may be of clinical significance, and it is possible that high vitamin E intakes may induce vitamin K deficiency.

Patient Groups at Increased Risk of Vitamin K Deficiency

Vitamin K deficiency (with respect to its coagulation function) is rare among healthy adults. The availability of vitamin K typically exceeds requirement and it is efficiently recycled. In addition, intestinal bacteria present in the large intestine synthesize a series of menaquinones, although it is unclear to what extent menaquinones can be absorbed and utilized since absorption of vitamin K is dependent on the availability of bile and pancreatic secretions which are largely absent this far along the gastrointestinal tract.

Any condition causing restriction in fat absorption (e.g., biliary obstruction, celiac disease, pancreatic insufficiency, cystic fibrosis, Crohn's disease, etc.) may confer acquired vitamin K deficiency. Conway et al. demonstrated that 70% of children with cystic fibrosis had a suboptimal vitamin K status based on a low serum phylloquinone, raised PIVKA-II, or both abnormalities.[23] In addition, vitamin K deficiency may develop in patients with poor dietary intake, hepatic dysfunction, diarrhea, or frequent antibiotic therapy.

The susceptibility of many patients with advanced cancer to vitamin K deficiency has been highlighted in studies. As an example, autologous bone marrow transplantation is linked with a rapid fall in circulatory concentrations of phylloquinone and the subsequent appearance of circulatory PIVKA-II within a few days.[24] In another study, very low serum phylloquinone concentrations were present in 22% of palliative care patients accompanied by an elevated PIVKA-II in 78% of patients.[25]

Vitamin K deficiency is also common in newborn infants as a consequence of very limited hepatic stores at birth, the low vitamin K content of human breast milk, and possibly impaired utilization in the early days of life. Many countries protect against vitamin K deficiency bleeding (VKDB) by routinely supplementing all infants with vitamin K after delivery. Classical VKDB occurs in the first week of life. Of particular concern is the late form of VKDB which presents at 3–8 weeks, often as debilitating or fatal intracranial bleeding.[26]

Genotypes Influencing Vitamin K Status

Point mutations in the gene that encodes GGCX lead to an autosomal recessive bleeding diathesis known as combined vitamin K-dependent coagulation factor deficiency type I (VKCFD).[27] For example, it was shown that the GGCX rs699664 SNP induces an increased carboxylase activity.[28] In a second autosomal recessive disorder, point mutations in the gene that encodes VKOR confer VKCFD type II.[29] Polymorphisms in the gene that encodes ApoE (i.e., E2, E3, and E4) and cytochrome P450 4F2 (CYP4F2) genes also have an impact on vitamin K status.[30, 31]

Analytical Methods
Historical Methods

Vitamin K was discovered during experiments on the cholesterol metabolism of chicks during 1928–1930[32]; the Danish biochemist Henrik Dam observed that chicks fed fat and cholesterol-free diets developed subdural or muscular hemorrhages at about the age of 3–4 weeks which were not prevented by dietary supplementation with cholesterol, lemon juice, yeast, cod-liver oil, or vitamins A, C, or D. Dam came to the conclusion that the missing fat-soluble factor was different from vitamins A, D, and E. He postulated the existence of a previously unidentified fat-soluble vitamin. He named this compound vitamin K. Not only was K the first letter of the alphabet not used to describe an existing or a suspected vitamin activity at the time, but it was also the first letter of the Danish and German word Koagulation. Shortly thereafter, Dam elucidated that vitamin K is abundant in liver and the green leaves of plants, and noted that the poor coagulability of plasma from bleeding chicks could not be reversed by the in vitro addition of vitamin K.

In 1939, Dam partially purified chick plasma prothrombin and demonstrated that its concentration is reduced in vitamin K-deficient chicks. The early methods for vitamin K status assessment therefore utilized chick bioassays.[33] Preventive and curative chick bioassays for vitamin K provided a means of establishing good dietary sources and led to the eventual isolation of the vitamin from alfalfa and putrefied fish meal. Almquist described a preventive bioassay[34] based on the occurrence of hemorrhage and simple measurement of the blood clotting time in which chicks were fed a basel diet known to induce vitamin K deficiency plus varying quantities of a test food substance. The curative bioassay worked on the premise that the clotting time of a chick suffering from vitamin K deficiency became normal in 3 days, when food containing sufficient vitamin K was given. The concentration of vitamin K was determined by the amount of clotting agent (a "watery" extract of hen lungs) required to cause formation of a clot in 180 s at 39°C when added to chick plasma diluted with an equal volume of Ringer's solution. The reciprocal (R_1) value of the amount of clotting agent required to clot a diseased plasma and that necessary to clot a normal ($R = 1$) plasma was calculated. When the R-values of a series of vitamin K-deficient chicks had been determined the animals received a weighed amount of a test substance in the form of tablets. On the fourth day the R-value was determined for a second time (R_2). The chicks were weighted every day during the 3-day period and the mean value calculated. From the R_2 value the amount of a test substance (test substance (mg)/chick body weight (g)/day) required to render the clotting power of diseased chicks normal was calculated by extrapolation. This amount was said to be one curative unit.

Laboratory Assessment of Vitamin K Status—Clinical Aspects

Vitamin K deficiency in the clinical setting is characterized by a bleeding tendency as a consequence of low concentrations of the coagulation factors II, VII, IX, and X. Prothrombin time (PT) based laboratory assays are insensitive and nonspecific indicators of vitamin K status and are suitable only for the diagnosis of acute episodes of vitamin K deficiency or antagonism. In studies utilizing diets containing approximately 10 µg/day or less of phylloquinone it was shown that prothrombin times (PTs) are not clinically useful indicators for vitamin K deficiency.[35–37] PTs are unable to reveal subclinical deficiency, which is prevalent in high-risk groups such as infants and those with malabsorption (e.g., biliary disease, pancreatic disease, cystic fibrosis, Crohn's disease, the elderly, short bowel). This is because for a detectable prolongation of the PT, the availability of fully γ-carboxylated prothrombin must fall by approximately 50%.

Laboratory Assessment of K Vitamins Status—Technical Aspects

Phylloquinone is the main circulating form of vitamin K and its concentration in human serum and plasma can be directly measured. Since circulating concentrations are extremely low (mean 0.5 µg/L, 1.1 nmol/L), sensitive detection methods are required for its quantification.

Several different methods and approaches for the determination of vitamin K status are available and between them an assessment of intake, absorption, and metabolism can be made.[38]

While vitamins A, D, and E are frequently measured in serum to evaluate status, the laboratory evaluation of vitamin K status and its homologs remains technically challenging and is performed by a relatively small number of specialist laboratories worldwide (<50 laboratory groups). Because the low vitamin K concentration in human blood (vitamin K is the least abundant and most lipophilic of the four fat-soluble vitamins) is set against the inherent complexity of the biological samples in which its measurement has clinical utility, the methods used must therefore be of high sensitivity and selectivity. In the majority of methods available, characteristic limitations for assessment of vitamin K status include intricate sample preparation, requirement for long chromatographic run times, and the need to chemically induce fluorescence to facilitate detection. Continued interest in the roles that vitamin K play beyond that of coagulation, and in particular in bone and vascular health has led to an increase in demand for robust laboratory assays capable of processing large numbers of requests.

The early physicochemical methods for the separation and detection of phylloquinone involved paper, column, and thin layer chromatography followed by quantification by colorimetry,[39] UV spectroscopy,[40, 41] or simple visual comparison with known amounts of phylloquinone on the chromatograms. These methods had not only sensitivity limitations but also problems of chromatographic resolution such as that in resolving phylloquinone from plastoquinone.[41]

Major challenges when analyzing vitamin K in biological materials include interferences by other lipids, especially VLDL, and lack of sensitivity necessary to quantitate vitamin K in the clinically relevant picogram range. However, the arrival of HPLC overcame these limitations and when coupled to fluorescence, electrochemical or tandem mass spectrometer detectors offered far superior selectivity and sensitivity.

Vitamin K status is currently most commonly determined through the measurement of phylloquinone in serum and plasma by HPLC with fluorescence detection[42, 43] and various modifications of this method.[44–47] Electrochemical detection has also been used, which provided enhanced sensitivity for endogenous concentrations of vitamin K[48, 49] and its urinary metabolites.[50]

Contemporary methods developed and applied in clinical laboratories are increasingly based on LC-MS/MS and only recently, the almost forgotten technique of supercritical fluid chromatography (SFC) has been applied for the first time as "Ultra Performance Convergence Chromatography" (UPC^2) which offers a powerful tool for the separation sciences community[51] and is characterized by high sensitivity and selectivity.

The development and validation of accurate and precise assays for the direct measurement of phylloquinone requires consideration of several key factors. Obstacles to success include the sensitivity of vitamin K to daylight and chemical properties that confer the need for special sample handling, e.g., the avoidance of exposing samples to alkaline conditions. All sample preparation steps should be carried out under subdued light, or samples should at least be protected from strong light by wrapping them tightly in the silver foil or using amber glassware.

Preanalytical Sample Preparation

Vitamin K is sensitive to light and alkaline conditions (it is essential to remove all traces of detergents from glassware) but is stable to temperature, oxygen, and resistant to acids. All procedures for vitamin K analysis should be performed in subdued light.

Sample Collection

Vitamin K is most frequently measured in serum and plasma. Venous blood (4–5 mL) should be collected from a peripheral vein into plain tubes (serum) or tubes containing a suitable anticoagulant such as EDTA or heparin (plasma). Samples should be stored at −20C. Whole blood collected into sealed vacuum tubes, protected from light can be stored at 4°C for 24–48 h.

Internal Standardization

The analysis of phylloquinone typically requires sample purification and multiple chromatographic separations—making the inclusion of an internal standard (IS) to account for losses in vitamin K during the process essential. The selection of an appropriate IS is confounded through the wide variety of matrix-related interferences that may occur while processing blood samples from a wide variety of subjects. An effective IS requires structural analogy; it is good practice to use a compound with the same analytical properties as the analyte of interest that is not endogenously present in the specimen being analyzed. In case of any chemical modification during chromatographic analysis (e.g., postcolumn zinc reduction—as discussed later in this chapter), the IS should express the same chemical behavior. Additionally, the IS compensates for losses during extraction, evaporation, and injection of the sample.

A selection of IS material, used in recent HPLC and LC analysis, is presented in Table 5.1. Internal standards for vitamin K analysis very often utilize analogs of vitamin K, which differ in the side chain in position 3 of the naphthoquinone. These compounds exhibit the same chemical properties but are sufficiently different to allow full resolution on a chromatographic column.

Stable isotope-labeled internal standards are widely used in LC-MS/MS analysis. Often the choice is between using deuterated or ^{13}C-enriched compounds, however the latter are more expensive with limited availability. It should be noted that a deuterium (^{2}H) labeled IS is at some risk of hydrogen-deuterium exchange. On the other hand, ^{13}C seems more stable in comparison to deuterated compounds, but sometimes will encounter ionization issues.[76] Isotopically labeled internal standards are characterized by the same extraction recovery, ionization response, and also similar chromatographic retention time to the compound of interest. Importantly, the IS should differ in mass by at least +4 to +7 AMU to show a signal that does not interfere with the analyte.

Table 5.1 Examples of Published Methods for Measurement of Vitamin K in Human Serum or Plasma

Author	Method	Column	Internal STD	PK	MK-4	MK-7	K₁O
				Lower limit of detection (LOD)			
Nannapaneni et al.[53]	UFLC-APCI-MS/MS	COSMOSIL packed column, 2.5 µm Cholester, 3.0 mm×75 mm	Trans-phytonadione-D7 cis-phytonadione-D7	0.1 µg/L			
Abro et al.[55]	LC-APCIMS	Chromolith performance RP-18e column 100×4.6 mm	Vitamin D2	0.17 µg/L			
Riphagen et al.[57]	LC–MS/MS	Phenyl-hexyl, column (particle size 3 µm, 2.0 mm i.d. ×100 mm	D7-labeled internal standards for PK, MK-4, and MK-7	0.14 nmol/L	0.14 nmol/L	4.4 nmol/L	
Nair et.al.[59]	LC–MS/MS	C30 column, 250×4.6 mm, 200A, 5 µm	D-labeled transphytonadione	0.65 µg/L for cis 0.652 µg/L for trans			
Karl et al.[61]	LC-APCI-MS	Kinetex C18, 2.6 µm, 150 mm×3.0 mm	K1-d7	30 pmol/g	30 pmol/g	5 pmol/g	
Gentili et al.[63]	LC–MS/MS	SupelcosilTMC18, 4.6 mm×50 mm; 5 µm	K1-d7, MK-4-d7, K1-2,3 epoxide-d7	0.087 µg/L	0.107 µg/L		0.107 µg/L
Ahmed et al.[65]	HPLC-CL	Capcell pack ODSUG 120 35 mm×1.5 mm i.d., 5 µm	2-Methyl-3-pentadecyl –1.4-naphthoquinone	0.03 ng/mL	0.04 ng/mL	0.10 ng/ mL	
Ducros et al.[67]	LC-APCI-MS/MS	Alltima C18 column 150 mm, 2.1 mm i.d., 3 µm	D4-ring vitamin K₁	36 ng/L			

Continued

Table 5.1 Examples of Published Methods for Measurement of Vitamin K in Human Serum or Plasma—cont'd

Author	Method	Column	Internal STD	PK	MK-4	MK-7	K₁O
Fu et al.[69]	LC-APCI/MS	ProntoSil C30 column 5µm, 250mm×4.6mm	Vitamin $K_{1(25)}$	0.2 nmol/L unlabelled, 0.5 nmol/L labeled			
Paroni et.al.[47]	HPLC-FL	Spherisorb Ultrasphere ODS Beckman column 125mm×4.6mm, 5µm	Proprietary vitamin K derivative	0.2 nmol/L.			
Song et al.[72]	HPLC-APCI-MS	Sepax GP-C8 column, 5µm, 250mm×4.6mm i.d.	teprenone	0.3 ng/mL			
Ahmed et al.[73]	HPLC-PO-CL	Develosil ODS UG-5 50mm×1.5mm, i.d.	2-methyl-3-pentadecyl-1,4-naphthoquinone	32 fmol	38 fmol	85 fmol	
Suhara et al.[74]	LC-APCI-MS/MS	Capcell PAK C18 UG120, 5µm; 4.6mm i.d.×250 mm	^{18}O-labeled vitamin K derivatives	50 pg/mL	40 pg/mL	80 pg/mL	
Kamao et al.[46]	HPLC-FL	CAPCELL, PAK C18 UG120 4.6mm×250mm, 5µm	Vitamin K analogs synthesized in house	2 pg	4 pg	4 pg	
Wang et al.[45]	HPLC-FL	Hypersil BDS-C18 3.2×150mm	MK-6	10–50 fmol			

Chromatography
Thin layer chromatography

Thin layer chromatography (TLC) is no longer used for the analysis of vitamin K although it is generally regarded as an analytical method suited to the screening, separation, and preliminary identification of compounds. Qualitative and sometimes quantitative analysis of chemical components in complex mixtures is possible. TLC was successively used for semiquantitative vitamin K analysis from the late 1960s until the middle of the 1980s. Because of its low speed, poor sensitivity, and incompatibility with online detection and quantitation methods this separation technique has been largely replaced for more reliable HPLC and LC-MS/MS methods. The relevant reviews can be found here.[70, 75, 77]

High performance liquid chromatography methods

Methods based on HPLC have been used for the analysis of vitamin K in serum/plasma since the late 1960s—albeit at pharmacological concentrations. Many improvements in technology and sample preparation have allowed this technique to become the first line choice for separation of vitamin K and its analogs in both research and diagnostic laboratories. The highly lipophilic and neutral character of the vitamin K is well suited to a variety of different HPLC modes (normal and reversed) coupled to a variety of different inline detectors. The methods based on HPLC provide two major advantages to vitamin K analysis in comparison to methods previously used. First, advances in column technology have given opportunities in achieving much better chromatographic resolution. Irregularly shaped microparticulate packings were used throughout the 1970s until spherical materials were developed and their manufacture perfected. Spherical packings are packed more homogeneously than their irregular predecessors, providing superior column efficiency and selectivity. Second, advances in the design of modern detectors such as UV, EC, FL, and MS/MS, and equally importantly the development of pumps capable of delivering a steady flow of mobile phase at high pressure, enabled the advance in vitamin K measurement.

Nevertheless, the low circulating concentrations of the vitamin K in a complex biological matrix, with coexistent lipids make this measurement analytically challenging. How these issues have been addressed in the modern HPLC systems and other liquid chromatography or SFC will be discussed in the following sections.

Adsorption chromatography (normal phase high performance liquid chromatography)

Typically, stationary phases for normal phase chromatography consist of columns packed with silica. Silica stationary phases provide poor selectivity for equipolar lipids which coelute with vitamin K, with the consequence of a relative poor separation from interfering lipids. Changing the organic modifier in the polar mobile phase may provide better selectivity, however options for this are limited. Hexane with 3%–10% diethyl ether or minimal amounts of acetonitrile, ethyl acetate, or diisopropyl ether (up to 1%) can be used for this purpose. The mobile phase is practically the only variable used to achieve any change in selectivity. Silica Sep-Pak cartridges are still widely used for a cleanup stage of serum samples prior to injection on the reversed-phase HPLC systems.

Reversed-phase high performance liquid chromatography

Reversed-phase HPLC is the most commonly used approach for the analysis of vitamin K. A great variety of packing materials are available allowing the use of eluents and additives to solve analytical challenges; additionally the choice of column packing material can be adjusted depending on the preferred method of detection. The separation of vitamin K homologs can be achieved using highly retentive octadecylsilyl (ODS) packing material, with column lengths between 50 and 300 mm, but most frequently 150–250 mm. The usual internal diameter of the columns is 2.1, 3, and 4.6 mm, with particle sizes of 3 μm or 5 μm.

Reversed-phase chromatography with fluorescence detection has been the most utilized technique for the separation of vitamin K. Reversed-phase HPLC is well suited for the injection of samples reconstituted in polar solvent, which is not ideal since vitamin K is very nonpolar—isopropanol or ethanol have both been used successfully. The introduction of nonpolar solvents into the system is prohibited since it would cause minimized retention and additionally lowered fluorescence yields of vitamin K.

In human blood, vitamin K is transported by lipoproteins—CM, remnants, and VLDL. When extracting vitamin K, effective denaturation of lipoproteins is necessary; however, there may be loss of analyte by occlusion by the precipitate. Several precipitation methods are available for protein isolation; these include acetone, ethanol, isopropanol, and chloroform/methanol. The most effective way to denature the vitamin K transport proteins is with ethanol (usually in a volume ratio 1:4). Subsequently, extraction of vitamin K is performed with hexane and then the sample extracts are purified using solid–phase extraction.

Example methods for vitamin K analysis with fluorescence detection

Langenberg and Tjaden[78, 79] were the first to use fluorescence detection coupled to HPLC for the determination of vitamin K in serum samples ultimately leading to the UV- and EC-based approaches being largely abandoned. Modifications by Haroon et al. and later by Davidson et al. led to significant improvements of this technique; its sensitivity and robustness, and until today this is the most commonly used methodology for measurement of phylloquinone.

Features of the method published by Haroon et al.[42]

- Sample volume requirements reduced to 0.5–1.0 mL. Previous assays had required >1 mL
- Phylloquinone extracted into acetonitrile
- A single liquid–liquid extraction step used
- Sample extract purified on silica, then treated with an acidic mixture of hexane/acetonitrile (1/4 by vol) containing 70 mmol/L of zinc chloride
- The lipid extract directly injected into reversed-phase HPLC column
- Phylloquinone quantified by comparison with the internal standard dihydro-vitamin K_1
- A lower limit of detection for phylloquinone of 50 pg/mL achieved

Features of method published by Davidson and Sadowski[43]

Davidson et al. [43] modified Haroon's method to gain sufficient sensitivity for the measurement of endogenous phylloquinone and the epoxide form. The authors achieved this using more recently developed small-bore analytical columns packed with $3 \mu m$ particle size material (BDS Hypersil (250×2.1 mm i.d.)) and a mobile phase consisting of dichloromethane-methanol (10:90, *v/v*) with 0.5% of an aqueous solution containing 2 M zinc chloride, 2 M glacial acetic acid, and 1 M sodium acetate. Oxygen was removed from the HPLC using a degasser to maximize reduction of the quinone to hydroquinone. The fluorescence detector was set at excitation 244 nm and emission 430 nm. A lower limit of detection of 15 pg/mL (33 pmol/L) was achievable.

Sample Preparation—Liquid-Liquid Extraction

- Pipette 0.5 mL of serum or plasma into a glass tube
- Add $20 \mu L$ of the internal standard $K_{1(25)}$ (2.0 pmol)
- Add 1 mL of ethanol and vortex to precipitate the plasma proteins
- Add 0.5 mL of H_2O and 3 mL of hexane and mix vigorously for 2 min prior to centrifugation at 1000*g* for 5 min
- Transfer top hexane layer to a new set of tubes
- Evaporate hexane extract to dryness at 50°C under oxygen-free nitrogen

Extract Purification by Solid-Phase Extraction

- Add $200 \mu L$ of 2-propanol to the dried sample extract and heat to 50°C
- Condition SPE C18 column with 3 mL of dichloromethane/methanol (20:80, *v/v*) then 3 mL of methanol and, finally 3 mL of deionized H_2O
- Apply dissolved sample extract directly onto the preconditioned SPE column
- Apply vacuum on the manifold for at least 2 min to dry the sorbent
- Wash the sorbent with 3 mL of H_2O/methanol (5:95, v/v) and with 3 mL of acetonitrile
- Elute the vitamin K fraction with 6 mL of dichloromethane/methanol (20:80, v/v)
- Evaporate eluent to dryness and reconstitute with $20 \mu L$ of dichloromethane immediately followed by $180 \mu L$ of methanol containing 10 mM zinc chloride, 5 mM acetic acid, and 5 mM sodium acetate
- Inject $100 \mu L$ of sample extract onto the system

Simultaneous determination of phylloquinone and vitamin K_1 2,3-epoxide

Blood concentrations of KO are increased in subjects taking warfarin and other vitamin K antagonists. Without vitamin K antagonist action the measurement of KO in serum and plasma is difficult since it is circulating at concentrations of less than a tenth of that of phylloquinone.[43] To determine phylloquinone and KO simultaneously Davidson et al ensured that residual oxygen was removed from the HPLC system and that the linear velocity of the mobile phase was optimized for the postcolumn reduction of KO. The reduction of KO to the hydroquinone requires two steps: the reduction to phylloquinone and then to hydroquinone. The two-step reduction, without the oxygen scrubber, was found to be incomplete leading to much lower sensitivity.

Method for determination of phylloquinone and vitamin K_1 2,3 epoxide

Modified Davidson et al method (as used by the chapter author)[43]

Phylloquinone and KO are extracted from serum or plasma into hexane after precipitation of proteins with ethanol and addition of an internal standard, MK-6. The hexane extract is subjected to a purification procedure to remove coextracted lipids along with contaminants using a silica SPE cartridge and facilitate injection onto a HPLC system. Final analysis is by reversed-phase HPLC using a column containing C18 modified silica. Phylloquinone, KO, and IS are detected following postcolumn chemical reduction (S·S column 2.1 (i.d) x 30 mm containing zinc dust, <10 μm, by fluorescence, (excitation 330 nm, emission 430 nm).

Sample preparation—liquid-liquid extraction

- Add 500 μL of serum/calibrators and 200 μL of MK-6 (3.6 ng/mL) to a glass test tube and vortex briefly
- Add 3.6 mL of ethanol to each tube and vortex for 1 min followed by 12 mL of hexane and mix for 10 min at 1000 g
- Transfer the upper organic layer to a new test tube and dry down under oxygen-free nitrogen at 50°C
- Reconstitute the extract in 2 mL of hexane

Extract purification by solid-phase extraction

- Reconstitute the lipid extract in 2 mL hexane and vortex mix
- Prime Sep-Pak cartridges with 4 mL of hexane and draw through under minimal pressure
- Load sample onto cartridge and draw through under minimal pressure
- Add 4 mL of hexane to sample tube and vortex mix
- Load onto Sep-Pak cartridge and elute under minimal pressure
- Add 2 mL of hexane to the cartridge and elute under minimal pressure
- Elute the vitamin K and internal standard fraction with 10 mL of 3% (v/v) diethyl ether in hexane
- Evaporate to dryness under a stream of oxygen-free nitrogen at 50°C
- Dissolve the extract in 50 μL ethanol
- Leave the samples resting on the bench protected from light for 10–15 min
- Inject the samples onto HPLC system

Example HPLC equipment

- Column BDS Hypersil (150×2.1 mm i.d., 3 μm)
- Zinc column[1]

---

[1] Take precautions while working with the zinc column as zinc particles in the column are being oxidized and are able to pass through the filter frits and enter the flow cell causing baseline drifts and anomalies.

- Column oven with temperature set at 50°C
- Degasser
- Fluorescence detector (excitation 330 nm, emission 430 nm)
- Injection volume: 10 µL
- Sampler temperature: 4°C

Mobile phase

- 0.5% zinc acid solution[2] in methanol
- Flow rate: 0.4 mL/min, isocratic

Lower limit of detection

- Phylloquinone = 0.05 µg/L (0.11 nmol/L)
- Vitamin K_1 2,3-epoxide = 0.12 µg/L (0.26 nmol/L)
- Run time: 22 min (while using the switching valve to divert late eluting chromatographic peaks to waste), 50 min (without the switching valve)

Supercritical fluid chromatography

Until recently instrumental and technological limitations prevented the widespread adoption of SFC. The application for vitamin K is based on the principle that under pressure, CO_2 occurs as a liquid that has unique properties as a solvent for chromatography. It is characterized by low viscosity and high diffusion coefficient which in turn make its performance something between liquid and gas chromatography. In addition, there are varieties of stationary phases available similar to those used in LC. A huge advantage of the system is the possibility of using the sub-2 µm particles in the column, which allow for higher speed (the higher linear velocity), thus offering much shorter separation time. For detection, SFC is coupled to the same panel of detectors as HPLC, including a UV spectrophotometer, FL detector, or MS/MS).[51]

The ACQUITY ultraperformance convergence chromatography system

An example of a commercially available SFC solution is the ACQUITY UPC^2 system which has been commercially available since 2012 from Waters (Milford, MA, USA). This instrument overcomes a number of the pitfalls reported in the previously designed SFC systems. Important features in the SFC system include the efficient cooling of the CO_2 pump heads and the design of a new dual stage back pressure regulator that is heated and insulated to avoid frost formation. In addition, the auxiliary injection valve has been designed to reduce pressure pulses from the injection sequence and reduce carryover, enabling repeatable and reproducible partial loop injections.

The UPC^2 system has similar selectivity to normal phase chromatography, compounds elute from the stationary phase according to its polarity whereas in reversed-phase chromatography polar analytes elute first. In convergence chromatography

---

[2] Zinc acid solution: mixture of 2 M $ZnCl$, 1 M sodium acetate, 2 M acetic acid.

polar analytes are retained and elute last. The UPC^2 platform offers minimized system and dwell volume, enabling users to maximize the separation power inherent to sub-2 μm particles.

Application for quantitative analysis of phylloquinone and vitamin K_1 2,3-epoxide

In the author's laboratory, a method has been developed using UPC^2 in tandem with a Xevo TQS micro to rapidly determine endogenous concentrations of phylloquinone and KO.[80] Baseline separation of phylloquinone, KO, and a deuterated form of vitamin K_1 (K_1–d_7) is achieved using an ACQUITY UPC^2 HSS C18 column in 2.4 min, with the total run time of 5.5 min under a gradient elution of a CO_2 and methanol mixture (1%–40% methanol), the makeup solvent was used (2% formic acid in methanol) to enhance ionization and increase sensitivity. Methanolic calibration curves exhibited linearity in the range of 0.1–100 μg/L for phylloquinone and for KO. The quantifier MRM transitions for K_1, KO, and K-d_7 were $451.4 \rightarrow 187.2$, $467.6 \rightarrow 307.5$, and $458.4 \rightarrow 194.1$ m/z, respectively. The reproducible injections of only 1 μL of serum extracts resulted in 0.1 μg/L LLOD for both K_1 and KO. The total run time of 5.5 min is four to five times faster than the typical run time achievable using HPLC or LC-MS/MS. The organic solvent consumption is <1 mL per injection resulting in the much reduced consumption of solvents in comparison to conventional liquid chromatography. This method is simple to perform and practical for routine phylloquinone and KO determination in serum with significant increases in throughput and decreases in operating costs.

Ultraperformance Convergence Chromatography Conditions

Instrumentation	ACQUITY UPC^2 With Xevo TQSmicro MS
Column	ACQUITY UPC^2 HSS C18 column, 1.8 μm, 2.1 mm × 100 mm
Mobile phase	Solvent A: CO_2
	Solvent B: methanol
Gradient	1%–40% Methanol
ABPR	2500 psi
Make up	2% Formic acid in methanol (0.35 mL/min)
Flow rate	1.5 mL/min
LLOQ	0.1 μg/L

Detection

Many different types of detectors have been used as the final step of analysis for the determination of vitamin K. Detection of vitamin K can take advantage of UV characteristics, fluorescence properties (hydroquinone form of vitamin K only), or electrochemical activity. Vitamin K and its homologs can also be detected with peroxyoxalate chemiluminescence coupled to HPLC systems (HPLC-PO-CL).[73] In recent years, it has been demonstrated that LC-MS/MS can be successfully applied to vitamin K analysis; therefore the number of published methods continues to grow (Table 5.1). The various mass spectrometer designs provide an impressive array of options for the analyst to choose from.

Ultraviolet detection

Ultraviolet detection coupled to HPLC was commonly used for vitamin K analysis during the 1970s. The methods suffered from relatively high limits of detection (500–1000 pg) and were not suitable for the detection of vitamin K at endogenous concentrations. Despite these limitations this approach is still useful when higher concentrations of vitamin K are expected in analyzed material. Vitamin K exhibits an UV spectrum that is characteristic of its naphthoquinone nucleus, with four distinctive peaks between 240 and 280 nm and much less absorption at around 320–330 nm. The maximum absorbance for phylloquinone in ethanol is E (1%, 1 cm) = 19,900 L/mol × cm at 248 nm.[81]

Fluorescence detection

Detectors that monitor the fluorescence emitted by analytes are generally more sensitive than photometric-based techniques and have been used successfully for the determination of vitamin K at endogenous concentrations since the 1990s. Vitamin K does not show native fluorescence, therefore the stable quinone form of the vitamin must be derivatized to the corresponding fluorescing hydroquinone. Inline pre and postcolumn reactors have been used to perform the derivatization which can be performed either chemically, wet chemically, and electrochemically. Example excitation and emission wavelengths are 329 and 430 nm, respectively, dependent on the detector configuration and lamp accordingly.

Generation of fluorescent hydroquinones

There are a number of methods available for the reduction of phylloquinone and menaquinones to their corresponding fluorescent hydroquinone. A reducing reagent can be added to the HPLC eluent or on a solid-phase reactor (pre or more commonly postcolumn). The most popular and effective technique exploits the reduction of vitamin K quinone by metallic zinc (200-mesh) packed into a short solid-phase postcolumn in the presence of zinc ions which have been added to the mobile phase.

Reduction on the zinc column:

$$Zn \rightarrow Zn^{2+} + 2e^-$$

$$2H^+ + 2e^- + K_1 \rightarrow K_1H_2$$

$$Zn + 2H^+ + K_1 \rightarrow Zn^{2+} + K_1H_2$$

The solid-phase postcolumn reactor is dry packed with zinc particles and placed between the main analytical column and the fluorescence detector. To the nonaqueous mobile phase (methanol) are added zinc chloride 2 M, 1 M sodium acetate, 2 M acetic acid. This allows efficient generation of hydrogen gas for the reduction of quinones. Many groups continue to use this approach with a variety of "in-house" adaptations.[43]

Electrochemical reduction

The electrochemical reduction of vitamin K quinones is achieved through a two-electron and two-proton transfer process. This chemical property was used successfully for the determination of vitamin K with fluorimetric detection by Langenberg et al.[78]

Electrochemical detection

In amperometric electrochemical detection, vitamin K quinones enter a flow cell, where they are reduced to the corresponding hydroquinones under a negative potential on an electrode. The main disadvantage of this type of detector is that the presence of even the smallest amounts of oxygen within the system causes high background currents and baseline drifts; moreover, serum samples led to periodical passivation of electrodes causing loss of sensitivity of the system.

For vitamin K analysis, a dual redox electrochemical detector system was often used as described by Canfield et al.[82] The principles of the method are based on the fact that the naphthoquinone nucleus is reduced to its naphthoquinol form on the first (upstream) electrode and then subsequently oxidized back to the respective quinone forms on the second electrode (downstream). A minimum potential of $-0.4\,V$ is necessary at the electrode for the reduction of phylloquinone (vitamin K 2,3-epoxide requires a lower potential of $-1.0\,V$), whereas on the downstream electrode a potential of $0.0\,V$ up to $+0.7\,V$ has been used.[49]

Mass spectrometry

Techniques based on LC-MS/MS are increasingly being developed and applied for vitamin K status determination. The analysis is reliant on three sequential events: ionization, mass-to-charge (m/z) ion separation, and ion detection. Liquid chromatography mass spectrometry-based method development for vitamin K has been technically challenging because of the low endogenous serum concentration, interfering serum lipids, and photo sensitivity. Awareness of the appropriate sample preparation and chromatographic separation prior to MS/MS detection for accurate measurement is crucial. It is also very important to choose the right ionization method on the LC-MS/MS system as this has a dramatic impact on the matrix effects and thereby assay accuracy and sensitivity. Methods using ESI and APCI ionization modes have both been described.[53, 83] It is thought that APCI provides superior ionization for vitamin K.

For phylloquinone, the parent ion ($m/z=451$, $[M+H]^+$) is fragmented to give a spectrum of daughter ion ($m/z=187$, $[M+H]^+-264$). This is achieved by elimination of $C_{19}H_{36}$ from the phytol side chain molecule. Analogically the other homologs and metabolites can be detected. It is recommended to monitor more than one fragmentation pattern (quantifier and qualifier) for better specificity of the method. A multiple reaction monitoring mode (MRM) is frequently used for acquiring the LC-MS/MS data.

Example Protocol for Sample Preparation

Sample preparation—liquid-liquid extraction

- Pipette $250\,\mu L$ of each sample into a glass tube
- Add $600\,\mu L$ of working internal standard solution using a multipipette dispenser to each tube, pulse vortex

- Using a multipipette dispenser, add 3 mL of hexane to each tube
- Carefully recap the tubes and mix the contents using a multitube vortex mixer for approximately 10 min
- Centrifuge the tubes (1000 g, 10 min)
- Transfer the supernatant (>2.5 mL) to 12 mL glass tubes using a glass Pasteur pipette
- Transfer each tube to an evaporator (e.g., XcelVap) in a fume hood, (50°C, 10 min, nitrogen pressure: start 18 psi—end 24 psi), evaporate to dryness under a stream of oxygen-free nitrogen)

Solid-Phase Extraction Cleanup Using Oasis PRiME HLB, Three-Step Protocol (Catch and Release)

- Reconstitute the samples in 1 mL of heptane
- Load 1 mL of sample extract onto the SPE cartridge
- Wash cartridge with 1 mL of 80% methanol
- Elute fraction of interest with 1 mL of heptane and collect into a vial
- Transfer each vial and evaporate to dryness in a fume hood, (settings 50°C, 10 min, nitrogen pressure: start 18 psi—end 24 psi)
- Reconstitute the samples in 50 μL of ethanol (wait 10 min to allow purified extract to dissolve)
- Transfer 50 μL of extract to a glass insert vial. Transfer vials to the LC-MS/MS autosampler
- 30 μL of extracted sample is injected

Liquid Chromatography Separation

Separation is achieved by means of an Agilent Eclipse Plus column with the dimensions 2.1×50 mm; packed with C18 with a particle size of 3.5 μm. The chromatographic mobile phases consist of 0.1% acetic acid (v/v) in water (eluent A) and 0.1% acetic acid (v/v) in methanol (eluent B). A column flow rate of 0.5 mL/min is used throughout the chromatographic run while the column temperature is maintained at 50°C. A fast gradient elution is then performed from 80% B to 100% B in 0.1 min, with an isocratic hold at 100% for 5.4 min to wash away late eluting components; the column is then equilibrated to baseline conditions.

Standardization

No international certified standard reference material is available for vitamin K. The international EQA provider KEQAS provides a noncertified standard reference material which is suitable as a control material for assigning values to in-house control materials and for validating analytical methods for measurement of phylloquinone in serum. The phylloquinone concentration of this material was assigned as a consensus value by the author's laboratory from 15 laboratories from 13 countries.

In-house calibrants for phylloquinone assays may be gravimetrically prepared in ethanol, serum, or serum-adjusted matrix. Preparation must be documented according to the internal procedures for the in-house calibrations and should be in line with ISO 15189 (2012) requirements.

Reference Range for Phylloquinone

In fasting adults: range: 0.17–0.68 µg/L (0.38–1.51 nmol/L) and nonfasting adults: range: 0.15–0.55 µg/L (0.33–3.44 nmol/L) Thane et al.[84, 85] There is little data available for the menaquinones.

External Quality Assurance Scheme

A quality assurance scheme for the determination of phylloquinone in serum, vitamin K external quality assurance scheme (KEQAS), has been available since 1996. The scheme provides a comprehensive help and advisory service in the development and validation of vitamin K assays. Aliquots of an untampered serum pool, ethanolic vitamin K standard spiked samples, and vitamin K depleted material as well as ethanolic phylloquinone standard solutions are regularly distributed to users of this service. The mean interlaboratory interrun CV for all laboratories shows CVs of 45% or less. This is consistent with the empirically derived Horwitz curve for the interlaboratory determination of analytes present at 1 ppb,[86] equivalent to the µg/L concentrations of phylloquinone determined in the EQA samples. The performance evaluation is undertaken by assessing the all laboratory trimmed mean which represents the target concentration. The target for results is the current standard deviation of proficiency testing (SDPT). The SDPT is calculated on the rolling 3-year basis and is responsive to changing trends in performance due to changes in, e.g., methodology.

The performance of such tests aims to assist in harmonization of phylloquinone analysis in order to improve the comparability of clinical and nutritional studies as well as to improve the quality of analysis. The scheme alerts laboratories if they report results that stray from the consensus of KEQAS members and provide support in rectifying analytical problems and to improve the quality of phylloquinone analysis through a Z scoring system.[87]

Functional Markers for Assessment of Vitamin K Status

Prothrombin time

Although coagulation type assays are widely used for the assessment of vitamin K status they offer only a limited, and relatively insensitive, overview and reflect the ability of the coagulation pathway to form fibrin rather than being specific for vitamin K. However, normalization of a previously prolonged PT after vitamin K administration is considered as diagnostic of vitamin K deficiency.

The PT only becomes prolonged once the concentration of prothrombin drops below 50% of normal.[15] Measurement of PT and the activated partial thromboplastin time (aPTT) can reflect an advanced stage of deficiency, but can be also suggestive of hepatic or hematologic diseases unrelated to vitamin K deficiency. Coagulation tests are used to assess the interference of vitamin K with oral anticoagulants and can give some limited indications on vitamin K status.

Urinary markers

Total urinary γ-carboxyglutamic acid

γ-Carboxyglutamic acid is synthesized posttranslationally and is excreted in urine by the kidney. As a result, urinary Gla excretion has been used as an indicator of vitamin K status in adults and reflects VKDP turnover and degradation. Urinary free Gla, which offers an overall assessment of the γ-carboxylation status of Gla proteins can be measured in the urine during a 24h collection (or concentration can be creatinine adjusted if spot urines are collected). The method is based on separation of o-phthalaldehyde derivatives by reversed-phase HPLC with fluorometric detection.[88, 89] There are no agreed cutoff values for urinary Gla concentration that would indicate vitamin K adequacy.[90]

Metabolites of vitamin K

In urine, the two major forms of vitamin K metabolites are the 5C- and 7C-aglycones. Their urinary excretion reflects dietary vitamin K intake and is a marker of total vitamin K status since all K-vitamers share the same metabolites.[50]

The urinary excretion of vitamin K metabolites has been studied in a young adult population [50] and in preterm and term neonates before and after vitamin K prophylaxis.[91] The study of vitamin K status in this vulnerable population has hitherto been problematic due to limitations in available techniques and reliance on invasive tests. Measurement of the 5C- and 7C-aglycones of vitamin K offers a novel noninvasive marker of total vitamin K status; however, similarly to urine Gla, the cutoff values for vitamin K adequacy have not been identified.[90]

The method for the measurement of urinary metabolites of vitamin K was published in 2005 by the author's colleagues and used HPLC with EC in the redox mode.[92] The preliminary cleanup process included removal of urinary salts by C18 SPE and deconjugation of metabolites with methanolic HCl. Then, the generated carboxylic acid aglycones were subject to cleanup procedure using silica SPE. The final analysis was by reversed-phase (C18) HPLC with a methanol-aqueous mobile phase. Metabolites were detected by amperometric, oxidative ECD of their quinol forms, which were generated by postcolumn coulometric reduction at an upstream electrode. The assay gave excellent linearity and high sensitivity with an on-column detection limit of <3.5 fmol (<1 pg).[92]

Example Protocol for Sample Preparation

Sample collection

Urine samples can be either spot urines, for which metabolite concentrations must be corrected for creatinine, or ideally 24h urine collections. Spot urines should be mid-stream samples collected in the morning.

When collecting 24h urine samples care should be taken to ensure that all the urine output during that period is collected and that the final volume is measured accurately. Samples should be wrapped in foil immediately to protect from light and frozen at −20°C for short-term storage or −70°C for long-term storage (>6 months).

Stage 1: Sample wash
- Prewash C18 SPE column with 1 mL methanol then with 1 mL deionized water
- Load 500 μL of urine sample on to column, elute under minimal pressure

- Wash with 1 mL deionized water under minimal pressure.
- Add 100 µL of 2-methyl-3-(7′-carboxy-heptyl)-1,4-naphthoquinone (methanolic Internal standard $c = 1.50$ µg/mL) to cartridge.
- Elute with 2 mL of methanol and collect the eluent.
- Evaporate to dryness under a stream of oxygen-free nitrogen at 40°C.

Stage 2: Deconjugation with hydrochloric acid
- Redissolve dried product in 1.1 mL methanolic HCl (1.5 mol/L) and vortex.
- Incubate for 16 h at room temperature

Stage 3: Bligh and Dyer liquid-liquid extraction
- Add 1.1 mL chloroform and 1 mL water to the extract dissolved in methanolic HCl and vortex for 30 s
- Centrifuge for 5 min
- Using a glass Pasteur pipette carefully remove and dispose of top layer
- Add 10 mL deionized water, replace cap, and invert several times
- Carefully remove all the aqueous top layer using a glass Pasteur pipette
- Evaporate to dryness under a stream of oxygen-free nitrogen at 40°C

Stage 4: Diazomethane production and methylation
- Add 20 mL potassium hydroxide (5 M) into a clean Duran
- Add 20 mL diethyl ether
- Using a spatula, carefully add methylating agent (MNNG[3]) until upper ethereal phase becomes a golden yellow color
- Add 300 µL of the upper yellow layer to the dried extracts from the previous stage, replace caps after each addition
- Gently agitate the samples
- Allow to react for 5 min
- Evaporate to dryness under a stream of oxygen-free nitrogen at 40°C

Stage 5: Solid-phase extraction
- Redissolve sample extracts in 2 mL hexane, vortex, heat gently (40°C), vortex again, and allow to cool.
- Assemble Sep-Pak cartridges with clean glass syringes on the manifold and waste tubes in the collection area.
- Prime Sep-Pak cartridge with 2 mL hexane.
- Pour the sample dissolved in hexane to the syringe on top of the Sep-Pak cartridge.
- Draw through under minimal pressure.
- Add 8 mL hexane to the tubes, replace cap and invert several times, add to the syringe, and draw through under minimal pressure.
- Replace waste tubes.
- Elute with 15% diethyl ether in hexane.
- Evaporate to dryness under a stream of oxygen-free nitrogen at 40°C.
- Redissolve in 100 µL methanol and inject onto HPLC system

---

[3] CAUTION. Take great care to comply with all safety information when using this very toxic and explosive reagent.

Protein induced by vitamin K absence

An inadequate supply of vitamin K prevents the optimal γ-carboxylation of VKDPs. These proteins lack functionality and are referred to as PIVKAs. The measurement of PIVKA-II (undercarboxylated factor II) is a functional indicator of hepatic vitamin K status, since PIVKA-II is not detectable at appreciable concentrations in the circulation of healthy adults. Undercarboxylated prothrombin species are specific for vitamin K deficiency and highly sensitive immunoassays are used for diagnosis of subclinical states of vitamin K deficiency.

Automated assay for undercarboxylated prothrombin

The Abbott ARCHITECT automated PIVKA-II assay is designed as a sandwich format immunoassay, using chemiluminescent paramagnetic microparticle analyzer (CMIA) based technology for quantitative determination of PIVKA-II species.[93] Two antibodies are utilized in the method. The first one is a recombinant murine monoclonal anti-PIVKA-II antibody 3C10 (Abbott Laboratories, IL, USA) which interacts with PIVKA-II and recognizes an epitope in PIVKA-II within the GLA domain, it is specific for amino acid glutamic residues at 13–27 positions in the N-terminal region (Fig. 5.4).

It was demonstrated that binding to 3C10 requires the Glu at positions 19–20, and additional Glu on position 25 resulted in increased binding capacity of the 3C10 antibody.[94] The second, the PIVKA-II microparticle complex is detected with an acridinium-labeled murine antiprothrombin monoclonal antibody MCA 1–8 (Abbott Laboratories) conjugate, which recognizes an epitope at the N-terminus in prothrombin (Fig. 5.4).

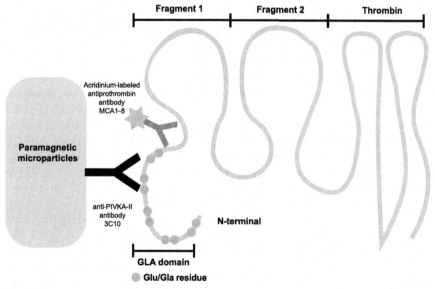

FIG. 5.4

Schematic illustration of the automated immunoassay ARCHITECHT PIVKA -II.

The automated assay is performed as follows: 30 μL of sample, 50 μL of assay buffer, and 50 μL of anti-PIVKA-II antibody-coated microparticles are combined and incubated for 18 min. After washing, 50 μL of acridinium-labeled antiprothrombin antibody conjugate is added and incubated for 4 min. Following an additional wash step, pretrigger and trigger solutions are added to the reaction mixture. The resulting chemiluminescent reaction is registered as relative light units (RLUs). The RLUs detected by the ARCHITECT optical system is related to the amount of PIVKA-II in the sample. Assay results can be obtained within 30 min after initial sample aspiration. The measuring range of this assay is 10–30,000 mAU/mL. Calibrators are prepared from PIVKA-II antigen diluted in buffer solution containing bovine serum albumin. The calibrator concentrations range from zero to 30,000 mAU/mL. PIVKA-II antigen for purpose of this assay was prepared from human prothrombin (Enzyme Research, IN, USA) by a thermal decarboxylation method described by Bajaj et al.[95] PIVKA-II value of the antigen was determined by comparison to the Picolumi PIVKA-II Assay (Eidia, Tokyo, Japan).

A linearity study was performed by Abbott Diagnostics based on guidance from the CLSI Document EP06-A.[96] The samples were prepared by dilution across the measuring interval with ARCHITECT PIVKA-II Calibrator A. The observed deviation from linearity ranged from −9% to 8% for samples from 20 mAU/mL to 30,000 mAU/mL. The observed deviation from linearity ranged from −1.27 mAU/mL to 0.55 mAU/mL for samples from LLOQ (5.06 mAU/mL) to 20 mAU/mL. Results >30000mAU/mL may be diluted using the Automated Dilution Protocol.

Undercarboxylated prothrombin determination by liquid chromatography mass spectrometry

Sohn et al.[97] used an MRM-MS approach to determine PIVKA-II for the diagnosis of hepatocellular carcinoma (HCC). They were able to quantify PIVKA-II at nanomolar concentrations (LLOQ, 0.49 nM (34.32 ng/mL); ULOQ, 1000.00 nM (70,037.00 ng/mL) in human serum samples. This method determines PIVKA-II due to the loss of their native quaternary, tertiary, and secondary structures. In brief, samples are subjected to immunodepletion, denaturation, chymotrypsin digestion, and desalting, and then chromatographed by LC. The signal was collected in the MRM-MS mode after electrospray ionization. In a clinical cohort of 250 samples (chronic hepatitis ($n=50$), liver cirrhosis ($n=50$), HCC ($n=50$), and recovery ($n=50$)), all AUROC values were higher than 0.7 and had diagnostic power P value <0.0001 compared with normal control ($n=50$). The correlation was $R=0.8335$ between established immunoassay and LC-MS/MS method.

Undercarboxylated prothrombin determination by manual enzyme linked immunoassay

The concentration of PIVKA-II in serum/plasma can be manually determined using a monoclonal antibody (MAb) to PIVKA-II in a sandwich ELISA format. The MAb (C4B6) used in the assay is conformation specific such that in the presence of calcium ions it binds only undercarboxylated species of prothrombin and does not cross-react with fully carboxylated native prothrombin. A 96-well plate is coated with the C4B6 MAb, diluted in carbonate–bicarbonate buffer and incubated overnight at 4°C. After incubation, the plate is washed and dilutions of PIVKA-II standards, unknowns, and QCs are added (in Tris-buffered-saline with Tween and calcium lactate) and incubated for 1 h. The plate is washed and a polyclonal antibody (rabbit antihuman prothrombin) added, the plate is then incubated for 30 min. Excess unbound antibody is removed by washing. A third HRP-conjugated antibody directed against rabbit immunoglobulin is then added and the plate incubated for 30 min. After a final wash sequence, ortho-phenylenediamine dihydrochloride (OPD) substrate solution is added. After approximately 10 min, the reaction is stopped with 2 M H_2SO_4 and the color change, which is proportional to the concentration of PIVKA-II measured at a wavelength of 490 nm using a spectrophotometer. The PIVKA-II serum/plasma standards used for calibration and the QC samples for monitoring assay reproducibility are derived from serum/plasma samples collected from subjects who are taking warfarin. The concentration of PIVKA-II in these serum/plasma standards is then measured using the Eitest Mono P-II micro cup EDO23 ELISA assay (Eisai Co Ltd., Tokyo). The concentrations of PIVKA-II by the Eisai assay are conventionally expressed as Arbitrary Units per ml (AU/ml). Belle et al[98] demonstrated that 1 AU of PIVKA-II in the C4B6 assay to be equivalent to 1 μg of purified multiple species of PIVKA-II measured by electrophoretic or ELISA techniques using polyclonal antibodies.[98] In the assay performed in the author's laboratory, the serum/plasma PIVKA-II standards were initially calibrated against the PIVKA-II standard provided in the Eisai kit, which had been widely used to assess vitamin K status in infants. Parallel assays of serum/plasma samples from newborns by both C4B6 and Eisai methods showed good agreement with respect to their relative quantification and specificity. The author's laboratory has found that once the PIVKA-II containing serum/plasma standards from warfarin-treated subjects had been calibrated against the Eisai assay, these standards are stable for several years when stored at −70°C. The threshold of detection of PIVKA-II is determined according to the upper value found in pools of vitamin K-replete healthy subjects (neonates or adults). The typical threshold of detection of 0.2 AU/mL represents a PIVKA-II concentration of approximately 200 ng/mL, which in turn represents 0.2% of total circulating prothrombin.

Sample collection

Venous blood (2 mL) should be collected from a peripheral vein into plain tubes (serum) or tubes containing a suitable anticoagulant such as EDTA or heparin (plasma). Samples may be stored up to 12 h at +4°C. If the length of time between sample collection and assay is to exceed 12 h, samples should be stored frozen under −20°C.

Reference range

Using the ARCHITECT assay, PIVKA-II was measured in 193 samples from healthy subjects in the European Union based on the guidance from CLSI document C28-A3c. A reference range of 17.36–50.90 mAU/mL was defined.

Using the manual PIVKA-II method the reference range was defined as ≤0.2 AU/mL.

Standardization

There is no international recognized method available or reference material for standardization of the PIVKA-II method.

Osteocalcin

OC is a small 49 amino acids long protein exclusively synthesized by osteoblasts and odontoblasts. In addition to its role in bone formation, OC plays a role in the regulation of energy metabolism and male fertility.[99] The process of γ-carboxylation leads to the activation of three specific glutamic acid residues in the mid molecular region of OC at position 17, 21, and 24.

The degree of OC carboxylation has been proposed as an indicator of the nutritional state of bone with respect to vitamin K.[100, 101] There is also evidence that circulating concentrations of undercarboxylated osteocalcin (ucOC) may be used as a sensitive marker of vitamin K inadequacy[31, 100, 101] or a poor diet.[102]

In the healthy adult population, approximately 30% of the circulating OC occurs in its undercarboxylated form,[103, 104] and increased vitamin K intake results in a rapid decline of ucOC,[8, 100] suggesting a state of subclinical vitamin K deficiency in healthy bone tissue. Because ucOC is detectable during sufficient intake of vitamin K to maintain coagulation, it is thought that ucOC may be a superior marker for detection of subclinical vitamin K deficiency in comparison to PIVKA-II.[31] On the other hand, some data indicates that ucOC does not exclusively depend on vitamin K status but also on vitamin D, suggesting that serum concentration of ucOC is not reliable for assessing vitamin K status.[105] Additionally ucOC concentrations have been reported to be increased both in postmenopausal women and in individuals who sustain hip fracture.[8, 106, 107]

Assays for ucOC are based on the higher affinity of native OC for hydroxyapatite[108] or barium sulfate[109] compared to carboxylated osteocalcin (cOC). Results of calculated ucOC should be corrected for the basal concentration of OC in the sample and be expressed as the percent of the total OC.[110]

The ratio of carboxylated (cOC) to not carboxylated osteocalcin (ucOC) is a measurement of vitamin K status in relation to bone health. cOC and ucOC can be quantified in serum with the Gla-OC and Glu-OC test kits from Takara (Shiga, Japan). Both kits are in vitro enzyme immunoassay based on a sandwich method that utilizes mouse monoclonal antibodies. Intra- and interassay precisions are 3.0%–4.8% and 0.7–2.4%, respectively, for the Gla-OC kit, whereas they are 4.5%–6.7% and 5.7%–9.9% for the Glu-OC kit.

Sample collection

Venous blood (2 mL) should be collected from a peripheral vein into plain tubes (serum) or tubes containing a suitable anticoagulant such as EDTA or heparin (plasma). Samples may be stored up to 12 h at 4°C. If the length of time between sample collection and assay is to exceed 12 h, samples should be stored frozen at −20°C or colder.

Reference range

A cutoff of 20% has been proposed by McKeown et al for %ucOC, to denote suboptimal vitamin K status.[7]

Matrix Gla Protein

MGP is one of the most studied VKDPs. The protein was first characterized by Price et al. who purified it from bovine bone matrix.[111] Many studies have shown MGP to be protective against vascular calcification by directly inhibiting calcium precipitation and crystallization.[112, 113] MGP, in contrast to OC, is expressed in many soft tissues including cartilage, lung, heart, kidney, and arteries. MGP-deficient mice have been shown to develop excessive and premature arterial calcifications, leading to death by rupture of the aorta before reaching the age of 2 months.[114]

MGP consists of 84 amino acids, its activation expressed through two posttranslational processes: the phosphorylation of three serine residues in positions 3, 6, and 9 (although the role of this phosphorylation process is still not well understood) and the carboxylation of five Glu.[115, 116] For this reason, different forms of MGP are circulating in plasma which may have, in fact, different physiological roles and functions.[113, 117]

There are methods available for the determination of the different forms of MGP: carboxylated, uncarboxylated and phosphorylated, and unphosphorylated.[112, 113, 118] The dephosphorylated and uncarboxylated (dp-ucMGP) variant is the only form associated with a response to vitamin K supplementation.[52, 54, 119] Additionally, studies on patients treated with vitamin K antagonist have shown increased concentrations of dp-ucMGP[113, 118] and inversely, lowered concentrations of dp-ucMGP have been shown when the treatment was stopped.[56]

The measurement of dp-ucMGP has been suggested to be a functional indicator of vitamin K status in tissues that utilize MGP.[31, 52] Also, it is possible that the measurement of dp-ucMGP reflects vitamin K status.[58, 118] Analogously to OC, the amount of circulating dp-ucMGP in plasma depends on the total amount of MGP available.[31]

Concentrations of circulating dp-ucMGP can be determined in plasma using sandwich ELISA techniques.[112] The first step of the assay uses a monoclonal antibody against the nonphosphorylated sequence 3–15. For the detection of dp-ucMGP, a second monoclonal antibody is used which is directed against the noncarboxylated sequence 35–49 in human MGP.

An automated platform for the measurement of dp-ucMGP has been introduced by IDS Plc, Boldon, UK. The assay uses a microtiter plate and is based on the method first developed by VitaK BV, Maastricht, The Netherlands.[60]

The IDS Automated Analyzer IDS-iSYS InaKtif MGP assay uses two highly specific monoclonal antibodies. One biotinylated antibody is coupled to magnetic particles. The second antibody is coupled to an acridinium ester derivative. Fifty microliters of sample is incubated with 20 µL of MP and the second antibody. After 60 min of incubation, followed by a washing step, triggers are added whereby the luminescence measured is directly proportional to the InaKtif MGP concentration present in the sample. The time to first result is 63 min. The throughput of the assay is 89 tests an hour. The analytical range is 200–10,000 pM.[60]

Sample collection

Blood sample should be collected into EDTA tube, separated within 1.5 h and kept frozen at −20 °C or lower. Samples may undergo up to three freeze–thaw cycles.

Standardization

The IDS InaKtif MGP assay is calibrated against an in-house standard prepared from horse serum using 2-point calibrations for master curve repositioning. The analytical range is 200–10,000 pM.[60]

Best Practice

In the author's laboratory, a recommended approach for assessing vitamin K status is to confirm deficiency in patients by measurements of serum phylloquinone in tandem with a functional marker such as PIVKA-II.[2] Circulatory concentrations of phylloquinone fall within a few days of poor dietary intake or a diminished ability to absorb the vitamin from food. Crucially PIVKA-II gives an indication of the cellular utilization of the vitamin. The determination of PIVKA-II gives an early warning of impaired coagulation, which when used in combination with phylloquinone measurements can reveal a chronic suboptimal status.[62] PIVKA-II also has the additional advantage of being a retrospective marker of vitamin K status. With a half-life of approximately 60 h, PIVKA-II remains elevated for a period of several days after a vitamin K deficiency has been treated where the administration of phylloquinone would mask evidence of deficiency through direct measurement of phylloquinone alone. In other scenarios, a patient may have circulatory phylloquinone concentrations within the desired reference range but detectable circulatory PIVKA-II. This indicates hepatic dysfunction—and therefore an impaired ability to use vitamin K, or a previous episode (within 5 days) of vitamin K deficiency. Conversely a low circulatory concentration of phylloquinone may be detected but no PIVKA-II. This may indicate precarious tissue stores that have yet to progress to a functional deficient state (with respect to the coagulation function of vitamin K). The increased abundance of PIVKA-II can signify disease unrelated to vitamin K deficiency, when there is impaired utilization of vitamin K, e.g., liver disease. In Japan, Korea, and Indonesia measurement of PIVKA-II has been approved for use as a marker of HCC.[64, 66, 68]

Conclusions

There is no single biomarker of vitamin K suitable for the assessment of vitamin K status. In this chapter, we reviewed the analytical methods available to support the assessment of vitamin K status. A practical approach is to asses both serum phylloquinone and PIVKA-II which represents tissue reserves combined with a functional marker of coagulation. Each of these methods has its own advantages and disadvantages (Table 5.2), which should be considered during interpretation of the results. Indirect methods, for example, urinary vitamin K metabolites show promise

Table 5.2 Advantages and Disadvantages of Methods for Vitamin K_1 and PIVKA-II

Vitamin K_1		PIVKA-II	
Advantages	**Disadvantages**	**Advantages**	**Disadvantages**
Direct measure of the vitamin concentration	Strongly influenced by lipid levels (especially in hypertriglyceridemic patients)	PIVKA-II is a functional marker	Relationship with liver function not always clear
Evidence that plasma levels of vitamin K_1 fall rapidly in parallel with fall in liver stores	Affected by recent food intake (suggest fasting levels)	Reflects vitamin K availability for carboxylation of prothrombin	Long half-life (~60 h)–may not be a measure of current vitamin K status
Evidence that plasma vitamin K_1 reflects dietary intake	Plasma levels influenced by genotype, e.g., ApoE	Very sensitive	Does not reflect vitamin K status of nonhepatic tissues
May reflect total body reserves	Reflects only one form of the vitamin (vitamin K_1)	Stable marker of vitamin K status able to detect past deficiency	
	Does not reflect menaquinone (vitamin K_2) reserves		

as markers of global vitamin K and hepatic function.[71] Other markers, circulating undercarboxylated forms of VKDPs in particular ucOC and dp-ucMGP have been proposed as biomarkers of vitamin K in respect for certain tissues; bone and vascular, respectively. Less commonly measured urinary free Gla may be a useful functional indicator too.

Summary of the Laboratory Test for Vitamin K Investigation

- Serum PIVKA II concentration is a useful and sensitive homeostatic marker of subclinical vitamin K deficiency, especially in combination with a static marker, the phylloquinone concentration in serum/plasma
- The direct quantification of phylloquinone concentration in plasma/serum together with PIVKA-II is probably the best approach for evaluation of vitamin K status.
- Circulating concentration of phylloquinone in serum/plasma is not a meaningful indicator for nutritional status because of its dependence on dietary intake within the last 24 h, it should be considered as a biomarker of short-term phylloquinone intake.
- Urinary vitamin K metabolites are reliable markers of global vitamin K status; however, this method is not widely available in clinical settings.
- The level of OC carboxylation, obtained by assaying ucOC and cOC, and therefore the ratio cOC/ucOC, which are relevant indicators of the nutritional state with respect to vitamin K.
- The measurement of dp-ucMGP has been proposed as a functional indicator of vitamin K status and vascular calcification

Acknowledgments

The authors gratefully acknowledge our colleagues from The Human Nutristasis Unit. We also thank Viapath Analytics and Guy's and St. Thomas' NHS Foundation Trust for their continued support.

References

1. Olson RE. Vitamin K. In: Shils M, Olson JA, Shike M, Ross A, editors. *Modern nutrition in health and disease*. 9th ed. Baltimore: Williams and Wilkins; 1999. p. 363–80.
2. Shearer MJ, Gorska R, Harrington DJ, Schurgers LJNP. Vitamin K. In: Herrmann W, Obeid R, editors. *Vitamins in the prevention of human diseases*. Berlin, New York: De Gruyter; 2011. p. 515–60.
3. Matschiner JT, Bell RG. Metabolism and vitamin K activity of cis phylloquinone in rats. *J Nutr* 1972;**102**:625–9.

4. Leenheer de AP, Lambert WE, Bocxlaer van JF. *Modern chromatographic analysis of vitamins*. Marcel Dekker; 2000.

5. Hirota Y, Tsugawa N, Nakagawa K, et al. Menadione (vitamin K3) is a catabolic product of oral phylloquinone (vitamin K1) in the intestine and a circulating precursor of tissue menaquinone-4 (vitamin K2) in rats. *J Biol Chem* 2013;**288**:33071–80.

6. Shearer MJ, Newman P. Metabolism and cell biology of vitamin K. *Thromb Haemost* 2008;**100**:530–47.

7. McKeown NM, Jacques PF, Gundberg CM, et al. Dietary and nondietary determinants of vitamin K biochemical measures in men and women. *J Nutr* 2002;**132**:1329–34.

8. Booth SL, Martini L, Peterson JW, Saltzman E, Dallal GE, Wood RJ. Dietary phylloquinone depletion and repletion in older women. *J Nutr* 2003;**133**:2565–9.

9. Thijssen HHW, Vervoort LMT, Schurgers LJ, Shearer MJ. Menadione is a metabolite of oral vitamin K. *Br J Nutr* 2006;**95**:260–6.

10. Shearer MJ, McBurney A, Barkhan P. Studies on the absorption and metabolism of phylloquinone (vitamin K1) in man. *Vitam Horm* 1974;**32**:513–42.

11. Furie B, Bouchard BA, Furie BC. Vitamin K-dependent biosynthesis of gamma-carboxyglutamic acid. *Blood* 1999;**93**:1798–808.

12. Usui Y, Tanimura H, Nishimura N, Kobayashi N, Okanoue T, Ozawa K. Vitamin K concentrations in the plasma and liver of surgical patients. *Am J Clin Nutr* 1990;**51**:846–52.

13. Oldenburg J, Marinova M, Müller-Reible C, Watzka M. The vitamin K cycle. *Vitam Horm* 2008;**78**:35–62.

14. Tie J-K, Jin D-Y, Straight DL, Stafford DW. Functional study of the vitamin K cycle in mammalian cells. *Blood* 2011;**117**:2967–74.

15. Suttie J. Vitamin K and human nutrition. *J Am Diet Assoc* 1992;**92**:589–90.

16. Last JA. The missing link: the story of Karl Paul link. *Toxicol Sci* 2002;**66**:4–6.

17. Vermeer C, Schurgers LJ. A comprehensive review of vitamin K and vitamin K antagonists. *Hematol Oncol Clin North Am* 2000;**14**:339–53.

18. Booth SL, Centurelli MA. Vitamin K: a practical guide to the dietary management of patients on warfarin. *Nutr Rev* 1999;**57**:288–96.

19. Chang C-H, Wang Y-W, Yeh Liu P-Y, Kao Yang Y-H. A practical approach to minimize the interaction of dietary vitamin K with warfarin. *J Clin Pharm Ther* 2014;**39**:56–60.

20. Traber MG. Vitamin E and K interactions—a 50-year-old problem. *Nutr Rev* 2008;**66**:624–9.

21. Pastori D, Carnevale R, Cangemi R, et al. Vitamin E serum levels and bleeding risk in patients receiving oral anticoagulant therapy: a retrospective cohort study. *J Am Heart Assoc* 2013;**2**:e000364.

22. Booth SL, Golly I, Sacheck JM, et al. Effect of vitamin E supplementation on vitamin K status in adults with normal coagulation status. *Am J Clin Nutr* 2004;**80**:143–8.

23. Conway SP, Morton AM, Oldroyd B, et al. Osteoporosis and osteopenia in adults and adolescents with cystic fibrosis: prevalence and associated factors. *Thorax* 2000;**55**:798–804.

24. Elston TN, Dudley JM, Shearer MJ, Schey SA. Vitamin K prophylaxis in high-dose chemotherapy. *Lancet* 1995;**345**:1245.

25. Harrington DJ, Western H, Seton-Jones C, Rangarajan S, Beynon T, Shearer MJ. A study of the prevalence of vitamin K deficiency in patients with cancer referred to a hospital palliative care team and its association with abnormal haemostasis. *J Clin Pathol* 2008;**61**:537–40.

26. Shearer MJ, Fu X, Booth SL. Vitamin K nutrition, metabolism, and requirements: current concepts and future research. *Adv Nutr* 2012;**3**:182–95.

27. McMillan CW, Roberts HR. Congenital combined deficiency of coagulation factors II, VII, IX and X. *N Engl J Med* 1966;**274**:1313–5.
28. Kinoshita H, Nakagawa K, Narusawa K, et al. A functional single nucleotide polymorphism in the vitamin-K-dependent gamma-glutamyl carboxylase gene (Arg325Gln) is associated with bone mineral density in elderly Japanese women. *Bone* 2007;**40**:451–6.
29. Weston BW, Monahan PE. Familial deficiency of vitamin K-dependent clotting factors. *Haemophilia* 2008;**14**:1209–13.
30. Dashti HS, Shea MK, Smith CE, et al. Meta-analysis of genome-wide association studies for circulating phylloquinone concentrations. *Am J Clin Nutr* 2014;**100**:1462–9.
31. Kyla Shea M, Booth SL. Concepts and controversies in evaluating vitamin K status in population-based studies. Multidisciplinary Digital Publishing Institute (MDPI), *Nutrients* 2016;**8**:8.
32. Dam H. *The discovery of vitamin K, its biological functions and therapeutical application.* Nobel Lecture; 1943.
33. Dam H, Glavind J. Determination of vitamin K by the curative blood-clotting method. *Biochem J* 1938;**32**:1018–23.
34. Almquist HJ, Mecchi E, Klose AA. Estimation of the anti-haemorrhagic vitamin. *Biochem J* 1938;**38**:1897–903.
35. Allison PM, Mummah-Schendel LL, Kindberg CG, Harms CS, Bang NU, Suttie JW. Effects of a vitamin K-deficient diet and antibiotics in normal human volunteers. *J Lab Clin Med* 1987;**110**:180–8.
36. Booth SL, O'Brien-Morse ME, Dallal GE, Davidson KW, Gundberg CM. Response of vitamin K status to different intakes and sources of phylloquinone-rich foods: comparison of younger and older adults. *Am J Clin Nutr* 1999;**70**:368–77.
37. Sokoll LJ, Booth SL, O'Brien ME, Davidson KW, Tsaioun KI, Sadowski JA. Changes in serum osteocalcin, plasma phylloquinone, and urinary gamma-carboxyglutamic acid in response to altered intakes of dietary phylloquinone in human subjects. *Am J Clin Nutr* 1997;**65**:779–84.
38. Potischman N. Biologic and methodologic issues for nutritional biomarkers. *J Nutr* 2003;**133**(Suppl):875S–80S.
39. Scudi JV, Buhs RP. On the colorimetric method for the determination of the K vitamins. *J Biol Chem* 1942;**143**:665–9.
40. Lefevere MF, Leenheer de AP, Claeys AE. High-performance liquid chromatographic assay of vitamin K in human serum. *J Chromatogr* 1979;**186**:749–62.
41. Shearer MJ. Assay of K vitamins in tissues by high-performance liquid chromatography with special reference to ultraviolet detection. *Methods Enzymol* 1986;**123**:235–51.
42. Haroon Y, Bacon DS, Sadowski JA. Liquid-chromatographic determination of vitamin K1 in plasma, with fluorometric detection. *Clin Chem* 1986;**32**:1925–9.
43. Davidson KW, Sadowski JA. Determination of vitamin K compounds in plasma or serum by high-performance liquid chromatography using postcolumn chemical reduction and fluorimetric detection. *Methods Enzymol* 1997;**282**:408–21.
44. Usui Y, Nishimura N, Kobayashi N, Okanoue T, Kimoto M, Ozawa K. Measurement of vitamin K in human liver by gradient elution high-performance liquid chromatography using platinum-black catalyst reduction and fluorimetric detection. *J Chromatogr* 1989;**489**:291–301.
45. Wang LY, Bates CJ, Yan L, Harrington DJ, Shearer MJ, Prentice A. Determination of phylloquinone (vitamin K1) in plasma and serum by HPLC with fluorescence detection. *Clin Chim Acta* 2004;**347**:199–207.

46. Kamao M, Suhara Y, Tsugawa N, Okano T. Determination of plasma vitamin K by high-performance liquid chromatography with fluorescence detection using vitamin K analogs as internal standards. *J Chromatogr B Anal Technol Biomed Life Sci* 2005;**816**:41–8.

47. Paroni R, Faioni EM, Razzari C, Fontana G, Cattaneo M. Determination of vitamin K1 in plasma by solid phase extraction and HPLC with fluorescence detection. *J Chromatogr B Anal Technol Biomed Life Sci* 2009;**877**:351–4.

48. Hart JP, Shearer MJ, Mccarthy PT. Enhanced sensitivity for the determination of endogenous Phylloquinone (vitamin K1) in plasma using high-performance liquid chromatography with dual-electrode electrochemical detection. *Analyst* 1985;**110**:1181–4.

49. McCarthy PT, Harrington DJ, Shearer MJ. Assay of phylloquinone in plasma by high-performance liquid chromatography with electrochemical detection. *Methods Enzymol* 1997;**282**:421–33.

50. Harrington DJ, Booth SL, Card DJ, Shearer MJ. Excretion of the urinary 5C- and 7C-aglycone metabolites of vitamin K by young adults responds to changes in dietary phylloquinone and dihydrophylloquinone intakes. *J Nutr* 2007;**137**:1763–8.

51. Nováková L, Perrenoud AG, Francois I, West C, Lesellier E, Guillarme D. Modern analytical supercritical fluid chromatography using columns packed with sub-2 µm particles: a tutorial. *Anal Chim Acta* 2014;**824**:18–35.

52. Westenfeld R, Krueger T, Schlieper G, et al. Effect of vitamin K2 supplementation on functional vitamin K deficiency in hemodialysis patients: a randomized trial. *Am J Kidney Dis* 2012;**59**:186–95.

53. Nannapaneni NK, Jalalpure SS, Muppavarapu R, Sirigiri SK, Kore P. A sensitive and rapid UFLC-APCI-MS/MS bioanalytical method for quantification of endogenous and exogenous vitamin K1 isomers in human plasma: development, validation and first application to a pharmacokinetic study. *Talanta* 2017;**164**:233–43.

54. Caluwé R, Pyfferoen L, De Boeck K, De Vriese AS. The effects of vitamin K supplementation and vitamin K antagonists on progression of vascular calcification: ongoing randomized controlled trials. *Clin Kidney J* 2016;**9**:273–9.

55. Abro K, Memon N, Bhanger MI, Abro S, Perveen S, Laghar AH. Determination of vitamins E, D3, and K1 in plasma by liquid chromatography in plasma by liquid chromatography-atmospheric pressure chemical ionization-mass spectrometry utilizing a monolithic column. *Anal Lett* 2013;**47**:14–24.

56. Delanaye P, Dubois BE, Lukas P, et al. Impact of stopping vitamin K antagonist therapy on concentrations of dephospho-uncarboxylated matrix Gla protein. *Clin Chem Lab Med* 2015;**53**:e191–3.

57. Riphagen IJ, van der Molen JC, van Faassen M, et al. Measurement of plasma vitamin K1 (phylloquinone) and K2 (menaquinones-4 and -7) using HPLC-tandem mass spectrometry. *Clin Chem Lab Med* 2016;**54**:1201–10.

58. van den Heuvel EG, van Schoor NM, Lips P, et al. Circulating uncarboxylated matrix Gla protein, a marker of vitamin K status, as a risk factor of cardiovascular disease. *Maturitas* 2014;**77**:137–41.

59. Nair SM, Sharma M, Karia D. Method validation of high performance liquid chromatography mass spectrometric method for the estimation of cis and trans isomers of phytonadione in human plasma using D-labeledtrans phytonadione. *J Sci Res* 2015;**4**:8.

60. Kasper D, Bougoussa M, Theuwissen E, Vermeer C. Assessment of vitamin K status by fully automated IDS-iSYS inaKtiv MGP. In: Presented at the European Calcified Tissue Society Congress ECTS, Bone Abstracts, Vol. 3; 2014. PP395.

61. Karl JP, Fu X, Dolnikowski GG, Saltzman E, Booth SL. Quantification of phylloquinone and menaquinones in feces, serum, and food by high-performance liquid chromatography-mass spectrometry. *J Chromatogr B Anal Technol Biomed Life Sci* 2014;**963**:128–33.
62. Shearer MJ. Vitamin K in parenteral nutrition. *Gastroenterology* 2009;**137**:S105–18.
63. Gentili A, Cafolla A, Gasperi T, et al. Rapid, high performance method for the determination of vitamin K 1, menaquinone-4 and vitamin K 1 2,3-epoxide in human serum and plasma using liquid chromatography-hybrid quadrupole linear ion trap mass spectrometry. *J Chromatogr A* 2014;**1338**:102–10.
64. Inagaki Y, Tang W, Makuuchi M, Hasegawa K, Sugawara Y, Kokudo N. Clinical and molecular insights into the hepatocellular carcinoma tumour marker des-γ-carboxyprothrombin. *Liver Int* 2011;**31**:22–35.
65. Ahmed S, Kishikawa N, Ohyama K, Imazato T, Ueki Y, Kuroda N. Selective chemiluminescence method for monitoring of vitamin K homologues in rheumatoid arthritis patients. *Talanta* 2011;**85**:230–6.
66. Wang X, Zhang W, Liu Y, et al. Diagnostic value of prothrombin induced by the absence of vitamin K or antagonist-II (PIVKA-II) for early stage HBV related hepatocellular carcinoma. *Infect Agent Cancer* 2017;**12**:47.
67. Ducros V, Pollicand M, Laporte F, Favier A. Quantitative determination of plasma vitamin K1 by high-performance liquid chromatography coupled to isotope dilution tandem mass spectrometry. *Anal Biochem* 2010;**401**:7–14.
68. Klimovich V, Voong K, Sherwood R, Harrington DJ. Clinical utility of PIVKA-II in the diagnosis of hepatocellular carcinoma. *Clin Lab Int* 2017;**41**:6–11.
69. Fu X, Peterson JW, Hdeib M, et al. Measurement of deuterium-labeled phylloquinone in plasma by high-performance liquid chromatography/mass spectrometry. *Anal Chem* 2009;**81**:5421–5.
70. Sherma J, Fried B. *Handbook of thin-layer chromatography*. Marcel Dekker; 2003.
71. Card DJ, Gorska R, Cutler J, Harrington DJ. Vitamin K metabolism: current knowledge and future research. *Mol Nutr Food Res* 2013;**58**:1590–600.
72. Song Q, Wen A, Ding L, Dai L, Yang L, Qi X. HPLC–APCI–MS for the determination of vitamin K 1 in human plasma: method and clinical application. *J Chromatogr B* 2008;**875**:541–5.
73. Ahmed S, Kishikawa N, Nakashima K, Kuroda N. Determination of vitamin K homologues by high-performance liquid chromatography with on-line photoreactor and peroxyoxalate chemiluminescence detection. *Anal Chim Acta* 2007;**591**:148–54.
74. Suhara Y, Kamao M, Tsugawa N, Okano T. Method for the determination of vitamin K homologues in human plasma using high-performance liquid chromatography-tandem mass spectrometry. *Anal Chem* 2005;**77**:757–63.
75. Leenheer de AP, Lambert WE, de MGM R. *Modern chromatographic analysis of the vitamins*. Marcel Dekker; 1985.
76. Vogeser M, Seger C. Pitfalls associated with the use of liquid chromatography–tandem mass spectrometry in the clinical laboratory. *Clin Chem* 2010;**56**:1234–44.
77. Sherma J, Fried B. Thin layer chromatographic analysis of biological samples. A review. *J Liq Chromatogr Relat Technol* 2005;**28**:2297–314.
78. Langenberg JP, Tjaden UR. Determination of (endogenous) vitamin K1 in human plasma by reversed-phase high-performance liquid chromatography using fluorometric detection after post-column electrochemical reduction. Comparison with ultraviolet, single and dual electrochemical detection. *J Chromatogr* 1984;**305**:61–72.

79. Langenberg JP, Tjaden UR. Improved method for the determination of vitamin K1 epoxide in human plasma with electrofluorimetric reaction detection. *J Chromatogr* 1984;**289**:377–85.

80. Gorska R, Liddicoat F, Harrington DJ. The determination of vitamin K1 and K1 2,3-epoxide in human serum/plasma by ultra performance convergence chromatographytm (UPC2) with LCMSMS. In: *Clinical chemistry and laboratory medicine Warsaw: the 4th joint EFLM-UEMS congress*. Warsaw, Poland: Laboratory Medicine at the Clinical Interface; 2016. p. 213–366.

81. Suttie JW. *Vitamin K in health and disease*. CRC Press; 2009.

82. Canfield LM, Hopkinson JM, Lima AF, et al. Quantitation of vitamin K in human milk. *Lipids* 1990;**25**:406–11.

83. Khaksari M, Mazzoleni LR, Ruan C, Kennedy RT, Minerick AR. Data representing two separate LC-MS methods for detection and quantification of water-soluble and fat-soluble vitamins in tears and blood serum. *Data Br* 2017;**11**:316–30. https://doi.org/10.1016/j.dib.2017.02.033.

84. Shearer MJ. Phylloquinone (vitamin K1) in serum or plasma by HPLC. In: Fizanda F, editor. *Nutritional status assessment: a manual for population studies, Nutritional status assessment: a manual for population studies*, London: Chapman & Hall; 1991. p. 214–20.

85. Thane CW, Wang LY, Coward WA. Plasma phylloquinone (vitamin K1) concentration and its relationship to intake in British adults aged 19–64 years. *Br J Nutr* 2006;**96**:1116–24.

86. Horwitz W. Evaluation of analytical methods used for regulation of foods and drugs. *Anal Chem* 1982;**54**:67–76.

87. Card DJ, Shearer MJ, Schurgers LJ, Harrington DJ. The external quality assurance of phylloquinone (vitamin K(1)) analysis in human serum. *Biomed Chromatogr* 2009;**23**:1276–82.

88. Haroon Y. Rapid assay for gamma-carboxyglutamic acid in urine and bone by precolumn derivatization and reversed-phase liquid chromatography. *Anal Biochem* 1984;**140**:343–8.

89. Sokoll LJ, Sadowski JA. Comparison of biochemical indexes for assessing vitamin K nutritional status in a healthy adult population. *Am J Clin Nutr* 1996;**63**:566–73.

90. Turck D, Bresson J, Burlingame B, et al. EFSA NDA Panel (EFSA Panel on Dietetic Products, Nutrition and Allergies), Opinion on the dietary reference values for vitamin K. *EFSA J* 2017;**15**:4780. 78 pp.

91. Harrington DJ, Clarke P, Card DJ, Mitchell SJ, Shearer MJ. Urinary excretion of vitamin K metabolites in term and preterm infants: relationship to vitamin K status and prophylaxis. *Pediatr Res* 2010;**68**:508–12.

92. Harrington DJ, Soper R, Edwards C, Savidge GF, Hodges SJ, Shearer MJ. Determination of the urinary aglycone metabolites of vitamin K by HPLC with redox-mode electrochemical detection. *J Lipid Res* 2005;**46**:1053–60.

93. Fujita K, Kinukawa H, Ohno K, Ito Y, Saegusa H, Yoshimura T. Development and evaluation of analytical performance of a fully automated chemiluminescent immunoassay for protein induced by vitamin K absence or antagonist II. *Clin Biochem* 2015;**48**:1330–6.

94. Kinukawa H, Shirakawa T, Yoshimura T. Epitope characterization of an anti-PIVKA-II antibody and evaluation of a fully automated chemiluminescent immunoassay for PIVKA-II. *Clin Biochem* 2015;**48**:1120–5.

95. Bajaj SP, Price PA, Russell WA. Decarboxylation of gamma-carboxyglutamic acid residues in human prothrombin. Stoichiometry of calcium binding to gamma-carboxyglutamic acid in prothrombin. *J Biol Chem* 1982;**257**:3726–31.

96. CLSI EP06-A. *Evaluation of the linearity of quantitative measurement procedures: a statistical approach*. Approved Guideline, 2003. Wayne, PA.

97. Sohn A, Kim H, Yu SJ, Yoon JH, Kim Y. A quantitative analytical method for PIVKA-II using multiple reaction monitoring-mass spectrometry for early diagnosis of hepatocellular carcinoma. *Anal Bioanal Chem* 2017;**409**:2829–38.

98. Belle M, Brebant R, Guinet R, Leclercq M. Production of a new monoclonal antibody specific to human des-gamma-carboxyprothrombin in the presence of calcium ions. Application to the development of a sensitive ELISA-test. *J Immunoassay* 1995;**16**:213–29.

99. Booth SL. Roles for vitamin K beyond coagulation. *Annu Rev Nutr* 2009;**29**:89–110.

100. Rehder DS, Gundberg CM, Booth SL, Borges CR. Gamma-carboxylation and fragmentation of osteocalcin in human serum defined by mass spectrometry. *Mol Cell Proteomics* 2015;**14**:1546–55.

101. Booth SL, Martini L, Peterson JW, et al. Human nutrition and metabolism dietary phylloquinone depletion and repletion in older women. *J Nutr* 2003;**133**:2565–9.

102. Herrmann W, Obeid R. *Vitamins in the prevention of human diseases*. De Gruyter; 2011.

103. Knapen MH, Nieuwenhuijzen Kruseman AC, Wouters RS, Vermeer C. Correlation of serum osteocalcin fractions with bone mineral density in women during the first 10 years after menopause. *Calcif Tissue Int* 1998;**63**:375–9.

104. Schurgers LJ, Shearer MJ, Hamulyák K, Stöcklin E, Vermeer C. Effect of vitamin K intake on the stability of oral anticoagulant treatment: dose-response relationships in healthy subjects. *Blood* 2004;**104**:2682–9.

105. Buranasinsup S, Bunyaratavej N. The intriguing correlation between undercarboxylated osteocalcin and vitamin D. *J Med Assoc Thail* 2015;**98**:S16–20.

106. Braam LAJLM, Knapen MHJ, Geusens P, et al. Vitamin K1 supplementation retards bone loss in postmenopausal women between 50 and 60 years of age. *Calcif Tissue Int* 2003;**73**:21–6.

107. Shearer MJ. The roles of vitamins D and K in bone health and osteoporosis prevention. *Proc Nutr Soc* 1997;**56**:915–37.

108. Price PA, Williamson MK, Lothringer JW. Origin of the vitamin K-dependent bone protein found in plasma and its clearance by kidney and bone. *J Biol Chem* 1981;**256**:12760–6.

109. Sokoll LJ, O'Brien ME, Camilo ME, Sadowski JA. Undercarboxylated osteocalcin and development of a method to determine vitamin K status. *Clin Chem* 1995;**41**:1121–8.

110. Gundberg CM, Nieman SD, Abrams S, Rosen H. Vitamin K status and bone health: an analysis of methods for determination of undercarboxylated osteocalcin. *J Clin Endocrinol Metab* 1998;**83**:3258–66.

111. Price PA, Urist MR, Otawara Y. Matrix Gla protein, a new γ-carboxyglutamic acid-containing protein which is associated with the organic matrix of bone. *Biochem Biophys Res Commun* 1983;**117**:765–71.

112. Schurgers LJ, Cranenburg ECM, Vermeer C. Matrix Gla-protein: the calcification inhibitor in need of vitamin K. *Thromb Haemost* 2008;**100**:593–603.

113. Cranenburg ECM, Koos R, Schurgers LJ, et al. Characterisation and potential diagnostic value of circulating matrix Gla protein (MGP) species. *Thromb Haemost* 2010;**104**:811–22.

114. Luo G, Ducy P, McKee MD, et al. Spontaneous calcification of arteries and cartilage in mice lacking matrix GLA protein. *Nature* 1997;**386**:78–81.

115. Murshed M, Schinke T, McKee MD, Karsenty G. Extracellular matrix mineralization is regulated locally; different roles of two gla-containing proteins. *J Cell Biol* 2004;**165**:625–30.
116. Schurgers LJ, Spronk HMH, Skepper JN, et al. Post-translational modifications regulate matrix Gla protein function: importance for inhibition of vascular smooth muscle cell calcification. *J Thromb Haemost* 2007;**5**:2503–11.
117. Schlieper G, Westenfeld R, Krüger T, et al. Circulating nonphosphorylated carboxylated matrix gla protein predicts survival in ESRD. *J Am Soc Nephrol* 2011;**22**:387–95.
118. Delanaye P, Krzesinski J-M, Warling X, et al. Dephosphorylated-uncarboxylated matrix Gla protein concentration is predictive of vitamin K status and is correlated with vascular calcification in a cohort of hemodialysis patients. *BMC Nephrol* 2014;**15**:145.
119. Cranenburg ECM, Vermeer C, Koos R, et al. The circulating inactive form of matrix Gla protein (ucMGP) as a biomarker for cardiovascular calcification. *J Vasc Res* 2008;**45**:427–36.

CHAPTER

Methods for assessment of Thiamine (Vitamin B$_1$)

Martin A Crook

Department of Clinical Biochemistry and Metabolic Medicine, Guys and St Thomas' Hospital, London, United Kingdom

Department of Clinical Biochemistry and Metabolic Medicine, University Hospital Lewisham, London, United Kingdom

Chapter Outline

---

[*] Drug or vitamin doses and requirements should always be confirmed by the readers' pharmacy and drug information unit. Authors accept no liability for accuracy of drug or vitamin statements.

Introduction

It was Eijkman, Nobel laureate in medicine, who first discovered thiamine while studying the cause of beriberi in Java.[1] This chapter looks at thiamine biochemistry and physiology. Clinical disorders associated with thiamine deficiency and the laboratory assessment of thiamine status will be discussed.

Structure and Function

Thiamine exists in multiple forms which differ in their phosphate groups. The forms that are found are the nonphosphorylated or free thiamine form, thiamine monophosphate (TMP), diphosphate (TPP), and thiamine triphosphate (TTP) (Fig. 6.1). These forms are collectively referred to as thiamine compounds in this chapter.[2–9]

After thiamine ingestion, free thiamine is converted to TPP via thiamine pyrophosphokinase and utilized as a cofactor for various metabolic pathways. Thiamine TPP can be further phosphorylated or dephosphorylated to form any of the thiamine compounds. Erythrocytes are by far the largest reservoir of TPP accounting for approximately 80% of total body stores. However, TMP and free thiamine are usually the only detectable thiamine compounds in extracellular fluid.[3–6]

Physiology

Human adults can store approximately 30 mg of thiamine, primarily in muscle tissue, liver, and kidneys. The vitamin is principally absorbed in the jejunum and ileum through an active process at low concentrations and by passive diffusion at high concentrations. In plasma, thiamine is largely bound to albumin with the vast majority in the circulation contained within erythrocytes (~90%). A daily thiamine intake of up to

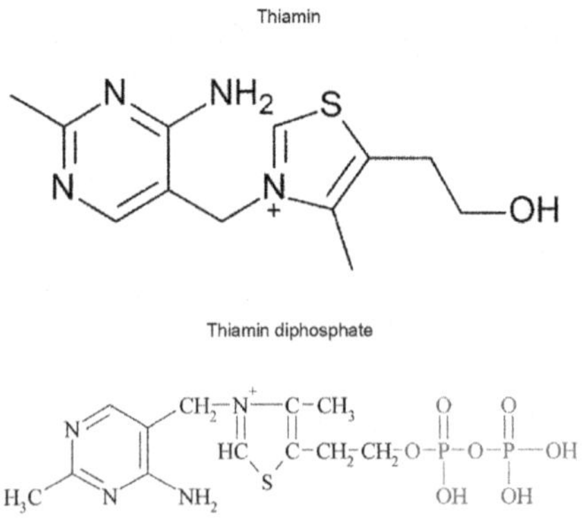

FIG. 6.1

Structures of thiamine and thiamine diphosphate.

3.4 mg results in a proportional increase in uptake by erythrocytes.[3–6] A specific binding protein called thiamine-binding protein (TBP) has been identified and plays an important role in tissue distribution. Thiamine's cellular transport is mediated by specific thiamine carrier transporters: human thiamine transporter-1 (hTHTR-1) which is the product of the SLC19A2 gene and hTHTR-2 which is the product of SLC19A3. Thiamine deficiency induces intestinal carrier-mediated uptake by increased expression of hTHTR-2.[7–9] Thiamine and its metabolites are usually renally excreted.

Sources

Thiamine biosynthesis occurs in bacteria, some protozoans, plants, and fungi. Humans cannot synthesize thiamine and are reliant on exogenous sources, which include many dietary components; wheat germ, oatmeal, and yeast are particularly rich sources of thiamine. Adequate amounts of thiamine are usually present in the diet, and deficiency is more common in alcoholics and in patients with anorexia nervosa and other under-nutrition states.[3–6] Humans generally only have body stores of thiamine capable of sustaining metabolic demand for approximately 2 weeks which can become depleted with poor dietary intake or more quickly if other clinical conditions are present.[3–6]

Functions

Thiamine TPP is produced by thiamine diphosphokinase and is an essential cofactor for the decarboxylation of 2-oxoacids, such as the conversion of pyruvate to acetyl coenzyme a and also other pathways including pyruvate dehydrogenase (PDH), α-ketogluterate dehydrogenase (KGDH), and branched-chain α-keto acid dehydrogenase (BCKDH), (Fig. 6.2). In thiamine deficiency, pyruvate cannot be metabolized and accumulates in the blood. Thiamine TPP is also an essential cofactor for transketolase in the pentose-phosphate pathway

Thiamine is essential for the optimal function of the nervous system and repair of myelin nerve sheaths. In turn magnesium is an important cofactor for thiamine-dependent enzymes.[2–9] In addition, other reputed noncofactor roles of thiamine compounds are shown within the oxidative stress response, gene regulation, cholinergic system, immune function, chloride channels, and neurotransmission.[2–6]

Clinical Application of Measurement

Thiamine deficiency may be more frequent than currently thought and some studies report a prevalence of 10%–20% in hospital patients. Deficiency of thiamine is usually because of poor intake and clinical manifestations are variable and complex. A well-described cause is excessive alcohol intake combined with high-carbohydrate and poor thiamine intake. One of the commonest causes worldwide is a diet high in unenriched white flour or rice particularly with diets high in raw fish which is high in thiaminase.[10–13]

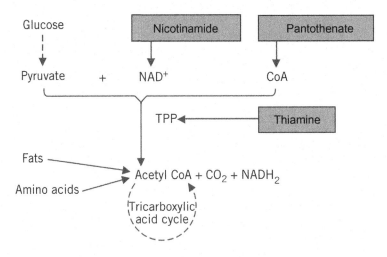

Role of B vitamins in formation of acetyl CoA

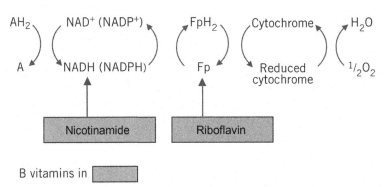

Role of B vitamins in electron transfer

FIG. 6.2

Biochemical pathways for B vitamins.

Fom Crook MA. In clinical biochemistry and metabolic medicine. 8th ed. Hodder/Arnold;, 2012. ISBN: 9781444144147.

Deficiency of thiamine causes beriberi, in which anorexia, wasting, neurological pathology (motor and sensory polyneuropathy, amnesia, encephalopathy called Wernicke-Korsakoff syndrome), and cardiac arrhythmias may occur. This form is called "dry" beriberi (shoshin) and is associated with low cardiac output. Conversely, in "wet" beriberi there is peripheral edema, sometimes associated with cardiac failure. Beriberi may be aggravated by a high-carbohydrate diet, possibly because this leads to an increased rate of glycolysis and therefore of pyruvate production. Other manifestations of thiamine deficiency are "gastrointestinal beriberi" displaying severe abdominal pain, mimicking mesenteric vascular occlusion and intestinal ischemia, with unexplained lactic acidosis. There is also a genetic form of beriberi in which thiamine is present but not appropriately intestinally absorbed.[14–22]

Thiamine deficiency is not confined to populations with poor nutritional intake. It may also occur in critically ill patients, especially those taking parenteral, or enteral nutritional support. Refeeding syndrome is another situation where thiamine deficiency can rapidly occur, associated with hypophosphatemia, hypomagnesemia, and hypokalemia. Here the insulin release after a period of prolonged starvation evokes movement of intracellular ions, such as phosphate, magnesium, and potassium into cells and utilization of thiamine occurs via carbohydrate pathways.[23, 24]

Thiamine deficiency states are not always recognized by clinicians because of a lack of familiarity with the condition, the paucity of "classic" signs and symptoms, as well as lack of readily available confirmatory laboratory tests. It is plausible that different patients may present with one or more of the varying manifestations of thiamine deficiency, depending on the genetic background.[25]

Subclinical thiamine deficiency has been observed in infancy and has implications on gross and fine motor function and balance skills in childhood, suggesting thiamine as having a crucial role in normal motor development in humans. Indeed, mitochondrial TPP transport is important for CNS development. Episodic encephalopathy can occur with deficiency of this vitamin and associated thiamine phosphokinase 1 (TPK1) mutations. Thiamine deficiency should be considered in children with diabetic ketoacidosis and encephalopathy.[26–43]

Thiamine-responsive megaloblastic anemia syndrome is characterized by megaloblastic anemia, progressive sensorineural hearing loss, and diabetes mellitus. This disorder occurs between infancy and adolescence.[26–43]

Biotin–thiamine-responsive basal ganglia disease is a rare severe neurogenetic disorder caused by pathogenic molecular variants in the thiamine transporter gene. Thiamine transporter genes SLC19A3 are important in a case of suspected and unexplained thiamine deficiency. This neurometabolic disease presents in early childhood with progressive neurodegeneration, such as seizures, dysphagia, coma, and death.[30–43]

Thiamine transporter SLC25A19 mutations cause Amish lethal microcephaly, which dramatically slows brain development and leads to alpha-ketoglutaric aciduria. Episodic encephalopathy-type thiamine metabolism dysfunction (OMIM 614458) is a very rare disorder due to TPK1 mutations that have recently been described. Furthermore, patients with senile dementia of the Alzheimer's type had lower levels plasma or RBC thiamine levels than matched controls without this condition.[36–45]

The binge drinking of alcohol by mothers during pregnancy compromises thiamine status and promotes fetal alcohol syndrome. Marchiafava–Bignami disease is characterized by primary degeneration of the corpus callosum associated with alcohol excess and low thiamine levels.[35–38,46–59] Thiamine deficiency has also recently been described in patients with postural tachycardia syndrome.[60]

Bariatric surgery-related, Wernicke–Korsakoff encephalopathy and residual memory impairment have been reported, and postoperative dysphagia following sleeve gastrectomy for morbid obesity is associated with low whole-blood thiamine concentrations, a strange irony that patients with excess intake of macronutrients show micronutrient deficiency.[42,43,48,61,62]

Interesting research suggests that patients with septic shock who were randomized to receive thiamine had a lower serum creatinine concentration and a lower rate of progression to renal impairment than patients randomized to placebo. Furthermore, thiamine deficiency occurs more often in those with cerebral malaria compared to those with uncomplicated malaria. Thiamine deficiency should also be considered in patients with unexplained lactic acidosis

Studies have shown a new hypertension-susceptibility locus, uncovering a previously unsuspected thiamine transporter whose genetic variants predicted several disturbances in cardiac and autonomic function. The thiamine transporter THTR2 gene expression is apparently down-regulated in breast carcinoma, which may predispose to resistance to apoptosis.[35–38,42,47,48,56–62]

Certain medications such as proton-pump inhibitors or Metformin are known to also cause thiamine deficiency. Causes of thiamine deficiency are summarized in Table 6.1.[35–38,42,47,48,56–62]

In summary, thiamine deficiency can result from poor intake, reduced gastrointestinal absorption, renal loss, or increased metabolic requirements; many cases are multifactorial

Table 6.1 Some Causes of Thiamine Deficiency

Poor dietary intake especially if "excessive intake" of carbohydrate-rich foods
Alcoholism
Refeeding syndrome
Anorexia nervosa
High dietary intake of thiaminases, e.g., betel nuts and raw fish
Malabsorption states
Folate deficiency
Drugs, e.g., proton-pump inhibitors, metformin, high-dose diuretics
HIV infection
Diabetes mellitus
Chronic vomiting
Bariatric surgery
Intensive care and critically ill patients
Severe sepsis
Hyperemesis gravidarum
Hemodialysis; chronic kidney disease
Thyrotoxicosis
Thiamine transporter-2 deficiency
Spinocerebellar ataxia type 2
Thiamine-responsive megaloblastic anemia syndrome (TRMA)

Deficiency Treatment

The daily oral requirement for thiamine is 1.2 mg. Mild deficiency is treated by oral medication, with 25–100 mg of thiamine given daily; while in severe deficiency a higher dose of 200–300 mg is used in divided doses. In severe thiamine deficiency intravenous thiamine preparations are necessary, particularly in patients at risk of Wernicke-Korsakoff syndrome. It is essential to remember that anaphylaxis may occasionally follow the intravenous injection of thiamine. The administration of intravenous thiamine is described in the MHRA/CHM advice https://www.gov.uk/drug-safety-update/pabrinex-allergic-reactions. Here intravenous administration of thiamine should be by cautious infusion over 30 min and it is essential that resuscitation facilities for treating anaphylaxis should be readily available. It is important to remember that severely thiamine-deficient mothers should avoid breast-feeding as the toxic methylglyoxal metabolite may be present in breast milk.[47,63,64]

Studies have not shown a difference in the time to resolution of Wernicke-Korsakoff encephalopathy clinical symptoms in relation to the dose of IV thiamine. Indeed, large-scale studies are required to determine the optimal dosing of thiamine for this condition.[47,63,64]

Analytical Methods

The laboratory determination of thiamine status is problematic and currently performed by specialist laboratories. Decreased plasma pyruvate and raised lactate concentrations may be clues as to the diagnosis of thiamine deficiency in the presence of a metabolic lactic acidosis.[3] However, serum thiamine concentrations may not always be a reliable marker of the body's thiamine status as this only represents a small proportion of the total body thiamine.

It is also possible to measure urinary excretion of thiamine and its metabolites. However, the diagnosis of thiamine deficiency is primarily a clinical one. The reader is recommended to see the outstanding review by Collie and colleagues which covers the topic in depth.[65]

Historical Methods

Historically the first assays with clinical utility for the assessment of thiamine status were based on microbiological methods. Status has also been estimated using erythrocyte transketolase activity, with and without added exogenous TPP. Reduced transketolase activity, when due to thiamine deficiency, will become normalized after the addition of the cofactor. An erythrocyte transketolase activation coefficient (ETKAC) is usually less than 1.25 in thiamine deficiency. Conversely, the difference is described as a percentage increase in transketolase activity, with an increase of around 25% or more indicating thiamine deficiency. However, the upper limit of the reference range is wide within the literature ranging from 15% to 40% of increased activity.[66–71] Both of these approaches provide indirect estimates of thiamine status. Transketolase testing estimates the functional activity of this enzyme in erythrocytes by measuring the rate that hexoses are produced from the TPP-dependent pentose phosphate pathway rather than thiamine abundance. The interpretation of transketolase assays is difficult in subjects with diabetes mellitus, hepatic disease, or certain anemias as well as any other disorder disturbing the synthesis of transketolase and thus may not reflect thiamine deficiency per se.[66–71]

Direct Assay Measurement of Thiamine Compounds

In comparison to indirect methods, serum, plasma, and urine can be used to measure other thiamine compounds besides TPP. This allows for the total thiamine status to be assessed. The developed assays use HPLC with detection using fluorescence or mass spectrometry with a limit of quantification for all thiamine compounds of ~3.0 nmol/L. Measuring plasma concentrations of thiamine forms is advantageous over erythrocyte indirect assay due to convenient sample handling, high throughput, and ability to develop low volume, multiplex assays.

Preanalytical

Blood samples should be collected following an overnight fast into tubes containing EDTA or citrate. Transportation should take place on a cool pack and protection from light should be provided to guard against degradation. Samples can be stored at ambient temperatures for a maximum of 24 h or whole blood can be frozen and stored at −20°C (or −70°C for long term storage). Thiamine standard compound solutions may be stable for one to five months.[65]

Analytical

Derivatives of thiamine are generated in some assays to enhance fluorescence intensity. These methods may utilize hazardous oxidizing chemicals such as cyanogen bromide and mercuric chloride. Less dangerous alternatives such as potassium ferrocyanide (or hexacyanoferrate) may be used but will reduce the achieved fluorescence intensity.

Derivatization

The derivatization may take place pre- or postchromatographic separation. Advantages of precolumn derivatization for thiamine analysis are sharper diagnostic peaks and enhanced resolution using equipment for a gradient elution. Conversely, the advantages of postcolumn derivatization thiamine methods include the prolonging of column life and preserved chromatographic performance, although a disadvantage is that a second pump to mix the effluent with the reprivatizing agent prior to the detector is required.[72–76]

Chromatographic separation

A variety of HPLC column stationary phases have been used successfully for thiamine analysis including amide, carbon, gel permeation, amide, and carbon/silica polymers. Mobile phases may consist of methanol, acetonitrile, potassium or sodium phosphate, hydrochloric acid, ammonium acetate. For fluorescence detection, excitation wavelength used are 240–375 nm and 430–443 nm for emission. Ion pairing has been utilized such as tetrabutylammonium hydroxide, sodium pentasulfonate 1-heptane sulfonic acid sodium salt. In addition, derivatization by various compounds such as potassium ferrocyanide, cyanogen bromide, sodium hydroxide which have all been used.[77–102] Several Liquid chromatography-mass spectrometry (LC-MS)/MS methods for thiamine and its products have also been developed and applied.[72,103–105]

Postanalytical

Reference ranges and interpretation

Currently, there are no standardized common reference intervals including those for gender, ethnicity, and age. Additionally, there is no consensus regarding decision limits to determine therapeutic concentrations for thiamine. This is clearly clinically problematic for the diagnosis of thiamine disorders. Indeed, published reference intervals for whole-blood TPP range from a lower limit of 63–105 nmol/L to an upper limit of 171–229 nmol/L.[65] Age-related reference ranges for thiamine should be established as the very elderly, i.e., those more than 75 years of age are known to have a reduced total thiamine compound concentration that can be about 25% lower than that of middle-aged individuals.[65]

Collie and coworkers reported intra- and interindividual biological variability for TPP concentrations in whole blood to be 4.8%–12.0%, respectively. The analytical goals for TPP assays of imprecision of 2.4%, bias of 3.2%, and total error of 7.2%.[106,107]

Standardization

No certified reference material or reference analytical method is available. This paucity of standardization and traceability confers a lack of reliability on calibration accuracy. The commutability of standards and calibrators is unknown.[108,109]

There is also a lack of robust internal standardization for thiamine assays. Thiamine and related compounds are usually unstable in alkaline environments, although their fluorescence intensity increases with assay alkalinity, with an optimum at pH 9.0. Also it is useful to remember column chromatographic performance and longevity can be reduced under highly alkaline conditions.

External Quality Assurance

Two external quality assessment schemes are available for the determination of thiamine in the form of whole-blood TPP, namely, The Royal College of Pathologists Australasia Quality Assurance Programs (RCPAQAP) and the Dutch Foundation for

Quality Assessment in Medical Laboratories (SKML). These EQA schemes incorporate a common lyophilized thiamine-containing material, purchased in bulk and distributed to the participating laboratories.[106,107] Generally it has been found that recent end-of-cycle reports from the RCPAQAP and SKML EQA vitamin programs indicate that probably no participating laboratory is routinely using indirect methodology.

As Jenco and colleagues summarize regarding thiamine analysis, currently there are indeed problems regarding paucity of automation, lack of assay reproducibility and standardization, with time-consuming pretreatment stages which automation will hopefully improve.[110] Furthermore, these tests are potentially expensive and time consuming and not readily available.

Summary

In summary, clinicians should be vigilant for thiamine deficiency that can present in various clinical scenarios and also insidiously (Table 6.1). The diagnosis of deficiency is largely a clinical one although prompt thiamine replacement may reduce morbidity and mortality. The assays for thiamine and related compounds consist of indirect and direct methods—the latter being the most favored by far. High performance liquid chromatography and more recently LC-MS/MS offer methods of thiamine analysis. Thiamine assay problems currently consist of lack of assay standardization, and paucity of reference material and calibrators. Also there are no agreed thiamine reference ranges and few quality control schemes for thiamine assays something which needs to be promptly resolved to aid better understanding of thiamine-associated clinical disorders.

References

1. Carpenter KJ. The discovery of thiamin. *Ann Nutr Metab* 2012;**61**:219–23.
2. Williams RR. The chemistry and biological significance of thiamin. *Science* 1938;**87**:559–63.
3. Crook MA. *In clinical biochemistry and metabolic medicine*. 8th ed. Hodder/Arnold; 2012ISBN: 9781444144147.
4. Singleton CK, Martin PR. Molecular mechanisms of thiamin utilization. *Curr Mol Med* 2001;**1**:197–207.
5. Lonsdale D. A review of the biochemistry, metabolism and clinical benefits of thiamin(e) and its derivatives. *Evid Based Complement Alternat Med* 2006;**3**:49–59.
6. Schattner A. An unlikely culprit—the many guises of thiamine deficiency. *Am J Emerg Med* 2013;**31**:. 635.e5–636.e5.
7. Manzetti S, Zhang J, Van Der Spoel D. Thiamin function, metabolism, uptake and transport. *Biochemistry* 2014;**53**:821–35.
8. Ganapathy V, Smith SB, Prasad PD. SLC19: the folate/thiamine transporter family. *Pflugers Arch-Eur J Physiol* 2004;**447**:641–6.
9. Diaz GA, Banikazemi M, Oishi K, Desnick RJ, Gelb BD. Mutations in a new gene encoding a thiamine transporter cause thiamine-responsive megaloblastic anaemia syndrome. *Nat Genet* 1999;**22**:309–12.

10. Dutta B, Huang W, Molero M, Kekuda R, Leibach FH, Devoe LD, et al. Cloning of the human thiamine transporter, a member of the folate transporter family. *J Biol Chem* 1999;**274**:31925–9.

11. Powers JS, Zimmer J, Meurer K, Manske E, Collins JC, Greene HL. Direct assay of vitamins B1, B2, and B6 in hospitalized patients: relationship to level of intake. *J Parenter Enter Nutr* 1993;**17**:315–6.

12. Wilkinson TJ, Hanger HC, George PM, Sainsbury R. Is thiamine deficiency in elderly people related to age or co-morbidity? *Age Ageing* 2000;**29**:111–6.

13. Jamieson CP, Obeid OA, Powell-Tuck J. The thiamin, riboflavin and pyridoxine status of patients on emergency admission to hospital. *Clin Nutr* 1999;**18**:87–91.

14. Mataix J, Aranda P, Sánchez C, Montellano MA, Planells E, Llopis J. Assessment of thiamin (vitamin B1) and riboflavin (vitamin B2) status in an adult Mediterranean population. *Br J Nutr* 2003;**90**:661–6.

15. Kittanamongkolchai W, Leeaphorn N, Srivali N, Cheungpasitporn W. Beriberi in a dialysis patient: do we need more thiamine? *Am J Emerg Med* 2013;**31**:753.

16. Chitra S, Lath KV. Wernicke's encephalopathy with visual loss in a patient with hyperemesis gravidarum. *J Assoc Physicians India* 2012;**60**:53–6.

17. Ward KE, Happel KI. An eating disorder leading to wet beriberi heart failure in a 30-year-old woman. *Am J Emerg Med* 2013;**31**:. 460.e5–466.e5.

18. Bonucchi J, Hassan I, Policeni B, Kaboli P. Thyrotoxicosis associated with Wernicke's encephalopathy. *J Gen Intern Med* 2008;**23**:106–9.

19. Kimura H, Takeda K, Muto Y, Mukai H, Furusho M, Nakashita S, et al. Development of Wernicke's encephalopathy during initiation of hemodialysis in an elderly non-alcoholic patient. *Clin Nephrol* 2012;**78**:487–91.

20. Oguz SS, Ergenekon E, Tümer L, Koç E. A rare case of severe lactic acidosis in a preterm infant: lack of thiamine during total parenteral nutrition. *J Pediatr Endocrinol Metab* 2011;**24**:843–5.

21. Mosnier E, Niemetzky F, Stroot J, Pommier de Santi V, Brousse P, Guarmit B, et al. A large outbreak of thiamine deficiency among illegal gold miners in French Guiana. *Am J Trop Med Hyg* 2017;**96**:1248–52.

22. Ozawa H, Homma Y, Arisawa H, et al. Severe metabolic acidosis and heart failure due to thiamine deficiency. *Nutrition* 2001;**17**:351–2.

23. Crook MA, Hally V, Panteli JV. The importance of the refeeding syndrome. *Nutrition* 2001;**17**:632–7.

24. Stanga Z, Brunner A, Leuenberger M, Grimble RF, Shenkin A, Allison SP, et al. Nutrition in clinical practice—the refeeding syndrome. *Eur J Clin Nutr* 2008;**62**:687–94.

25. Sriram K, Manzanares W, Joseph K. Thiamine supplementation in nutrition support. *Nutr Clin Pract* 2012;**27**:41–50.

26. Costantini A, Pala MI, Colangeli M, Savelli S. Thiamine and spinocerebellar ataxia type 2. *BMJ Case Rep* 2013;https://doi.org/10.1136/bcr-2012-007302. pii: bcr2012007302.

27. Serrano M, Rebollo M, Depienne C, Rastetter A, Fernández-Álvarez E, Muchart J, et al. Reversible generalized dystonia and encephalopathy from thiamine transporter 2 deficiency. *Mov Disord* 2012;**27**:1295–8.

28. Yilmaz Agladioglu S, Aycan Z, Bas VN, Peltek Kendirci HN, Onder A. Thiamine-responsive megaloblastic anemia syndrome: a novel mutation. *Genet Couns* 2012;**23**:149–56.

29. Pérez-Dueñas B, Serrano M, Rebollo M, Muchart J, Gargallo E, Dupuits C, et al. Reversible lactic acidosis in a newborn with thiamine transporter-2 deficiency. *Pediatrics* 2013;**131**:e1670–5.

30. Larkin JR, Zhang F, Godfrey L, Molostvov G, Zehnder D, Rabbani N, et al. Glucose-induced down regulation of thiamine transporters in the kidney proximal tubular epithelium produces thiamine insufficiency in diabetes. *PLoS One* 2012;**7**:e53175.
31. Pichler H, Zeitlhofer P, Dworzak MN, Diakos C, Haas OA, Kager L. Thiamine-responsive megaloblastic anemia (TRMA) in an Austrian boy with compound heterozygous SLC19A2 mutations. *Eur J Pediatr* 2012;**171**:1711–5.
32. Scharfe C, Hauschild M, Klopstock T, Janssen AJ, Heidemann PH, Meitinger T, et al. A novel mutation in the thiamine responsive megaloblastic anaemia gene SLC19A2 in a patient with deficiency of respiratory chain complex I. *J Med Genet* 2000;**37**:669–73.
33. Fleming JC, Tartaglini E, Steinkamp MP, Schorderet DF, Cohen N, Neufeld EJ. The gene mutated in thiamine responsive anaemia with diabetes and deafness (TRMA) encodes a functional thiamine transporter. *Nat Genet* 1999;**22**:305–8.
34. Labay V, Raz T, Baron D, Mandel H, Williams H, Barrett T, et al. Mutations in SLC19A2 cause thiamine responsive megaloblastic anaemia associated with diabetes mellitus and deafness. *Nat Genet* 1999;**22**:300–4.
35. Piekutowska-Abramczuk D, Jurkiewicz E, Rokicki D, Ciara E, Trubicka J, Iwanicka-Pronicka K, et al. Neuropathological characteristics of the brain in two patients with SLC19A3 mutations related to the biotin-thiamine-responsive basal ganglia disease. *Folia Neuropathol* 2017;**55**:146–53.
36. Whitford W, Hawkins I, Glamuzina E, Wilson F, Marshall A, Ashton F, et al. Compound heterozygous SLC19A3 mutations further refine the critical promoter region for biotin-thiamine-responsive basal ganglia disease. *Cold Spring Harb Mol Case Stud* 2017;**3**:https://doi.org/10.1101/mcs.a001909. Jul 10. pii: mcs.a001909. [Epub ahead of print].
37. McCann A, Midttun Ø, Whitfield KC, Kroeun H, Borath M, Sophonneary P, et al. Comparable performance characteristics of plasma thiamine and erythrocyte thiamine diphosphate in response to thiamine fortification in rural Cambodian women. *Nutrients* 2017;**9**:https://doi.org/10.3390/nu9070676. pii: E676.
38. Dean RK, Subedi R, Gill D, Nat A. Consideration of alternative causes of lactic acidosis: thiamine deficiency in malignancy. *Am J Emerg Med* 2017;https://doi.org/10.1016/j.ajem.2017.05.016. May 15. pii: S0735–6757(17)30380-7. [Epub ahead of print].
39. Hamel J, Logigian EL. Acute nutritional axonal neuropathy. *Muscle Nerve* 2017;https://doi.org/10.1002/mus.25702. [Epub ahead of print].
40. Chiurazzi C, Cioffi I, De Caprio C, De Filippo E, Marra M, Sammarco R, et al. Adequacy of nutrient intake in women with restrictive anorexia nervosa. *Nutrition* 2017;**38**:80–4. https://doi.org/10.1016/j.nut.2017.02.004. [Epub 2017 Feb 24].
41. Bâ A. Alcohol and thiamine deficiency trigger differential mitochondrial transition pore opening mediating cellular death. *Apoptosis* 2017;**22**:741–52.
42. Lawton AW, Frisard NE. Visual loss, retinal hemorrhages, and optic disc edema resulting from thiamine deficiency following bariatric surgery complicated by prolonged vomiting. *Ochsner J* 2017;**17**:112–4.
43. Fernandes LMP, Bezerra FR, Monteiro MC, Silva ML, de Oliveira FR, Lima RR, et al. Thiamine deficiency, oxidative metabolic pathways and ethanol-induced neurotoxicity: how poor nutrition contributes to the alcoholic syndrome, as Marchiafava-Bignami disease. *Eur J Clin Nutr* 2017;**71**:580–6.
44. Moskowitz A, Andersen LW, Cocchi MN, Karlsson M, Patel PV, Donnino MW. Thiamine as a renal protective agent in septic shock. A secondary analysis of a randomized, double-blind, placebo-controlled trial. *Ann Am Thorac Soc* 2017;**14**:737–41.

45. Harel Y, Zuk L, Guindy M, Nakar O, Lotan D, Fattal-Valevski A. The effect of subclinical infantile thiamine deficiency on motor function in preschool children. *Matern Child Nutr* 2017;https://doi.org/10.1111/mcn.12397. [Epub ahead of print].

46. Masood U, Sharma A, Nijjar S, Sitaraman K. B-cell lymphoma, thiamine deficiency, and lactic acidosis. *Proc (Baylor Univ Med Cent)* 2017;**30**:69–70.

47. Potter K, Wu J, Lauzon J, Ho J. Beta cell function and clinical course in three siblings with thiamine-responsive megaloblastic anemia (TRMA) treated with thiamine supplementation. *J Pediatr Endocrinol Metab* 2017;**30**:241–6.

48. Nath A, Yewale S, Tran T, Brebbia JS, Shope TR, Koch TR. Dysphagia after vertical sleeve gastrectomy: evaluation of risk factors and assessment of endoscopic intervention. *World J Gastroenterol* 2016;**22**:10371–9.

49. Banka S, de Goede C, Yue WW, Morris AA, von Bremen B, Chandler KE, et al. Expanding the clinical and molecular spectrum of thiamine pyrophosphokinase deficiency: a treatable neurological disorder caused by TPK1 mutations. *Mol Genet Metab* 2014;**113**:301–6.

50. Zhang K, Huentelman MJ, Rao F, Sun EI, Corneveaux JJ, Schork AJ, et alInternational Consortium for Blood Pressure Genome-Wide Association Studies, Schork NJ, Eskin E, Nievergelt CM, Saier Jr. MH, O'Connor DT. Genetic implication of a novel thiamine transporter in human hypertension. *J Am Coll Cardiol* 2014;**63**:1542–55.

51. Mayr JA, Freisinger P, Schlachter K, Rolinski B, Zimmermann FA, Scheffner T, et al. Thiamine pyrophosphokinase deficiency in encephalopathic children with defects in the pyruvate oxidation pathway. *Am J Hum Genet* 2009;**89**:806–12.

52. Lindhurst MJ, Fiermonte G, Song S, Struys E, De Leonardis F, Schwartzberg PL, et al. Knockout of Slc25a19 causes mitochondrial thiamine pyrophosphate depletion, embryonic lethality, CNS malformations, and anemia. *Proc Natl Acad Sci U S A* 2006;**103**:15927–32.

53. Clark JA, Burny I, Sarnaik AP, Audhya TK. Acute thiamine deficiency in diabetic ketoacidosis: diagnosis and management. *Pediatr Crit Care Med* 2006;**7**:595–9.

54. Liu S, Huang H, Lu X, Golinski M, Comesse S, Watt D, Grossman RB, et al. Down-regulation of thiamine transporter THTR2 gene expression in breast cancer and its association with resistance to apoptosis. *Mol Cancer Res* 2003;**1**:665–73.

55. Shamir R, Dagan O, Abramovitch D, Abramovitch T, Vidne BA, Dinari G. Thiamine deficiency in children with congenital heart disease before and after corrective surgery. *J Parenter Enter Nutr* 2000;**24**:154–8.

56. Krishna S, Taylor AM, Supanaranond W, Pukrittayakamee S, ter Kuile F, Tawfiq KM, Holloway PA, White NJ. Thiamine deficiency and malaria in adults from southeast Asia. *Lancet* 1999;**353**:546–9.

57. Gold M, Chen MF, Johnson K. Plasma and red blood cell thiamine deficiency in patients with dementia of the Alzheimer's type. *Arch Neurol* 1995;**52**:1081–6.

58. Oishi K, Diaz GA. Thiamine-responsive megaloblastic anemia syndrome. In: Pagon RA, Adam MP, Ardinger HH, Wallace SE, Amemiya A, LJH B, Bird TD, editors. *GeneReviews®*. Seattle, WA: University of Washington; 1993–2017. [Internet]. Seattle (WA). 2003 Oct 24 [updated 2017 May 4].

59. Blitshteyn S. Vitamin B1 deficiency in patients with postural tachycardia syndrome (POTS). *Neurol Res* 2017;**39**:685–8.

60. Gasquoine PG. A case of bariatric surgery-related Wernicke-Korsakoff syndrome with persisting anterograde amnesia. *Arch Clin Neuropsychol* 2017;**20**:1–8.

61. Valentino D, Sriram K, Shankar P. Update on micronutrients in bariatric surgery. *Curr Opin Clin Nutr Metab Care* 2011;**14**:635–41.

62. Scarano V, Milone M, Di Minno MN, Panariello G, Bertogliatti S, Terracciano M, et al. Late micronutrient deficiency and neurological dysfunction after laparoscopic sleeve gastrectomy: a case report. *Eur J Clin Nutr* 2012;**66**:645–7.

63. Alim U, Bates D, Langevin A, Werry D, Dersch-Mills D, Herman RJ, et al. Thiamine prescribing practices for adult patients admitted to an internal medicine service. *Can J Hosp Pharm* 2017;**70**:179–87.

64. Nishimoto A, Usery J, Winton JC, Twilla J. High-dose parenteral thiamine in treatment of Wernicke's encephalopathy: case series and review of the literature. *In Vivo* 2017;**31**:121–4.

65. Collie JTB, Greaves RF, Jones OAH, Lam Q, Eastwood GM, Bellomo R. Vitamin B1 in critically ill patients: needs and challenges. *Clin Chem Lab Med* 2017;**55**:1652–68.

66. Ridyard HN. Quantitative assay of aneurin (vitamin B1). *Nature* 1946;**157**:301.

67. Ballard CW, Ballard EJ. The colorimetric determination of aneurine by Auerbach's method. *J Pharm Pharmacol* 1949;**1**:330–3.

68. Dawburn MC. The determination of thiamine in white bread by the thiochrome method. *Aust J Exp Biol Med Sci* 1949;**27**:207–14.

69. Teeri AE. A new fluorometric determination of thiamine. *J Biol Chem* 1952;**196**:547–50.

70. Bessey OA, Lowry OH, Davis EB. The measurement of thiamine in urine. *J Biol Chem* 1952;**195**:453–8.

71. Villeneuva JR. A new method for the microbiological assay of thiamin (vitamin B1). *Nature* 1955;**176**:465.

72. Lopez-Flores J, Fernandez-De Cordova ML, Molina-Diaz A. Implementation of flow-through solid phase spectroscopic transduction with photochemically induced fluorescence: determination of thiamine. *Anal Chim Acta* 2005;**535**:161–8.

73. Heydari R, Elyasi NS. Ion-pair cloud-point extraction: a new method for the determination of water-soluble vitamins in plasma and urine. *J Sep Sci* 2014;**37**:2724–31.

74. Ihara H, Matsumoto T, Shino Y, Hashizume N. Assay values for thiamine or thiamine phosphate esters in whole blood do not depend on the anticoagulant used. *J Clin Lab Anal* 2005;**19**:205–8.

75. Talwar D, Davidson H, Cooney J, St JO'Reilly D. Vitamin B1 status assessed by direct measurement of thiamin pyrophosphate in erythrocytes or whole blood by HPLC: comparison with erythrocyte transketolase activation assay. *Clin Chem* 2000;**46**:704–10.

76. Tashirova OA, Ramenskaya GV, Vlasov AM, Khaitov MR. Development and validation of an LC/MS methods for quantitative determination of thiamine in blood plasma. *Pharm Chem J* 2013;**46**:742–4.

77. El-Sabban MZ, EL-Maghrabi MS. A microfluorometric method for the assay of vitamin B1. *J Egypt Med Assoc* 1957;**40**:831–8.

78. Mickelsen O, Yamamoto S. Methods for the determination of thiamine. *Methods Biochem Anal* 1958;**6**:191–257.

79. Deibel RH, Evans JB, Niven Jr. CF. Microbiological assay for thiamin using lactobacillus viridescens. *J Bacteriol* 1957;**74**:818–21.

80. Maciasr FM. Improved medium for assay of thiamine with lactobacillus fermenti. *Appl Microbiol* 1957;**5**:249–52.

81. Hoad KE, Johnson LA, Woollard GA, Walmsley TA, Briscoe S, Jolly LM, et al. Vitamin B1 and B6 method harmonization: comparison of performance between laboratories enrolled in the RCPA quality assurance program. *Clin Biochem* 2013;**46**:772–6.

82. Icke GC, Nicol DJ. Thiamin status in pregnancy as determined by direct microbiological assay. *Int J Vitam Nutr Res* 1994;**64**:33–5.

83. Olkowski AA, Gooneratne SR. Microbiological methods of thiamine measurement in biological material. *Int J Vitam Nutr Res* 1992;**62**:34–42.

84. Lonsdale D, Shamberger RJ. Red cell transketolase as an indicator of nutritional deficiency. *Am J Clin Nutr* 1980;**33**:205–11.

85. Gibson GE, Ksiezak-Reding H, Sheu KF. Correlation of enzymatic, metabolic, and behavioural deficits in thiamine deficiency and its reversal. *Neurochem Res* 2004;**9**:803–14.

86. Mount JN, Heduan E, Herd C, Jupp R, Kearney E, Marsh A. Adaptation of coenzyme stimulation assays for the nutritional assessment of vitamins B1, B2 and B6 using the Cobas Bio centrifugal analyser. *Ann Clin Biochem* 1987;**24**:41–6.

87. Dancy M, Evans G, Gaitonde MK, Maxwell JD. Blood thiamine and thiamine phosphate ester concentrations in alcoholic and non-alcoholic liver diseases. *Br Med J (Clin Res Ed)* 1984;**289**:79–82.

88. Kirk JR. Automated analysis of thiamine, ascorbic acid, and vitamin A. *J Assoc Off Anal Chem* 1977;**60**:1234–7.

89. Williams DG. Methods for the estimation of three vitamin dependent red cell enzymes. *Clin Biochem* 1976;**9**:252–5.

90. Walker MC, Carpenter BE, Cooper EL. Simultaneous determination of niacinamide, pyridoxine, riboflavin, and thiamine in multivitamin products by high-pressure liquid chromatography. *J Pharm Sci* 1981;**70**:99–101.

91. Li K. Simultaneous determination of nicotinamide, pyridoxine hydrochloride, thiamine mononitrate and riboflavin in multivitamin with minerals tablets by reversed-phase ion-pair high performance liquid chromatography. *Biomed Chromatogr* 2002;**16**:504–7.

92. Such V, Traveset J, Gonzalo R, Gelpí E. Stability assays of aged pharmaceutical formulas for thiamine and pyridoxine by high performance thin-layer chromatography and derivative ultraviolet spectrometry. *Anal Chem* 1980;**52**:412–9.

93. Berg TM, Behagel HA. Semiautomated method for microbiological vitamin assays. *Appl Microbiol* 1972;**23**:531–42.

94. Smeets EH, Muller H. A NADH-dependent transketolase assay in erythrocyte hemolysates. *Clin Chem Acta* 1971;**33**:379–86.

95. Nordentoft M, Timm S, Hasselbalch E, Roesen A, Gammeltoft S, Hemmingsen R. Thiamine pyrophosphate effect and erythrocyte transketolase activity during severe alcohol withdrawal syndrome. *Acta Psychiatr Scand* 1993;**88**:80–4.

96. Lynch PL, Young IS. Determination of thiamine by high-performance liquid chromatography. *J Chromatogr A* 2000;**881**:267–84.

97. Vinas P, Lopez-Erroz C, Balsalobre N, Hernandez-Cordoba M. Comparison of ion-pair and amide-based column reverse phase liquid chromatography for the separation of thiamine related compounds. *J Chromatogr B* 2001;**757**:301–8.

98. Losa R, Sierra MI, Fernandez A, Blanco D, Buesa JM. Determination of thiamine and its phosphorylated forms in human plasma, erythrocytes and urine by HPLC and fluorescence detection: a preliminary study on cancer patients. *J Pharm Biomed Anal* 2005;**37**:1025–9.

99. Korner RW, Vierzig A, Roth B, Muller C. Determination of thiamin diphosphate in whole blood samples by high-performance liquid chromatography-a method suitable for pediatric diagnostics. *J Chromatogr B* 2009;**877**:1882–6.

100. Bettendorff L, Peeters M, Jouan C, Wins P, Schoffeniels E. Determination of thiamin and its phosphate esters in cultured neurons and astrocytes using an ion-pair reversed-phase high performance liquid chromatographic method. *Anal Biochem* 1991;**198**:52–9.

101. Chatzimichalakis PF, Samanidou VF, Verpoorte R, Papadoyannis IN. Development of a validated HPLC method for the determination of B-complex vitamins in pharmaceuticals and biological fluids after solid phase extraction. *J Sep Sci* 2004;**27**:1181–8.

102. Basiri B, Sutton JM, Hanberry BS, Zastre JA, Bartlett MG. Ion pair liquid chromatography method for the determination of thiamine (vitamin B1) homeostasis. *Biomed Chromatogr* 2016;**30**:35–41.

103. Puts J, de Groot M, Haex M, Jakobs B. Simultaneous determination of underivatized vitamin B1 and B6 in whole blood by reversed phase ultra high performance liquid chromatography tandem mass spectrometry. *PLoS One* 2015;**10**:e0132018.

104. Bohrer D, Do Nascimento PC, Ramirez AG, Medndonca JK, de Carvahlo LM, Promblum SC. Determination of thiamine in blood serum and urine by high-performance liquid chromatography with direct injection and post-column derivatization. *Microchem J* 2004;**78**:71–6.

105. Mancinelli R, Ceccanti M, Guiducci MS, Sasso GF, Sebastiani G, Attilia ML, et al. Simultaneous liquid chromatographic assessment of thiamine, thiamine monophosphate and thiamine diphosphate in human erythrocytes: a study on alcoholics. *J Chromatogr B* 2003;**789**:355–63.

106. SKML. End of cycle vitamin B1 reports 20151 and 20152. In: *Dutch foundation for quality assessment in medical laboratories*. 2016.

107. RCPA Quality Assurance Programs. *Thiamine pyrophosphate end of cycle 33 report*. http://www.rcpaqap.com.au/chempath; 2016. Available from:.

108. Talwar DK, Azharuddin MK, Williamson C, Teoh YP, McMillan DC, O'Reilly DS. Biological variation of vitamins in blood of healthy individuals. *Clin Chem* 2005;**51**:2145–50.

109. Sandberg S, Fraser CG, Horvath AR, Jansen R, Jones G, Oosterhuis W, et al. Defining analytical performance specifications: consensus statement from the 1st strategic conference of the European Federation of Clinical Chemistry and Laboratory Medicine. *Clin Chem Lab Med* 2015;**53**:833–5.

110. Jenčo J, Krčmová LK, Solichová D, Solich P. Recent trends in determination of thiamine and its derivatives in clinical practice. *J Chromatogr A* 2017;**1510**:1–12.

Further Reading

111. Touisni N, Charmantray F, Hélaine V, Hecquet L, Mousty C. An efficient amperometric transketolase assay: towards inhibitor screening. *Biosens Bioelectron* 2014;**62**:90–6.

112. Roser RL, Andrist AH, Harrington WH, Naito HK, Lonsdale D. Determination of urinary thiamine by high-pressure liquid chromatography utilizing the thiochrome fluorescent method. *J Chromatogr* 1978;**146**:43–53.

113. Ihara H, Matsumoto T, Kakinoki T, Shino Y, Hashimoto R, Hashizume N. Estimation of vitamin B1 excretion in 24-hr urine by assay of first-morning urine. *J Clin Lab Anal* 2008;**22**:291–4.

114. Lynch PL, Trimble ER, Young IS. High-performance liquid chromatographic determination of thiamine diphosphate in erythrocytes using internal standard methodology. *J Chromatogr B* 1997;**70**:120–3.

Methods for assessment of Vitamin B$_2$

7

Roy A. Sherwood,

King's College London, London, United Kingdom

Chapter Outline

Structure and Function

Structure

Riboflavin (vitamin B$_2$) was initially isolated as a yellow-green pigment from milk whey in 1872; though it was not until the 20th century, when research on accessory factors began, that scientists started to investigate the nature of this pigment further. Riboflavin is fluorescent under ultraviolet light and in pure form is a yellow crystalline solid that is poorly soluble in comparison to the other B vitamins.

Laboratory Assessment of Vitamin Status. https://doi.org/10.1016/B978-0-12-813050-6.00007-3

Function

Although riboflavin (Fig 7.1A) itself exerts minimal enzymatic activity, it is the precursor to the essential coenzymes flavin mononucleotide (FMN), (Fig 7.1B) and flavin adenine dinucleotide (FAD), (Fig. 7.1C).

Flavin mononucleotide and FAD are called flavocoenzymes and the enzymes that require their presence to function are termed flavoproteins. Riboflavin cannot be synthesized in humans and is not stored in the body so ongoing adequate dietary intake is required to maintain the cellular concentrations of FMN and FAD at an appropriate level.

Flavin mononucleotide and FAD participate in a number of critical oxidation-reduction reactions that are involved in amino acid metabolism, fatty acid synthesis, DNA repair, and in the activation of folate (Chapter 11) and pyridoxine (Chapter 9).[1] Flavin mononucleotide is a coenzyme in complex I of the electron transport chain while FAD is involved in complex II. Flavin adenine dinucleotide is required by the

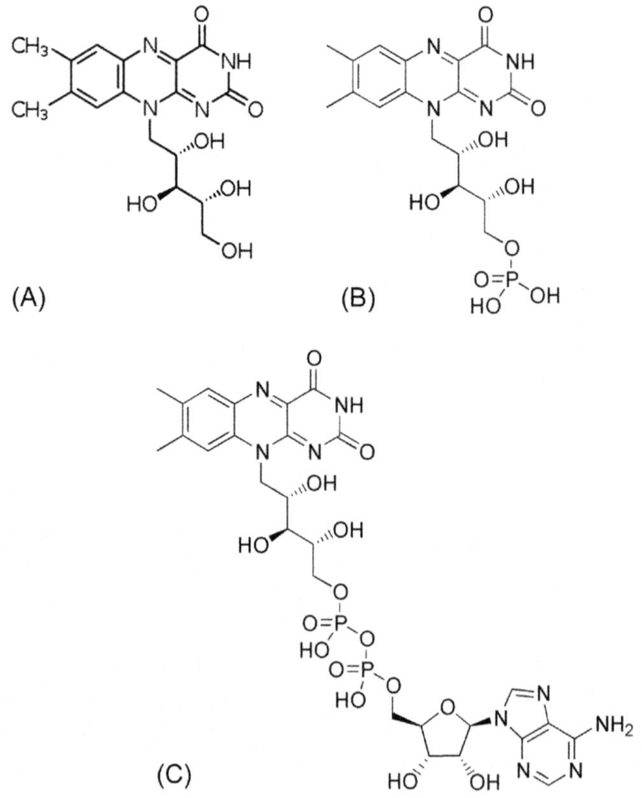

FIG. 7.1

Chemical structures for (A) riboflavin, (B) flavin mononucleotide (FMN), and (C) flavin adenine dinucleotide (FAD).

dehydrogenases that act on pyruvate, α-ketoglutarate, and the branched chain amino acids. The conversion of retinol to retinoic acid (Chapter 2) is catalyzed by cytosolic retinal dehydrogenase, which is an FAD-dependent enzyme. The reduction of the oxidized form of glutathione to its reduced form by glutathione reductase is FAD dependent. Flavin adenine dinucleotide is the most common form of the flavocoenzymes at a cellular level.

Free riboflavin in the circulation is bound to albumin and to certain immunoglobulins that can also bind FMN and FAD.[2] At the tissue level, the flavocoenzymes are bound to various enzymes, for example, FAD is bound to succinic dehydrogenase.[3] Vitamin B_2 concentrations are fourfold to fivefold higher in erythrocytes when compared to plasma.

Dietary Intake

Milk and dairy products are the predominant source of riboflavin in typical diets in the developed world. Cereals, meat (particularly liver and kidney), fatty fish, and green vegetables are also sources. In many countries certain breads and cereals are routinely fortified with riboflavin and cereals are now estimated to account for >20% of dietary riboflavin in the United Kingdom. Current recommended dietary intakes of riboflavin are consistent across Europe and the United States ranging from 0.3 mg/d in infants to 1.3 mg/d in adults.[4, 5] An additional 0.3–0.5 mg/d is recommended during pregnancy or lactation. Placental transfer of riboflavin involves binding proteins that are specific to pregnancy and leads to concentrations in fetal circulation that are fourfold higher than material plasma.

Riboflavin is heat stable so cooking has no effect on bioavailability but prolonged exposure to light can lead to a degree of degradation.

Riboflavin absorption is an active process facilitated by the secretion of bile salts. Most of the absorption occurs in the proximal small intestine via a carrier-mediated transport across the enterocytes.[6] This process is saturable at approximately 30 mg of riboflavin with little extra being absorbed beyond that.[7] Free riboflavin taken up from the enterocytes is converted to FMN by ATP-dependent phosphorylation by flavokinase and then to FAD by FAD-synthase. Urinary excretion is predominantly as free riboflavin with small amounts of other metabolites. In individuals taking large quantities of riboflavin supplements the free riboflavin can impart a significant yellow coloration to the urine.

Clinical Application of Measurement
Riboflavin Deficiency in Humans

Riboflavin deficiency is relatively uncommon in the developed world but in developing countries mild deficiency can be seen in up to 50% of the population. Riboflavin deficiency can be associated with inadequate dietary intake, malabsorptive conditions, for example, celiac disease, but is often seen in combination with a generalized B vitamin deficiency. Riboflavin deficiency (sometimes called ariboflavinosis) causes stomatitis of the mouth and tongue, cheilosis (chapped and fissured lips) and a scaly rash on the genitalia. It has also been associated with visual disturbances including night blindness, migraine headaches, mild anemia, and psychological effects including depression.[1] Treatment is by dietary modification or supplements; usually combined vitamin B supplements in view of the association of riboflavin deficiency with deficiencies of the other B vitamins.

Analytical Methods
Assessment of Riboflavin Status in Humans

A wide variety of assays have been described for the measurement of vitamin B$_2$ in pharmaceutical products and in food and drink. A much more limited range of techniques are available for the analysis of riboflavin in body fluids. The methods available for measurement of vitamin B$_2$ in various matrices were reviewed by Gul et al.[8]

The flavins can be measured in whole blood, serum, plasma, or urine. Samples for FAD analysis have been shown to be stable for several days at room temperature and for months at −20°C. The flavins exhibit varying degrees of degradation when exposed to light, so it is advisable that samples should be protected from direct exposure to light before and during analysis.

Microbiological Assays

One of the first methods described, in 1939, was a microbiological assay for riboflavin based on the observation that the amount of growth of *Lactobacillus casei* was directly proportional to the riboflavin concentration in the culture medium.[9]

Fluorometric Assays

Following on from the microbiological assay for riboflavin published in 1939[9] several methods for urinary riboflavin that utilized the native fluorescence of riboflavin were described.[10, 11] A number of modifications to these initial methods have subsequently been described.

Functional Assays

For many years the preferred method for assessing riboflavin status was a functional assay based on the measurement of the activity of glutathione reductase in erythrocytes rather than direct measurement of the vitamin.[12] It is still considered by some to give the best indication of tissue saturation vitamin B$_2$ and hence long-term status. The method is usually carried out with and without the in vitro addition of FAD to permit determination of the activity coefficient which is referred to as the Erythrocyte Glutathione Reductase Activation Coefficient (EGRAC). An EGRAC of >1.4 (i.e., an increase of >40% on addition of FAD) is considered indicative of riboflavin deficiency, an EGRAC between 1.2 and 1.4 suggests riboflavin status is suboptimal while an EGRAC <1.2 indicated adequate status.

Enzymatic Assay

An enzyme-linked ligand-sorbent assay (ELLSA) has been reported that is capable of detecting riboflavin in human plasma and urine. Conjugates obtained by coupling 3-carboxymethylriboflavin with carbodiimide to bovine serum albumin were adsorbed on to a microtiter plate. Detection was carried out spectrophotometrically using a horseradish peroxide label. The detection limit was quoted as 0.8 pmol of riboflavin but reported CVs were around 20%.[13]

Capillary Electrophoretic Methods

A method for the measurement of riboflavin, FMN, and FAD based on micellar electrokinetic capillary chromatography with laser-induced fluorescence (LIF) detection was reported in 1999.[14] Plasma samples were pretreated with trichloroacetic acid followed by solid-phase extraction using reversed-phase columns. Recoveries of 90%–103% were reported with CVs of between 4% and 12%. The limit of detection was quoted as being below physiological concentrations. A method for the measurement of riboflavin in urine using CE-LIF was published in 2003.[15] The authors suggested that as FAD and FMN fluoresce at the same wavelength (467 nm) as riboflavin they could also be detected by the method, although they did not present any data to support this assertion. A further method using CE-LIF was reported in 2007.[16] While the method was primarily intended for measurement of riboflavin in beverages, including green tea, the authors did analyze human urine samples. The limit of detection was reported as 3 nmol/L. A simple method for assay of riboflavin in urine was reported that used capillary electrochromatography with a methacrylate-based monolithic column.[17]

Chromatographic Methods

Most HPLC methods have been developed for the relatively higher concentrations of riboflavin found in pharmaceutical preparations, foodstuffs, and urine.[18, 19] Methods have, however, been reported that can measure the lower concentrations found in whole blood,[20–23] serum,[21, 22, 24] or plasma.[25–29] The majority of these methods involve a sample preparation step with protein precipitation, typically with trichloroacetic acid, followed by reversed-phase chromatography with C18 columns. Most methods use fluorimetric detection utilizing the native fluorescence of the flavins. A commercially available assay for the flavins using HPLC with fluorescence detection is produced by Chromsystems (Germany). A representative chromatogram from the Chromsystems method for the flavins is shown in Fig. 7.2.

Methods using HPLC with mass spectrophotometric detection were first described for urinary riboflavin. A method for the simultaneous detection of vitamins B_2 and

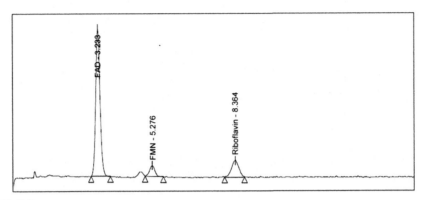

FIG. 7.2

Example reversed-phase HPLC with fluorimetric detection showing the separation of flavin adenine dinucleotide (FAD), flavin mononucleotide (FMN), and riboflavin.

B$_{12}$ (Chapter 12) using matrix-assisted laser desorption ionization time-of-flight mass spectrometry (MALDI-TOF MS) was described in 2009.[30] A method for urine riboflavin alone was subsequently reported using LC-MS/MS that had a detection limit of 30 nmol/L.[31] More recently a method for the simultaneous assay of vitamins B$_2$ and B$_6$ (Chapter 9) using LC-MS/MS has been described with detection using positive ion mode.[32] The analysis time was 8 min. Limits of detection were in the range 0.1–0.4 nmol/L and within- and between-day CVs were 3%–20% and 6%–22%, respectively. Plasma concentrations (median; 10–90th percentiles) in 94 healthy subjects were riboflavin 10.3 (4.9–38.4) nmol/L, FAD 63.1 (49.9–84.8) nmol/L, and FMN 7.5 (3.3–13.4) nmol/L.

Standardization

Certified pharmaceutical secondary standards are commercially available from a number of suppliers to support the determination of riboflavin in pharmaceutical formulations and human urine samples using HPLC.

External Quality Assurance

An external quality assurance scheme is operated by Instand (Germany) http://instand-ev.de/en/eqas.html for the analysis of riboflavin in serum and for FAD, FMN, and riboflavin in whole blood.

References

1. Powers HJ. Riboflavin (vitamin B-2) and health. *Am J Clin Nutr* 2003;**77**:1352–60.
2. Innis WS, McCormick DB, Merrill Jr. AH. Variations in riboflavin binding by human plasma: identification of immunoglobulins as the major proteins responsible. *Biochem Med* 1986;**34**:151–65.

3. Singer TP, Kenney WC. Biochemistry of covalently-bound flavins. *Vitam Horm* 1974;**32**:1–45.

4. Institute of Medicine. *Dietary reference intakes for thiamin, riboflavin, niacin, vitamin B6, folate, vitamin B12, pantothenic acid, biotin and choline*. Washington, DC: The National Acadamies Press; 1998. p. 87–122.

5. EU. *Overview on dietary reference values for the EU population as derived by the EFSA panel on dietetic products, nutrition and allergies*. European Union; 2017.

6. Jusko WJ, Levy G. Absorption, metabolism and excretion of riboflavin 5'-phosphate in man. *J Pharm Sci* 1967;**56**:58–62.

7. Zempleni J, Galloway JR, McCormick DB. Pharmacokinetics of orally and intravenously administered riboflavin in healthy humans. *Am J Clin Nutr* 1996;**63**:54–66.

8. Gul W, Anwar Z, Qadeer K, Perveen S, Ahmad I. Methods of analysis of riboflavin (vitamin B2): a review. *J Pharm Pharm Sci* 2014;**2**:10–21.

9. Snell EE, Strong FM. A microbiological assay for riboflavin. *Ind Eng Chem Anal Ed* 1939;**11**:346–50.

10. Najjar VA. The fluorometric determination of riboflavin in urine and other biological fluids. *J Biol Chem* 1941;**141**:355–64.

11. Morell DB, Slater EC. The fluorometric determination of riboflavin in urine. *Biochem J* 1946;**40**:652–7.

12. Gibson RS. *Riboflavin in principles of nutritional*. assessment 2nd ed. Oxford: Oxford University Press; 2005.

13. Kozik A. Microtitre-plate enzyme-linked ligand-sorbent assay of riboflavin (vitamin B2) in human plasma and urine. *Analyst* 1996;**121**:333–7.

14. Hustad S, Ueland PM, Schneede J. Quantification of riboflavin, flavin mononucleotide, and flavin adenine dinucleotide in human plasma by capillary electrophoresis and laser-induced fluorescence detection. *Clin Chem* 1999;**45**:862–8.

15. Su A-K, Lin C-H. Determination of riboflavin in urine by capillary electrophoresis—blue light emitting diode-induced fluorescence detection combined with a stacking technique. *J Chromatogr B* 2003;**785**:39–46.

16. Hu L, Yand X, Wang C, Yuan H, Xiao D. Determination of riboflavin in urine and beverages by capillary electrophoresis with in-column optical fibre laser-induced fluorescence detection. *J Chromatogr B* 2007;**856**:245–51.

17. Wei X, Qi L, Qiao J, Yao C, Wang F, Chen Y. Assay of vitamin B in urine by capillary electrochromatography with a methacrylate-based monolithic column. *Electrophoresis* 2010;**31**:3227–32.

18. Nielsen P. Flavins. In: De Leenheer AP, Lambert WE, Nelis HF, editors. *Modern chromatographic analysis of vitamins*. New York: Marcel Dekker; 1992. p. 355–98.

19. Smith MD. Rapid method for determination of riboflavin in urine by high-performance liquid chromatography. *J Chromatogr B Biomed Sci Appl* 1980;**182**:285–91.

20. Floridi A, Palmerini CA, Fini C, Pupita M, Fidanza F. High performance liquid chromatographic analysis of flavin adenine dinucleotide in whole blood. *Int J Vitam Nutr Res* 1985;**55**:187–91.

21. Bötticher B, Bötticher D. A new HPLC-method for the simultaneous determination of B1-, B2- and B6-vitamers in serum and whole blood. *Int J Vitam Nutr Res* 1987;**57**:273–8.

22. Ohkawa H. A simple method for micro-determination of flavins in whole blood by high-performance liquid chromatography. *Biochem Int* 1982;**4**:187–94.

23. Speek AJ, van Schaik F, Schrijver J, Schreurs WH. Determination of the B2 vitamer flavin adenine dinucleotide in whole blood by high-performance liquid chromatography with fluorometric detection. *J Chromatogr* 1982;**228**:311–6.

24. Lambert WE, Cammaert PM, de Leenheer AP. Liquid-chromatographic measurement of riboflavin in serum and urine with iroriboflavin as internal standard. *Clin Chem* 1985;**31**:1371–3.

25. Capo-chichi CD, Gueant JL, Feillet F, Namour F, Vidailhet M. Analysis of riboflavin and riboflavin cofactor levels in plasma by high-performance liquid chromatography. *J Chromatogr B Biomed Sci Appl* 2000;**739**:219–24.

26. Pietta P, Calatroni A, Rava A. Hydrolysis of riboflavin nucleotides in plasma monitored by high-performance liquid chromatography. *J Chromatgr* 1982;**229**:445–9.

27. Lopez-Anays A, Mayersohn M. Quantification of riboflavin, riboflavin 5'-phosphate and flavin adenine dinucleotide by high-performance liquid chromatography. *J Chromatogr* 1987;**423**:105–13.

28. Zempleni J. Determination of riboflavin and flavocoenzymes in human blood plasma by high-performance liquid chromatography. *Ann Nutr Metab* 1995;**39**:224–6.

29. Petteys BJ, Frank EL. Rapid determination of vitamin B2 (riboflavin) in plasma by HPLC. *Clin Chim Acta* 2011;**412**:38–43.

30. Mandal SM, Mandal M, Ghosh AK, Dey S. Rapid determination of vitamin B2 and B12 in human urine by isocratic lquid chromatography. *Anal Chim Acta* 2009;**640**:110–3.

31. Bishop AM, Fernandez C, Whitehead Jr. RD, Morales AP, Barr DB, Wilder LC, Baker SE. Quantification of riboflavin in human urine using high performance liquid chromatography-tandem mass spectrometry. *J Chromtogr B Abalyt Technol Biomed Life Sci* 2011;**879**:1823–6.

32. Midttun O, Hustad S, Solheim E, Schneede J, Ueland PE. Multianalyte quantification of vitamin B6 and B2 species in the nanomolar range in human plasma by liquid chromatography-tandem mass spectrometry. *Clin Chem* 2005;**51**:1206–16.

Methods for assessment of pantothenic acid (Vitamin B₅)

Rick Huisjes, David J. Card

Nutristasis Unit, Viapath, St. Thomas' Hospital, London, United Kingdom

Chapter Outline

Structure and Function

Vitamin B₅ (pantothenic acid) is an essential micronutrient required for the synthesis of coenzyme A (CoA), (Fig. 8.1). Coenzyme A is the key factor in the first step of the tricarboxylic acid (TCA) cycle, responsible for transferring the acetyl group from pyruvate oxidation to oxaloacetate yielding citrate. Coenzyme A is also a critical cofactor in fatty acid metabolism.

Pantothenic acid (or the anion pantothenate) is formed by pantothenate synthase, which catalyzes an ATP-dependent condensation reaction between β-alanine and pantoic acid. The reaction occurs in almost all plants and microorganisms hence the Greek-derived prefix 'panto,' meaning everywhere.[1]

Pantothenic acid is water soluble (log $P=-1.69$) and negatively charged under physiological conditions. At low pH (pH = 2–3) it is neutrally charged.[2,3] Pantothenic acid is relatively unstable and chemical decomposition occurs rapidly in acidic and basic conditions. Moreover, the rate of pantothenic acid degradation is increased by heat.[3] There is no evidence that the stability of pantothenic acid is influenced by light exposure. The biophysical properties of vitamin B₅ are shown in Table 8.1.

Laboratory Assessment of Vitamin Status. https://doi.org/10.1016/B978-0-12-813050-6.00008-5

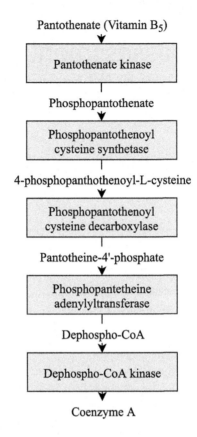

FIG. 8.1

Coenzyme A (CoA) synthesis from vitamin B₅. (A) pantothenate kinase (rate-limiting); (B) phosphopantothenoyl cysteine synthetase; (C) phosphopantothenoyl cysteine decarboxylase; (D) phosphopantetheine adenylyltransferase; (E) dephospho-CoA kinase.

Image adapted from Basu SS, Mesaros C, Gelhaus SL, Blair IA. Stable isotope labeling by essential nutrients in cell culture for preparation of labeled coenzyme A and its thioesters. Anal Chem 2011;83:1363–9.

Table 8.1 Biophysical Properties of Vitamin B₅

Molecular weight	219.24 g/mol
Estimated log *P*	−1.69
Solubility	Freely soluble in water, ethyl acetate, dioxane, glacial acetic acid Moderately soluble in ether, amyl alcohol Practical insoluble in benzene and chloroform

Dietary Intake

Humans are dependent on endogenous sources of pantothenic acid to satisfy metabolic demand. Since the vitamin is widely distributed in a variety of foods, deficiency is very rare. Good sources of pantothenic acid include dried mushrooms, eggs, beef, and chicken.

Absorption, Transport, Kinetics, and Storage

Pantothenic acid is absorbed by a sodium-dependent transporter that is present in both the small and large intestines—an uptake mechanism that competes with biotin.[4] Excess pantothenic acid is not stored and is predominately excreted unmodified in urine. A proportion of pantothenic acid is excreted in urine as 4'-phosphopantothenate.[5] No data is available for distribution volume and elimination halftime. However, as pantothenic acid is highly water soluble, it is expected to have a low distribution volume.

Clinical Application of Measurement

Pantothenic acid deficiency has historically been associated with the severe clinical presentation seen in burning feet syndrome,[6] which was first described in Japanese prison camps during the Second World War.[6, 7] In patients with severe malnutrition, it is likely that other concomitant vitamin deficiencies contribute to the clinical presentation. Symptoms of pantothenic acid deficiency are thought to be related to impaired CoA synthesis, i.e., fatigue, apathy, and irritability. Case reports are notably absent from the literature, although deficient states have been induced and studied experimentally in human volunteers.[8]

Pantothenic acid deficiency can be induced experimentally by blocking CoA synthesis using a pantothenic acid kinase antagonist such as Ω-methylpantothenate.[9] In healthy volunteers this led to reduced mobility and gastrointestinal complaints (i.e., gastric burning, abdominal cramps, and diarrhea). One subject developed paresthesia (tingling in the legs) and a burning sensation on the soles of the feet. Pantothenic acid deficiency can be accompanied by hypokalemia and can result in disturbances of T-waves of the electrocardiogram of the heart. In controlled animal models, isolated deficiency has been shown to result in convulsions, damage to the adrenal gland, anemia, hypoglycemia, hyperventilation, and tachycardia.[10] Therefore in suspected cases of pantothenic acid deficiency deranged electrolytes and adrenocorticotrophic hormone concentrations may be seen.[9] Coenzyme A can be considered to be the active, functional form of pantothenic acid,[11] therefore in cases of suspected deficiency low blood concentrations of CoA may be present. Although low concentrations of CoA could also result from genetic polymorphisms in pantothenate kinase, phosphopantothenoyl cysteine synthetase, phosphopantothenoyl cysteine decarboxylase, phosphopantetheine adenylyltransferase, or dephospho-CoA kinase.[11, 12]

There is no evidence of toxicity associated with pantothenic acid in the literature.

Analytical Methods

Pantothenic acid can be measured in whole blood, red cells, plasma, serum, and urine. Whole blood and urine are the sample matrixes that have proven most informative. Reference values of 1.57–2.66 μmol/L have been reported for whole blood concentrations of pantothenic acid.[13] Whole blood pantothenic acid concentrations are strongly correlated with dietary intake[13, 14] making it the preferred matrix over serum/plasma. Also, serum/plasma concentrations of pantothenic acid are lower than in whole blood.[13]

Pantothenic acid can also be measured in urine. Urine concentrations (in μmol/day, postcorrection for 24-h urine volume) were found to correlate well the dietary intake of pantothenic acid.[15] An increased intake of 42 μmol/day resulted in 14.6 μmol/day urinary excretion of pantothenic acid, indicating it is either metabolized or excreted via nonrenal mechanisms.

Historical Methods

The first method to become available to quantify pantothenic acid was the microbiological assay[16] which was first described by Wright and Skeggs in 1944.[17] Pantothenic acid is essential for the growth of *Lactobacillus plantarum* and the assay is specific enough that pantothenic acid cannot be replaced by β-alanine.[18] By using a standard curve of pantothenic acid in culture media with other nutrients, the growth of the microorganism can be measured.[19] The microbiological assay is a very long-standing method and it is still considered to be the gold standard and currently used to compare the validity and accuracy of new tests.[20] The method is considered to be very robust and no specialist equipment is required. The disadvantages of this technique are that the stimulation of pantothenic acid to the growth of *L. plantarum* is not linear, but exponential and subject to carrying capacity.[19] Therefore multiple sample dilutions are necessary for high concentration samples and calibration standards. Moreover, biological fluids (such as urine, plasma, saliva) contain many compounds other than pantothenic acid which may interfere with the rate of bacterial growth. The sensitivity and specificity of the microbiological assay of vitamin B$_5$ is, however, considered to be suitable to diagnose pantothenic acid deficiency in the clinical setting.

Best Practice

Many articles have been published for the determination of pantothenic acid in pharmaceuticals and foods. These include the simultaneous analysis of water-soluble vitamins in various matrixes, such as described by Martin et al.[21] and Sim et al.[22]

The determination of pantothenic acid in blood and urine has been less studied and has proved to be complex. The two main approaches for the detection and quantification of pantothenic acid in biological fluids are HPLC and ELISA.

High Performance Liquid Chromatography

A robust and cost-effective method for the measurement of pantothenic acid in biological fluids is HPLC.[23] Samples require purification prior to analysis and detection by UV is hampered by relatively low UV absorbance at wavelengths above 240 nm.[23] Moreover, coextraction of other water-soluble vitamins can result in interfering chromatographic peaks and a loss of sensitivity. Recently, Takahaski et al.[20] developed an improved method to quantify pantothenic acid in urine by reversed phase HPLC and UV detection. Firstly, pantothenic acid is separated by reversed phase HPLC column. Subsequently, sodium hydroxide (NaOH) is pumped inline toward the sample mixture leading to decomposition of pantothenic acid to pantoic acid and β-alanine. After decomposition, ortho-phthalaldehyde (OPA) and 3-mercaptopropionic acid (3-MPA) are added to derivatize pantoic acid and β-alanine and to obtain a fluorescent derivative. Although highly sensitive and specific for pantothenic acid (limit of detection = 3 pmol from an on-column 20-μL injection volume), the method requires multiple steps to purify, decompose, and derivatize the vitamin. Although all steps occur inline, there is an increased risk of analytical error due to incomplete derivatization or blockages. The OPA is used in this method to derivatize β-alanine[24] and aids the fluorescence detection of pantothenic acid (excitation/emission = 340 nm/455 nm).

Recently, Zhang et al. employed LC-MS/MS for the analysis of pantothenic acid in plasma.[25] The method they developed could simultaneously analyze seven biomarkers of insulin resistance and among these biomarkers is pantothenic acid. A reversed phase column with a linear mobile phase gradient moving from 100% formic acid/water/NH_4Cl (0.025:1000:0.001, v/v/wt) toward a mixture with 40% methanol/acetonitrile/NH_4Cl (2000,1000:0.001, v/v/wt). The method described by Zhang et al. showed good linearity, accuracy, and precision. The method is very robust and suitable for use in the clinical setting; these benefits are set against the relative difficultly in implementing and the requirement for expensive equipment.

Enzyme-Linked Immunosorbent Assay

Enzyme-linked immunosorbent methods for the measurements of pantothenic acid are available.[26] They include purification and derivatization steps because pantothenic acid is not immunogenic and therefore not directly detectible by antibodies. To confer immunogenicity, the vitamin needs to be covalently linked with bromoacetyl bromide (Ace) and adipoyl dichloride (Adi). After linkage, B_5-Ace or B_5-Adi complexes are coupled to carrier proteins, such as bovine serum albumin (BSA), swine thyroglobulin (Tyr), Keyhole Limpet Hemocyanin (KLH), or human serum albumin (HSA). The immunogens BSA-Adi-B_5, Thyro-Adi-B_5, and BSA-Ace-B_5 have previously been injected in to rabbits and the antibodies generated isolated by immunochromatography. Enzyme-linked immunosorbent method plates coated with antibodies directed to BSA-Adi-B_5 have been found to be optimal. After addition of the samples and standards, goat anti-rabbit IgG-horseradish peroxidase is added, and the reaction initiated with 3,3′,5,5′-tetramethylbenzidine and halted with H_2SO_4. The optical density of the plates is determined at 450 nm. There is an inverse relationship between pantothenic acid concentration in samples and optical density.

Standardization

No standard reference material is available at this time. External quality assurance is provided for measurement of vitamin B$_5$ in lyophilized serum by Instand.

References

1. Coxon KM, Chakauya E, Ottenhof HH, Whitney HM, Blundell TL, Abell C, et al. Pantothenate biosynthesis in higher plants. *Biochem Soc Trans* 2005;**33**:743–6.
2. Brayfield A., 2014. Martindale: the complete drug reference.
3. O'Neil MJ, Royal Society of Chemistry (Great Britain). *The Merck index: an encyclopedia of chemicals, drugs, and biologicals.* Royal Society of Chemistry (Great Britain); 2013.
4. Said HM, Ortiz A, McCloud E, Dyer D, Moyer MP, Rubin S. Biotin uptake by human colonic epithelial NCM460 cells: a carrier-mediated process shared with pantothenic acid. *Am J Physiol* 1998;**275**:C1365–71.
5. Friedrich W. *Vitamins.* De Gruyter; 1988.
6. Bible SW, Lionel ND, Dunuwille R, Perera G. Pantothenol and the burning feet syndrome. *Br J Nutr* 1957;**11**:434–9.
7. Smith DA, Woodruff MFA. Report on deficiency diseases in Japanese prison camps. *Spec Rep Ser Med Res Counc (G B)* 1951;**274**:1–209.
8. Gropper SS, Smith JL. *Advanced nutrition and human metabolism.* Cengage Learning; 2012.
9. Hodges RE, Ohlson MA, Bean WB. Pantothenic acid deficiency in Man1. *J Clin Invest* 1958;**37**:1642–57.
10. Erdman JW, MacDonald I, Zeisel SH. *Present knowledge in nutrition.* International Life Sciences Institute; 2012.
11. Tsuchiya Y, Pham U, Gout I. Methods for measuring CoA and CoA derivatives in biological samples. *Biochem Soc Trans* 2014;**42**:1107–11.
12. Basu SS, Mesaros C, Gelhaus SL, Blair IA. Stable isotope labeling by essential nutrients in cell culture for preparation of labeled coenzyme a and its thioesters. *Anal Chem* 2011;**83**:1363–9.
13. Institute of Medicine (US). *Standing committee on the scientific evaluation of dietary reference intakes and its panel on folate OBV and C. pantothenic acid.* 1998.
14. Pearson PB. The pantothenic acid content of the blood of mammalia. *J Biol Chem* 1941;**140**:423–6.
15. Fukuwatari T, Shibata K. Urinary water-soluble vitamins and their metabolite contents as nutritional markers for evaluating vitamin intakes in young Japanese women. *J Nutr Sci Vitaminol (Tokyo)* 2008;**54**:223–9.
16. Hewitt W, Vincent S. *Theory and application of microbiological assay.* Academic Press; 1989.
17. Wright LD, Skeggs HR. The growth factor requirements of certain streptococci. *J Bacteriol* 1944;**48**:117–8.
18. Stanbery SR, Robbins S. *The effect of pantothenic acid on the growth of various genera of bacteria.* 1939.
19. Rowatt E. The relation of pantothenic acid to acetylcholine formation by a strain of Lactobacillus plantarum. *J Gen Microbiol* 1948;**2**:25–30.

20. Takahashi K, Fukuwatari T, Shibata K. Fluorometric determination of pantothenic acid in human urine by isocratic reversed-phase ion-pair high-performance liquid chromatography with post-column derivatization. *J Chromatogr B* 2009;**877**:2168–72.
21. Martin F, Campos-Giménez E. Pantothenic acid (vitamin B$_5$) in infant formula and adult/pediatric nutritional formula by ultra-high pressure liquid chromatography/tandem mass spectrometry method: collaborative study, final action 2012.16. *J AOAC Int* 2015;**98**:1697–701.
22. Sim H-J, Kim B, Lee J. A systematic approach for the determination of B-group vitamins in multivitamin dietary supplements by high-performance liquid chromatography with diode-array detection and mass spectrometry. *J AOAC Int* 2016;**99**:1223–32.
23. Hudson TS, Subramanian S, Allen RJ. Determination of pantothenic acid, biotin, and vitamin B12 in nutritional products. J Assoc Off Anal Chem, 1984;**67**:994–8.
24. Iijima S, Sato Y, Bounoshita M, Miyaji T, Tognarelli DJ, Saito M. Optimization of an online post-column derivatization system for ultra high-performance liquid chromatography (UHPLC) and its applications to analysis of biogenic amines. *Anal Sci* 2013;**29**:539–45.
25. Zhang Q, Ford LA, Goodman KD, Freed TA, Hauser DM, Conner JK, Vroom KET, Toal DR. LC-MS/MS method for quantitation of seven biomarkers in human plasma for the assessment of insulin resistance and impaired glucose tolerance. *J Chromatogr B Analyt Technol Biomed Life Sci* 2016;**1038**:101–8.
26. Gonthier A, Boullanger P, Fayol V, Hartmann DJ. Development of an ELISA for pantothenic acid (vitamin B5) for application in the nutrition and biological fields. J Immunoassay, 1998;19:**167**–94.

Further Reading

27. Heudi O, Fontannaz P. Determination of vitamin B5 in human urine by high-performance liquid chromatography coupled with mass spectrometry. *J Sep Sci* 2005;**28**:669–72.

Methods for assessment of Vitamin B$_6$

Roy A. Sherwood,

King's College London, London, United Kingdom

Chapter Outline

Structure and Function

Vitamin B$_6$ comprises six compounds based on a 2-methyl-3-hydroxypyridine structure with differing subunits at positions C4 and C5 that are interconvertible. These are pyridoxine (PYN), pyridoxamine (PYM), and pyridoxal (PYL) and their phosphorylated derivatives pyridoxine-5-phosphate (PNP), pyridoxamine-5-phosphate (PMP), and pyridoxal-5-phosphate (PLP) (Fig. 9.1). PLP is a cofactor in >150 different enzyme reactions including transamination and decarboxylation enzyme systems. Many PLP-dependent enzymes are involved in amino acid metabolism and they are also involved in organic acid, glucose, and lipid pathways.[1] PLP-dependent enzymes play an important role in the metabolism of neurotransmitters including

Laboratory Assessment of Vitamin Status. https://doi.org/10.1016/B978-0-12-813050-6.00009-7

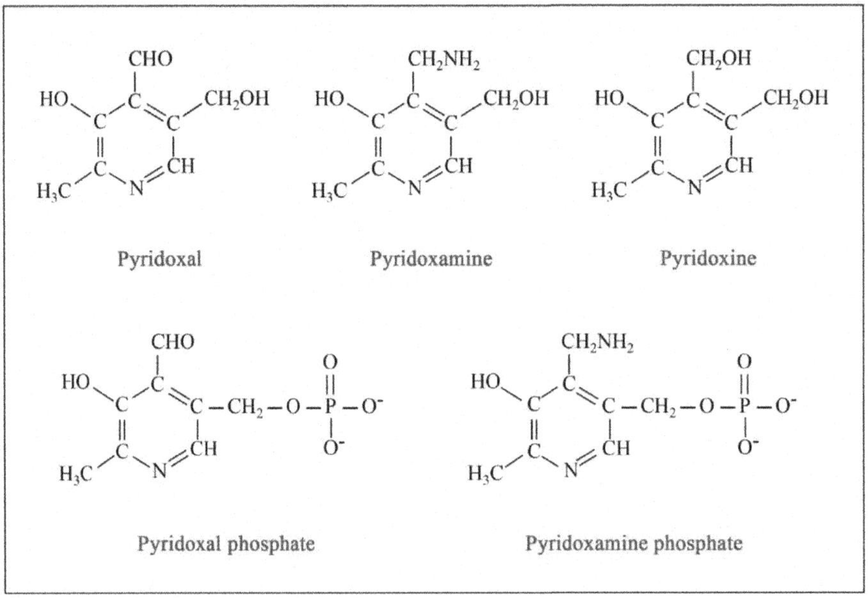

FIG. 9.1

Structures of the forms of vitamin B$_6$.

serotonin, dopamine, and γ-aminobutyric acid (GABA). PLP is synthesized in the liver from ingested PYN in a closely regulated process.[2]

In plasma, vitamin B$_6$ exists as PLP (~60%–70%), PYL (~30%), and a small amount of 4-pyridoxic acid (PA) that is the predominant form of the vitamin excreted in urine.[3] In plasma, PLP is strongly bound to albumin[4] with free PLP being converted to PYL by tissue nonspecific alkaline phosphatase (ALP).

Dietary Intake

Vitamin B$_6$ is present in most foods and therefore dietary deficiency alone is uncommon in developed countries. The recommended daily intake for adults is 1.5–1.7 mg/day, although higher intakes are recommended in those over 65 years of age.[5]

Clinical Applications of Measurement

The symptoms of vitamin B$_6$ deficiency include peripheral neuropathy, pellagra-like syndrome with seborrheic dermatitis and glossitis. Prolonged deficiency can lead to depression, confusion, and in severe cases causes abnormalities in EEG signals and seizures. Vitamin B$_6$ deficiency is often associated with deficiencies in other B vitamins and can be due to overall poor nutrition caused by a chaotic lifestyle, e.g., in alcoholic subjects.[6] Suboptimal vitamin B_6 status has also been reported in oral contraceptive users, smokers, and patients with celiac disease or diabetes.[6] Secondary vitamin B_6 deficiency can occur in treatments with drugs that interact with PLP, e.g., isoniazid[5] or inborn errors in vitamin B_6 salvage pathways or when mutations result in the accumulation of intermediates that react with PLP.[7] Vitamin B_6 deficiency has been reported to be associated with an increased risk of cardiovascular disease,[8] stroke,[9] and cancer.[10] Low plasma PLP has been reported in a number of diseases associated with inflammation including rheumatoid arthritis,[11] inflammatory bowel disease,[12] diabetes,[13] and deep vein thrombosis.[14] Plasma PLP has an inverse relationship to C-reactive protein and other acute phase reactants.[15]

Indices of Vitamin B_6 Status

The assessment of vitamin B_6 status can be undertaken using either direct indices, e.g., PLP, PYL, or PA or indirect markers that reflect the effect of deficiency on the various pathways involving PLP-dependent enzymes, e.g., kynurenines, transaminases, and various amino acids.

Direct indices of vitamin B_6 status

Plasma PLP is the most commonly used direct index of vitamin B_6 status.[16] Plasma PLP exhibits a positive relationship with vitamin B_6 intake[16] and increases 10-fold following supplementation[17] responding within 1–2 weeks to either depletion[18] or repletion.[3] Plasma PLP decreases within hours following carbohydrate ingestion[19] and is inversely related to HbA1c and fasting glucose.[20] This suggests that fasting samples are preferable to minimize within-person variability in food intake. There is a good reproducibility within an individual over years allowing a single assessment of vitamin B_6 status.[21] PLP is relatively stable at $4°C$[22] or at $-40°C$[23] but declines in samples kept at room temperature[23] or exposed to light.[24] Plasma should therefore be separated as soon as possible and stored frozen.

Plasma PLP is bound to albumin[4, 11, 25] and low results can be seen in states where PLP binding to albumin may be impaired, e.g., critical illness.[26] Plasma PLP falls in the second trimester of pregnancy normalizing within weeks postpartum.[27] Because of the high protein binding of PLP in plasma deproteinization of samples during laboratory analysis is essential. Plasma PLP is inversely associated with tissue nonspecific ALP.[11, 28] In children with hypophosphatasia, PLP is markedly increased[28, 29] whereas in rickets it is decreased in proportion to the increase in ALP.[30]

PLP can be detected in cerebrospinal fluid (CSF) although PYL is the predominant form of vitamin B_6 in CSF. CSF vitamin B_6 measurements are most useful when studying inborn errors of vitamin B_6 metabolism.[31] CSF PLP decreases with age until approximately 2 years of age[32] and correlates well in both normal children and adults with plasma PLP.

Erythrocyte PLP has been proposed as a more responsive marker of vitamin B_6 status than plasma PLP.[33] Erythrocyte PLP rises markedly within 1 h of PYN intake falling back over 4–5 h.[34] There are good correlations between erythrocyte PLP and plasma PLP, PYL, and PA and with urinary PA.[16] Unlike plasma PLP, erythrocyte PLP is unaffected by inflammation or ALP activity.[16] PLP binds to hemoglobin with high affinity, but with different affinities to hemoglobin variants. Sickle cell hemoglobin (HbS) has a higher affinity for PP than hemoglobin A and, therefore, sickle patients have a low plasma PLP and high erythrocyte PLP.[35]

Analytical Methods

Pyridoxal-5-Phosphate

The most frequently used analytical methodology for PLP is based on HPLC following derivatization and using fluorescence detection.[36–48] PLP is converted to a semicarbazone using precolumn derivatization, separated by reversed phase or cation exchange chromatography, and detected by fluorescence at alkaline pH. More recent methods have utilized LC-MS/MS to measure PLP in whole blood, plasma, or CSF that obviates the need for derivatization.[49–51] An UHPLC-MS/MS method for simultaneous detection of vitamins B$_1$ (Chapter 6) and B$_6$ in whole blood has been described that only requires protein precipitation with perchloric acid for sample pretreatment.[52] Deuterated PLP was used as the internal standard. Reversed phase chromatography with a C18 column was used and separation could be achieved in 2.5 min. Imprecision was ~10% and the limit of quantitation was 25.9 nmol/L. A good correlation was obtained with a conventional HPLC method.

Pyridoxal

Plasma PYL is the transport form of vitamin B$_6$ and has been proposed as an alternative marker of vitamin B$_6$ status.[3] Plasma PYL shows a good correlation with PLP,[17, 53] responds more than PLP to supplementation,[17, 54, 55] but less than PLP to dietary intake.[16] The decline in PLP in samples kept at room temperature is due to dephosphorylation to PYL, which therefore increases with suboptimal storage conditions.[23] PYL appears to be more stable in plasma prepared using EDTA than heparin or citrate,[23] but like PLP is sensitive to light exposure.[24] To improve within-person reproducibility it has been suggested that total vitamin B$_6$-aldehyde (PLP + PYL) may be the best marker of vitamin B$_6$ status[56] as it is less affected by changes in plasma albumin or ALP.

High performance liquid chromatography-based methods using reversed phase chromatography following chlorite or semicarbazide derivatization have been described most of which measure both PLP and PYL.[46, 48] Liquid chromatography mass spectrometry-based methods have also been described.[57]

4-Pyridoxic Acid

4-PA is the primary catabolite of vitamin B$_6$ formed in the liver. It is not protein bound in plasma and has a high renal clearance rate.[58] 4-PA is strongly correlated with both PLP and PYL. Plasma PA responds within 1–2 weeks to any changes in vitamin B$_6$ intake[16, 59] and increases about 50-fold following PYN supplementation.[17] In contrast to PLP and PYL, PA is stable in EDTA, citrate, and heparin plasma at room temperature or in frozen samples[23] and is unaffected by exposure to light.[24] It is not related to either plasma albumin or ALP. Significant increases in plasma PA are seen in critically ill patients, particularly those with renal impairment.[26] Plasma PA can be measured alongside PLP and PYL by HPLC[46, 60] or LC-MS/MS.[57]

4-PA represents >90% of vitamin B$_6$ excreted in urine with a small amount of PYL and PYM.[61, 62] Urinary PA increases rapidly following PYN supplements[63] and therefore is a marker of recent vitamin B$_6$ intake.[3] There appears to be no influence from age,[64] pregnancy,[65] alcohol, or oral contraceptives.[66–68] An excretion rate of >3 µmol/day has been proposed to indicate adequate vitamin B$_6$ status.[3] Disadvantages of PA as a marker of vitamin B$_6$ status are sensitivity to renal impairment and the presence of a circadian rhythm[55] that can result in a large within-person reproducibility.

Indirect Functional Indices of Vitamin B$_6$ Status
Transaminases

The transaminases, aspartic acid transaminase (AST; EC 2.6.1.1), and alanine transaminase (ALT; EC 2.6.1.2) are PLP dependent. While the amounts of AST and ALT in plasma reflect release from damaged hepatocytes, muscle cells, or red blood cells, measurement of AST and ALT in erythrocytes (Erythrocyte AST (EAST) and Erythrocyte ALT (EALT) respectively) has been used to assess vitamin B$_6$ status in either basal samples or after *in vitro* addition of PYN.[6] The results from the latter measurements are reported as activation coefficients (EAST-AC and EALT-AC for AST and ALT respectively). A higher AC reflects a lower vitamin B$_6$ status. EAST and EALT and their respective ACs are considered to be long-term indicators of vitamin B$_6$ status as they are related to the lifespan of the red cells.[3] It has been suggested that the transaminases require higher doses of vitamin B$_6$ to become normal compared to plasma PLP[69]; however, another study found that the transaminases, plasma or erythrocyte PLP, and urinary PA all responded similarly.[70] An advantage of the transaminases is that they are not related to albumin or ALP[71] but are affected by impaired renal function. Further disadvantages are that fresh whole blood is required (frozen samples cannot be used) and that the hemoglobin concentration affects the results.

Kynurenines

The amino acid tryptophan is catabolized principally by the kynurenine pathway. Two of the enzymes in this pathway are PLP dependent; kynurenine transaminase (KAT) and kynureninase (KYN). KYN activity is reduced with low dietary vitamin B$_6$ intake and is more responsive than KAT to vitamin B$_6$ deficiency.[72, 73] Assays are available to measure a range of kynurenines including kynurenic acid (KA), xanthurenic acid (XA), 3-hydroxykynurenine (HK), and 3-hydroxyanthranilic acid (HAA).[57, 74] 3-HK is the only kynurenine that is increased in plasma from vitamin B$_6$-deficient subjects, probably reflecting that its clearance not its formation involves PLP-dependent enzymes. There is a nonlinear inverse relationship between plasma HK and PLP.[3] Plasma HK decreases following PYN supplementation.[75] Plasma HK has a positive association with the concentration of tryptophan and inflammatory markers and a negative correlation with renal function.[76] A slight increase in HK with a slight decrease in KA and HAA occur in moderate vitamin B$_6$ deficiency, becoming more pronounced in more severe deficiencies. XA is only increased in significant deficiency but is measured as part of the tryptophan loading test (see later). The kynurenines are stable in plasma at $-80°C$, but HK and HAA decrease in plasma, whole blood, and serum at higher temperatures.[23]

Kynurenine metabolite ratios have been suggested to improve their capability as markers of vitamin B$_6$ status. HK/XA and HK/HAA identify low PLP better than HK alone.[76] Measurements of urinary HK and XA have been used to evaluate vitamin B$_6$ status.[3] The most established test is measurement of increases of urinary XA after an oral dose of tryptophan (2 g). Excretion of <65 umol/24 h of XA has been suggested to indicate adequate vitamin B$_6$ status.[3] An alternative using a 5 g tryptophan load and a 6 h urine collection has been used with an excretion of XA of >25 mg proposed to be indicative of vitamin B$_6$ deficiency.[62] Although the tryptophan load test has been in use for many years, a number of limitations must be taken into account. A number of factors have been shown to affect the concentration of tryptophan metabolites in urine including inflammation,[77] pregnancy,[78] and drugs including oral contraceptives and isoniazid.[78]

Plasma Amino Acids

Vitamin B$_6$ deficiency has been shown to affect the metabolism of amino acids related to one-carbon metabolism including serine, glycine, and cystathionine.[60, 79, 80] Glycine and serine are metabolized by the PLP-dependent enzyme serine hydroxymethyltransferase which is present in both the mitochondria and cytoplasm of many cells. Glycine is also metabolized by a PLP-dependent enzyme glycine decarboxylase that is part of the glycine cleavage system. Increased plasma concentrations of both glycine and serine have been reported that may be the result of reduced activity of the earlier enzymes in the presence of suboptimal vitamin B$_6$ status. Of these enzymes glycine decarboxylase appears to be the most important enzyme in producing increases in plasma glycine and serine.[80]

Cystathionine is a component of the transsulfuration pathway. Its plasma concentration is increased by >50% in individuals with marginal vitamin B$_6$ deficiency.[79] The mechanism for this increase appears to be the reduced activity of the enzyme cystathionine γ-lyase in the liver, while the activity of the enzyme involved in the formation of cystathionine, cystathionine β-synthase is not dependent on PLP.

Standardization
Standard Reference Materials

The United States National Institute of Standards and Technology (NIST) has developed a standard reference material, SRM 3950 for vitamin B_6 in serum that has two certified concentrations for PLP and indicative values for PA. SRM 1950 (Metabolites in Human Plasma) has certified values for various amino acids and nine vitamins including PLP. Standard reference materials for the other markers are not currently available.

The EQA provider Instand provides an assurance program for PLP. For more information about this scheme, visit http://www.instandev.de

Conclusions

The assessment of vitamin B_6 status can involve direct or functional markers. Historically, the tryptophan load test was considered to be one of the best means of assessing vitamin B_6 status but this has now been superseded by measurement of PLP in plasma or whole blood due to the development of LC-MC/MS methods for PLP assay.

References

1. Eliot AC, Kirsch JF. Pyridoxal phosphate enzymes: mechanistic, structural, and evolutionary considerations. *Annu Rev Biochem* 2004;**73**:383–415.
2. Merrill AH, Horiike K, McCormick DB. Evidence for the regulation of pyridoxal 5-phosphate formation by the liver by pyridoxamine (pyridoxine) 5-phosphate oxidase. *Biochem Biophys Res Commun* 1978;**83**:984–90.
3. Leklem JE. Vitamin B-6: a status report. *J Nutr* 1990;**120**(Suppl. 11):1503–7.
4. Lemeng L, Brashear RE, Li TK. Pyridoxal 5′-phosphate in plasma: source, protein-binding, and cellular transport. *J Lab Clin Med* 1974;**84**:334–43.
5. Bender DA. Vitamin B6: beyond adequacy. *J Evid Based Complementary Altern Med* 2011;**16**:29–39.
6. Spinneker A, Sola R, Mennen V, Castillo MJ, Pietrzik K, et al. Vitamin B-6 status, deficiency and its consequence—an overview. *Nutr Hosp* 2007;**22**:7–24.
7. Clayton PT. B6-responsive disorders: a model of vitamin deficiency. *J Inherit Metab Dis* 2006;**29**:317–26.
8. Lotto V, Choi SW, Friso S. Vitamin B-6: a challenging link between nutrition and inflammation in CVD. *Br J Nutr* 2011;**106**:183–95.
9. Kelly PJ, Kistler JP, Shih VE, Mandell R, Atassi N, et al. Inflammation, homocysteine, and vitamin B6 status after ischemic stroke. *Stroke* 2004;**35**:12–5.
10. Johansson M, Relton C, Uelund PM, Vollset SE, Midttun O, et al. Serum B vitamin levels and risk of lung cancer. *JAMA* 2010;**303**:2377–85.
11. Chiang EP, Smith DE, Selhub J, Dallal G, Wang YC, Roubenoff R. Inflammation causes tissue-specific depletion of vitamin B6. *Arthritis Res Ther* 2005;**7**:R1254–62.
12. Selhub J, Byun A, Liu Z, Mason JB, Bronson TR, Crott JW. Dietary vitamin B6 intake modulates colonic inflammation in the IL10−/− model of inflammatory bowel disease. *J Nutr Biochem* 2013;**24**:2138–43.

13. Friedman AN, Hunsicker LG, Selhub J, Bostom AG. Clinical and nutritional correlates of C-reactive protein in type 2 diabetic nephropathy. *Atherosclerosis* 2004;**172**:121–5.

14. Saibeni S, Cattaneo M, Vecchi M, Zignetti ML, Lecchi A, et al. Low vitamin B6 plasma levels, a risk factor for thrombosis, in inflammatory bowel disease: role of inflammation and correlation with acute phase reactants. *Am J Gastroenterol* 2003;**98**:112–7.

15. Sakakeeny L, Roubenoff R, Obin M, Fontes JD, Benjamin EJ, et al. Plasma pyrodoxal-5-phosphate is inversely associated with systemic markers of inflammation in a population of U.S. adults. *J Nutr* 2012;**142**:1280–5.

16. Hansen CM, Shultz TD, Kwak HK, Memon HS, Leklem JE. Assessment of vitamin B-6 status in young women consuming a controlled diet containing four levels of vitamin B-6 provides an estimated average requirement and recommended dietary allowance. *J Nutr* 2001;**131**:1777–86.

17. Bor MV, Refsum H, Bisp MR, Bleie O, Schneede J, et al. Plasma vitamin B6 vitamers before and after vitamin B6 treatment: a randomized placebo-controlled study. *Clin Chem* 2003;**49**:155–61.

18. Lu CC. Vitamin B6, blood PLP level, and risk of colorectal cancer. *JAMA* 2010;**303**:2251–2.

19. Leklem JE, Hollenbeck CB. Acute ingestion of glucose decreases plasma pyridoxal 5'-phosphate and total vitamin B-6 concentration. *Am J Clin Nutr* 1990;**51**:832–6.

20. Shen J, Lai CQ, Mattei J, Ordovas JM, Tucker KL. Associated on vitamin B-6 status with inflammation, oxidative stress, and chronic inflammatory conditions: the Boston Puerto Rican Health Study. *Am J Clin Nutr* 2010;**91**:337–42.

21. Cope EL, Shrubsole MJ, Cohen SS, Cai Q, Wu J, et al. Intraindividual variation in one-carbon metabolism plasma biomarkers. *Cancer Epidemiol Biomark Prev* 2013;**22**:1894–9.

22. Midttun O, Townsend MK, Nygård O, Tworoger SS, Brennan P, et al. Most blood biomarkers related to vitamin status, one-carbon metabolism, and the kynurenine pathway show adequate preanalytical stability and within-person reproducibility to allow assessment of exposure or nutritional status in healthy women and cardiovascular patients. *J Nutr* 2014;**144**:784–90.

23. Hustad S, Eussen S, Midttun O, Ulvik A, van de Kant PM, et al. Kinetic modelling of storage effects on biomarkers related to B vitamin status and one-carbon metabolism. *Clin Chem* 2012;**58**:401–10.

24. van der Ham M, Albersen M, de Koning TJ, Visser G, Middendorp A, et al. Quantification of vitamin B6 vitamers in human cerebrospinal fluid by ultra performance liquid chromatography-tandem mass spectrometry. *Anal Chim Acta* 2012;**712**:108–14.

25. Vasilaki AT, McMillan DC, Kinsells J, Duncan A, O'Reilly DS, et al. Relation between pyridoxal and pyridoxal phosphate concentrations in plasma, red cells, and white cells in patients with critical illness. *Am J Clin Nutr* 2008;**88**:140–6.

26. Huang YC, Chang HH, Huang SC, Cheng CH, Lee BJ, et al. Plasma pyridoxal 5'-phosphate is a significant indicator of immune response in the mechanically ventilated critically ill. *Nutrition* 2005;**21**:779–85.

27. Shibata K, Fukuwatari T, Sasaki S, Sano M, Susuki K, et al. Urinary excretion levels of water soluble vitamins in pregnant and lactating women in Japan. *J Nutr Sci Vitaminol (Tokyo)* 2013;**59**:178–86.

28. Iqbal SJ, Brain A, Reynolds TM, Penny M, Volland S. Relationship between serum alkaline phosphatase and pyridoxal 5'-phosphate levels in hypophosphatasia. *Clin Sci (Lond)* 1998;**94**:203–6.

29. Whyte MP, Mahuren JD, Vrabel LA, Coburn SZP. Markedly increased circulating pyridoxal-5'-phosphate levels in hypophosphatasia. Alkaline phosphatase acts in vitamin B6 metabolism. *J Clin Invest* 1985;**76**:752–6.

30. Reynolds RD, Lorenc RS, Wieczorek E, Pronicka E. Extremely low serum pyridoxal 5'-phosphate in children with familial hypophosphatemic rickets. *Am J Clin Nutr* 1991;**53**:698–701.

31. Plecko B, Struys EA, Jacobs C. Vitamin B6-dependent and responsive disorders. In: Blau N, Duran M, Gibson KM, Vici CD, editors. *Physician's guide to the diagnosis, treatment, and follow-up of inherited metabolic diseases*. Berlin: Spinger-Verlag; 2014. p. 179–90.

32. Footitt EJ, Heales SJ, Mills PB, Allen GF, Oppenheim M, Clayton PT. Pyridoxal 5'-phosphate in cerebrospinal fluid; factors affecting concentration. *J Inherit Metab Dis* 2011;**34**:529–38.

33. Quasim T, McMillan DC, Talwar D, Vasilaki A, O'Reilly DS, Kinsella J. The relationship between plasma and red cell B-vitamin concentration in critically-ill patients. *Clin Nutr* 2005;**24**:956–60.

34. Reynolds RD. Biochemical methods for status assessment. In: Raiten DL, editor. *Vitamin B-6 metabolism in pregnancy, lactation and infancy*. Boca Raton, FL: CRC Press; 1995. p. 41–59.

35. Natta CL, Reynolds RD. Apparent vitamin B6 deficiency in sickle cell anemia. *Am J Clin Nutr* 1984;**40**:235–9.

36. Schrijver J, Speek AJ, Schreurs WHP. Semi-automated fluorometric determination of pyridoxal-5'-phosphate (vitamin B6) in whole blood by high-performance liquid chromatography (HPLC). *Int J Vitam Nutr Res* 1981;**51**:216–22.

37. Coburn SP, Mahuren JD. A versatile cation-exchange procedure of measuring the seven major forms of vitamin B6 in biological sample. *Anal Biochem* 1983;**129**:310–7.

38. Sampson DA, O'Connor DK. Analysis of B6 vitamers and pyridoxic acid in plasma, tissues and urine using high performance liquid chromatography. *Nutr Res* 1989;**9**:259–72.

39. Edwards P, Liu PKS, Rose AA. Simple liquid-chromatographic method for measuring vitamin B6 compounds in plasma. *Clin Chem* 1989;**35**:241–5.

40. Qureshi S, Huang H. Determination of B6 vitamers in serum by simple isocratic high performance liquid chromatography. *J Liq Chromatogr* 1990;**13**:191–201.

41. Sharma SK, Dakshinamurti K. Determination of vitamin B6 vitamers and pyridoxic acid in biological samples. *J Chromatogr* 1992;**578**:45–51.

42. Reynolds TM, Brain A. A simple internally-standardised isocratic HPLC assay for vitamin B6 in human serum. *J Liq Chromatogr* 1992;**15**:897–914.

43. Kimura M, Kanehira K, Yokoi K. Highly sensitive and simple liquid chromatographic determination in plasma of B6 vitamers, especially pyridoxal 5'-phosphate. *J Chromatogr A* 1996;**722**:295–301.

44. Bates CJ, Pentieva KD, Matthews N, Macdonald A. A simple, sensitive and reproducible assay for pyridoxal 5'-phosphate and 4-pyridoxic acid in human plasma. *Clin Chim Acta* 1999;**280**:101–11.

45. Deitrick CL, Katholi RE, Huddleston DJ, Hardiek K, Burrus L. Clinical adaptation of a high-performance liquid chromatographic method for the assay of pyridoxal 5'-phosphate in human plasma. *J Chromatogr B Biomed Sci Appl* 2001;**751**:383–7.

46. Bisp MR, Vakur Bor M, Heinsvig EM, Kall MA, Nexo E. Determination of vitamin B6 vitamers and pyridoxic acid in plasma: development and evaluation of a high-performance liquid chromatographic assay. *Anal Biochem* 2002;**305**:82–9.

47. Talwar D, Quasim T, McMillan DC, Kinsella J, Williamson C, O'Reilly DS. Optimisation and validation of a sensitive high performance liquid chromatographic assay for routine measurement of pyridoxal 5-phosphate in human plasma and red cells using pre-column semicarbazide derivatisation. *J Chromatogr B Anal Technol Biomed Life Sci* 2003;**792**:333–43.

48. Rybak ME, Pfeiffer CM. Clinical analysis of vitamin B(6): determination of pyridoxal 5′-phosphate and 4-pyridoxic acid in human serum by reversed-phase high-performance liquid chromatography with chlorite postcolumn derivatisation. *Anal Biochem* 2004;**333**:336–44.

49. Van Zeist BD, De Jonge R. A stable isotope dilution LC-ESI-MS/MS method for the quantification of pyridoxal-5′-phosphate in whole blood. *J Chromatogr* 2012;**903**:134–41.

50. Midttun O, Hustad S, Solheim E, Schneed J, Ueland PM. Multianalyte quantification of vitamin B6 and B2 species in the nanomolar range in human plasma by liquid chromatography-tandem mass spectrometry. *Clin Chem* 2005;**51**:1206–16.

51. Footitt EJ, Calyton PT, Mills K, Heales SJ, Neergheen V, Oppenheim M, et al. Measurement of plasma B6 vitamer profiles in children with inborn errors of vitamin B6 metabolism using an LC-MS/MS method. *J Inherit Metab Dis* 2012;**36**:139–45.

52. Puts J, de Groot M, Haex M, Jakobs B. Simultaneous determination of underivatized vitamin B1 and B6 in whole blood by reversed phase ultra high performance liquid chromatography tandem mass spectrometry. *PLoS ONE* 2015;**10**:e0132018.

53. Midttun O, Hustad S, Schnnede J, Vollser SE, Ueland PM. Plasma vitanin B-6 forms and their relation to transsulfuration metabolites in a large, population-based study. *Am J Clin Nutr* 2007;**86**:131–8.

54. Thakker KM, Sitren HS, Gregory III JF, Schmidt GL, Baumgartner TG. Dosage forms and formulation effects on the bioavailability of vitamin E, riboflavin and vitamin B-6 from multivitamin preparations. *Am J Clin Nutr* 1987;**45**:1472–9.

55. Zempleni J. Pharmacokinetics of vitamin B6 supplements in humans. *J Am Coll Nutr* 1995;**14**:579–86.

56. Barnard HC, de Kock JJ, Pentieva WJ, McNulty H, Seljeflot I, Strandjord RE. A new perspective in the assessment of vitamin B-6 nutritional status during pregnancy in humans. *J Nutr* 1987;**117**:1303–6.

57. Midttun O, Hustad S, Ueland PM. Quantitative profiling of biomarkers related to B-vitamin status, tryptophan metabolism and inflammation in human serum by liquid chromatography/tandem mass spectrometry. *Rapid Commun Mass Spectrom* 2009;**23**:1371.

58. Zempleni J, Kübler W. Metabolism of vitamin B6 by human kidney. *Nutr Res* 1995;**15**:187–92.

59. Bates CJ, Pentieva KD, Prentice A, Mansoor MA, Finch S. Plasma pyridoxal phosphate and pyridoxic acid and their relationship to plasma homocysteine in a representative sample of British men and women aged 65 years and over. *Br J Nutr* 1999;**81**:191–201.

60. David SR, Quinlivan EP, Stacpoole PW, Gregory JF. Plasma glutathione and cystathionine concentrations are elevated but cysteine flux is unchanged by dietary vitamin B-6 restriction in young men and women. *J Nutr* 2006;**81**:373–8.

61. Rose RC, McCorrmick DB, Li TK, Lumeng L, Haddad JG, Spector R. Transport and metabolism of vitamins. *Fed Proc* 1971;**45**:30–9.

62. Sauberlich HE, Canham JE, Baker EM, Raica N, Herman YF. Biochemical assessment of the nutritional status of vitamin B6 in the human. *Am J Clin Nutr* 1972;**25**:629–42.

63. Lui A, Lemeng L, Aronoff GR, Li TK. Relationship between body store of vitamin B6 and plasma pyridoxal clearance: metabolic balance studies in humans. *J Lab Clin Med* 1985;**106**:491–7.

64. Kant AK, Moser-Veillon PB, Reynolds RD. Effect of age on changes in plasma, erythrocyte, and urinary B-6 vitamers after an oral vitamin B-6 load. *Am J Clin Nutr* 1988;**48**:1284–90.

65. Trumbo PR, Wang JW. Vitamin B-6 status indices are lower in pregnant than in nonpregnant women but urinary excretion of 4-pyridoxic acid does not differ. *J Nutr* 1993;**123**:2137–41.

66. Bitsch R. Vitamin B6. *Int J Vitam Nutr Res* 1993;**63**:278–82.

67. Brown RR, Rose DP, Leklem JE, Linkswiler H, Anand R. Urinary 4-pyridoxic acid, plasma pyridoxal phosphate, and erythrocyte aminotransferases levels in oral contraceptive users receiving controlled intakes of vitamin B6. *Am J Clin Nutr* 1975;**28**:10–9.

68. Leklem JE, Brown RR, Rose DP, Linkswiler HM. Vitamin B6 requirements of women using oral contraceptives. *Am J Clin Nutr* 1975;**28**:535–41.

69. Hansen CM, Leklem JE, Miller LT. Changes in vitamin B-6 status indicators of women fed a constant protein diet with varying levels of vitamin B-6. *Am J Clin Nutr* 1997;**66**:1379–87.

70. Huang YC, Chen W, Evans MA, Mitchell ME, Schultz TD. Vitamin B-6 requirement and status assessment of young women fed a high-protein diet with various levels of vitamin B-6. *Am J Clin Nutr* 1998;**67**:208–20.

71. Brussard JH, Löwik MR, van den Berg H, Brants HA, Bemelmans W. Dietary and other determinants of vitamin B6 parameters. *Eur J Clin Nutr* 1997;**51**(Suppl. 3):S39–45.

72. Okada M, Miyamoto E, Nishida T, Tomida T, Shibuya M. Effect of vitamin B6 nutrition and diabetes on vitamin B6 metabolism. *J Nutr Biochem* 1997;**8**:44–8.

73. van de Kamp JL, Smolen A. Response of kynurenine pathway enzymes to pregnancy and dietary level of vitamin B-6. *Pharmacol Biochem Behav* 1995;**51**:753–8.

74. Zheng X, Kang A, Dai C, Liang Y, Xie T, et al. Quantitative analysis of neurochemical panel in rat brain and plasma by liquid chromatography-tandem mass spectrometry. *Anal Chem* 2012;**84**:10044–51.

75. Theofylaktopoulou D, Ulvik A, Midttun O, Ueland PM, Vollset SE, et al. Vitamins B2 and B6 as determinants of kynurenines and related markers of interferon-γ-mediated immune activation in the community-based Hordaland Health Study. *Br J Nutr* 2014;**112**:1065–72.

76. Ulvik A, Theofylaktopoulou D, Midttun O, Nygård O, Eussen SJ, et al. Substrate product ratios of enzymes in the kynurenine pathway measured in plasma as indicators of functional vitamin B-6 status. *Am J Clin Nutr* 2013;**98**:934–40.

77. Pedersen ER, Svingen GF, Schartum-Hansen H, Ueland PM, Ebbing M, et al. Urinary excretion of kynurenine and tryptophan, cardiovascular events, and mortality after elective coronary angiography. *Eur Heart J* 2013;**34**:2689–96.

78. Rose DP, Braidman IP. Excretion of tryptophan metabolites as affected by pregnancy, contraceptive steroids, and steroid hormones. *Am J Clin Nutr* 1971;**24**:673–83.

79. da Silva VR, Rios-Avila L, Lamers Y, Ralat MA, Midttun O, et al. Metabolite profile analysis reveals functional effects of 28-day vitamin B-6 restriction on one-carbon metabolism and tryptophan catabolic pathways in healthy men and women. *J Nutr* 2013;**143**:1719–27.

80. Lamers Y, Williamson J, Ralat M, Quinlivan EP, Gilbert LR, et al. Moderate dietary vitamin B-6 restriction raises plasma glycine and cystathionine concentrations whilst minimally affecting the rates of glycine turnover and glycine cleavage in healthy men and women. *J Nutr* 2009;**139**:452–60.

Further Reading

81. Gregory JF, Kirk JR. Determination of urinary 4-pyridoxic acid using high performance liquid chromatography. *Am J Clin Nutr* 1979;**32**:879–83.

Methods for assessment of biotin (Vitamin B$_7$)

10

Rachel S. Carling*, Charles Turner†

Biochemical Sciences, Viapath, St Thomas' Hospital, London, United Kingdom*
WellChild Laboratory, Evelina London Children's Hospital, Guy's & St Thomas' NHS Foundation Trust, London, United Kingdom†

Chapter Outline

Laboratory Assessment of Vitamin Status. https://doi.org/10.1016/B978-0-12-813050-6.00010-3

Introduction

Vitamin B$_7$ (biotin) is a water-soluble B-complex vitamin, formerly known as Bios, coenzyme R, vitamin H, and protective factor X. The discovery of biotin cannot be attributed to one individual and instead represents the collective progress made by several investigators in the early part of the 20th century. In 1901, Wildiers hypothesized that yeast required an organic substance for normal growth and named this substance "Bios," after the Greek word meaning "that by which life is sustained".[1] In 1916, Bateman reported rats becoming ill when fed a diet rich in raw egg[2] and in 1927 a condition called "egg white injury" was described by Boas,[3] in which rats fed a diet with dried egg as the protein source developed alopecia, dermatitis, spastic gait, and ultimately death in four to six weeks. Boas also identified a substance present in yeast and raw liver, protective factor X, which could cure egg white injury. The discovery of coenzyme R was reported in 1933 by Allison, Hoover, and Burk who described a substance isolated from egg yolk which was essential for the growth of a Rhizobium, a nitrogen-fixing bacterium found in leguminous plants.[4] Less than a decade later, György also described a substance which could prevent egg white injury and named it vitamin H.[5] However, it was not until 1940, that bios, vitamin H, protective factor X, and coenzyme R were confirmed as chemically identical.[6] Further work then established that egg white injury occurred because dietary biotin binds tightly to avidin, a protein found in raw egg white, preventing absorption.[7]

Structure and Function

Biotin (hexahydro-2-oxo-1H-thieno[3,4-*d*]imidazole-4-pentanoic acid, C$_{18}$H$_{16}$N$_2$O$_3$S) has a molar mass of 244.309 g/mol. It is derived from urea, valeric acid, and thiophene; a ureido ring is fused with a tetrahydrothiophene ring which has a valeric acid substituent attached to one of its carbon atoms. Biotin is a colorless, crystalline material which is stable to air at room temperature but may be destroyed by UV light. The pKa of biotin is 4.5 so at physiological pH it is found predominantly in the anionic, deprotonated form. It is slightly soluble in water (220 mg/L at 25°C) and alcohol. There are eight stereoisomers of biotin, only one of which is biologically active, D(+)biotin.

Mammals cannot synthesize biotin and rely upon dietary intake from plant and microbial sources. Plant cells synthesize biotin endogenously as do microorganisms such as yeast and certain bacteria. Biotin is present in many foods but at a lower concentration than the other water-soluble vitamins.[8] Rich dietary sources of biotin include liver, kidneys, heart, pancreas, egg yolk, poultry, yeast, and cow's milk. Smaller quantities are found in plants, mainly in seeds.[9] Dietary sources of biotin exist in the free and protein-bound form, with the relative ratio of the two being dependent upon the given source. Thus the bioavailability of biotin varies depending upon the dietary source; in wheat 5% is bioavailable, in corn it is 100%.[8]

Another potentially important source of biotin for humans is enteric synthesis by gut bacteria in the large intestine, the majority of which is present in the unbound/absorbable form.[10, 11] While it is well established that biotin is produced by, and absorbed from, the large intestine, the significance of this contribution to the overall body pool has not yet been elucidated.[12] However, recent evidence indicates that the intestinal microbial synthesis of biotin does contribute to human nutrition and the identification of a specific biotin uptake mechanism, the Sodium-dependent Multi Vitamin Transporter (SMVT) further supports this.[13]

Biotin Requirements and Recommended Daily Allowance

Recommendations for the dietary intake of biotin are simply a reflection of the general population intake in healthy individuals. Formal estimates of biotin requirements in normal individuals have not been made.[14, 15] The EU recommended daily allowance (RDA) of biotin has been set at 50 µg.[16] This is based upon guidance from the joint FAO/WHO Expert consultation on recommended allowances of nutrients for food labeling purposes held in Helsinki in 1998 (FAO/WHO 1998). In the United States, the Food and Nutrition Board (FNB) at the National Academies of Sciences, Engineering, and Medicine found the evidence insufficient to establish a formal RDA or estimated average requirement (EAR) for biotin and have therefore published an Adequate Intake (AI): intake at this level is assumed to ensure nutritional adequacy.[17] The US reference AIs are shown in Table 10.1. Evidence suggests oral supplements of biotin are 100% bioavailable.[18]

Functions of Biotin

Biotin is found in small quantities in all living cells and has several important biochemical functions. In humans it is the coenzyme for four carboxylases: pyruvate carboxylase (PC), propionyl CoA carboxylase (PCC), multiple CoA carboxylase (MCC), and acetylCoA carboxylase I and II (ACC1, ACCII). The first three are

Table 10.1 Dietary Reference Adequate Intakes for Biotin

Age (Years)	Daily Requirement (µg)
0–6 months	5
7–12 months	6
1–3	8
4–8	12
9–13	20
14–18	25
>19	30

mitochondrial enzymes, the latter exists in two genetically distinct forms, one of which is found in the mitochondria, the other in the cytosol. The carboxylases are converted from the inactive apo forms to the active holo forms by holocarboxylase synthetase (HLCS: EC 6.2.4.10), in a two-step, ATP-dependent reaction in which a biotin molecule is covalently bound to the lysine residue in the carboxylases. Holocarboxylase synthetase is present in the nuclei, mitochondria, and cytoplasm. Biotinylated proteins are proteolytically degraded to biocytin (biotinyl-L-lysine) and biotinylated oligopeptides that can be subsequently cleaved by biotinidase (EC 3.5.1.12), recycling biotin. The four carboxylase enzymes are involved in carbon chain elongation steps in gluconeogenesis, fatty acid synthesis, propionate metabolism, and leucine catabolism, respectively.[19–22]

Pyruvate Carboxylase

Pyruvate carboxylase is found in the mitochondria and has a key role in gluconeogenesis. It catalyzes the carboxylation of pyruvate, producing oxaloacetate, an intermediate in the tricarboxylic acid cycle, which is converted to glucose in the liver, kidney, and other gluconeogenic tissues. Reduced PC activity causes lactic acidosis.

Propionyl CoA Carboxylase

Propionyl CoA carboxylase catalyzes the carboxylation of propionyl-CoA to form D-methylmalonyl-CoA, which is racemized to the L-isomer, prior to isomerization to succinylCoA and entry into the tricarboxylic acid cycle. Reduced PCC activity causes accumulation of 3-hydroxypropionic acid and 2-methylcitric acid.

Multiple CoA Carboxylase

Multiple CoA carboxylase catalyzes the catabolism of leucine, a branched chain amino acid. Reduced activity of this enzyme and metabolism via an alternative metabolic pathway results in the production of 3-hydroxyisovaleric acid (3HIVA) and 3-methylcrotonylglycine.

Acetyl CoA Carboxylase I and II

AcetylCoA carboxylase I and II have a role in the generation of malonyl CoA. While both enzymes catalyze the binding of bicarbonate to acetyl CoA to form malonyl CoA, *AccI* is found in the cytosol, where it has a key role in fatty acid synthesis, and ACCII is found in the mitochondria where it is involved in the regulation of fatty acid oxidation.

Biotin has also been shown to play a role in histone modification, gene expression, and immune function. Chromatin is a complex of macromolecules found in cells consisting of DNA, protein, and RNA. The chief protein components of chromatin are the histones, which have an important role in the folding of DNA. Recent studies have demonstrated that biotin is attached to the histones via an amide bond and to date, eleven different biotinylation sites have been identified in human histones.[23] Histone biotinylation is dependent upon the dietary supply of biotin. Both biotinidase and HLCS catalyze histone biotinylation, with the action of the latter thought to be the more important. Functions of histone biotinylation include gene expression and DNA repair. Knockdown of HLCS or biotinidase in *Drosophila melanogaster* decreases histone biotinylation resulting in abnormal gene expression patterns and a phenotype with decreased life span and heat resistance. Biotinylation of histones is a reversible process. Although little is known about the mechanism for modification at this time, it has been postulated that biotinidase may act as a catalyst here too.

Biotin has a key function in gene expression with >2000 biotin-dependent genes having been identified in human liver and lymphoid cells. These genes are part of specific gene clusters based on signaling pathways, chromosomal location, cellular localization of gene products, biological function, and molecular function. Bisnorbiotin has also been shown to affect gene expression leading to the hypothesis that biotin metabolites may display biotin-like activity in humans. Other studies have indicated that biotin is involved in normal immune functions such as antibody production, macrophage function, and the differentiation of T and B lymphocytes and biotin has also been linked to cell proliferation.[24, 25] Further information is outside the scope of this chapter and the reader is referred to other articles on this topic.[26, 27]

Biotin Homeostasis

Under normal homeostasis, the amount of biotin in the body is regulated by dietary intake, biotinidase, HLCS, and the SMVT. A disturbance in one or more of these regulatory mechanisms may result in biotin deficiency. It has also been hypothesized that the monocarboxylate transporter 1 (MCT1) may have a role in biotin homeostasis.

Holocarboxylase synthetase S catalyzes the attachment of biotin to lysine residues of apocarboxylases and histones. The biotin-dependent carboxylases have relatively short half-lives of between 1 and 8 days. Proteolytic degradation of the holocarboxylases results in the production of biocytin and biotinylated peptides. Free biotin is liberated from these breakdown products by the action of biotinidase, effectively recycling endogenous biotin for holocarboxylase synthesis. Likewise, biotinidase also releases free biotin from dietary biocytin and biotinylated peptides, thus recycling dietary biotin. Biotinidase is predominantly synthesized in the liver and secreted into the bloodstream; it is also present in pancreatic fluid, intestinal secretions, brush border membranes, and intestinal flora.

The SMVT is a high affinity transporter responsible for the intestinal absorption of free biotin, the renal reabsorption and transport across cell membranes in the liver and peripheral tissues. As mammals are unable to synthesize biotin de novo, biotin must be derived from exogenous sources and a high affinity transport mechanism is required to ensure cells have sufficient biotin to maintain activity of the biotin-dependent carboxylases. The SMVT utilizes a transmembrane sodium ion gradient and membrane potential to drive cellular uptake of biotin, pantothenic acid (Chapter 8), folate (Chapter 11), and lipoic acid. Pantothenic acid is key to the metabolism of carbohydrate, fat, and protein; it is required for the biosynthesis of coenzyme A and acyl carrier proteins. Lipoate is a potent antioxidant with a role in the redox cycling of other antioxidants such as vitamins C (Chapter 13) and E (Chapter 4) and regulating intracellular concentrations of glutathione. In biotin deficiency states, intestinal uptake of biotin is upregulated and there is a concomitant increase in the level of the SMVT protein. Conversely, biotin supplementation appears to have the opposite effect.

When extracellular concentrations of biotin exceed 25 μmol/L, passive diffusion across the cell membrane may also occur. The SMVT is also responsible for uptake into liver, peripheral tissue, and renal reabsorption. The exit of biotin from the enterocytes is carrier mediated.

It has been hypothesized that monocarboxylate transporter 1 (MCT1) may mediate the uptake of biotin (which is a monocarboxylic acid) into human cells and recent evidence confirms this theory is correct for lymphoid cells.[28] Monocarboxylate transporter 1 facilitates the acquisition of biotin into Peripheral Blood Mononuclear Cells (PBMC) using a proton-dependent counter transport mechanism. As a result, uptake of biotin is inhibited by monocarboxylic compounds such as lactate, pyruvate, 3-hydroxybutyrate, and acetate.

Biotin Metabolism

Before absorption can take place, dietary biotin, which is covalently bound to lysine residues in proteins, must be digested. Biotin-containing proteins are digested by gastrointestinal proteases and peptidases, releasing biocytin and biotinylated peptides.[29] These are then cleaved by biotinidase, which is present in pancreatic fluid, intestinal secretions, brush border membranes, and intestinal flora, releasing free biotin.

Liberation of free biotin is an important step as it is in this form that biotin is most efficiently absorbed and optimally bioavailable. Intestinal uptake of free biotin occurs primarily in the proximal small intestine in a process mediated by the SMVT. The SMVT transports biotin against a concentration gradient and is the rate-limiting step in movement across the intestinal epithelium.

The intestinal absorption of free biotin is inhibited by structural analogs of biotin with a free carboxyl group at the valeric acid side chain, for example, desthiobiotin, whereas analogs with a blocked carboxyl group, e.g., biocytin, do not. Biotin exits the intestinal epithelial cells via a sodium-independent carrier-mediated mechanism.

Bacterially synthesized biotin is found predominantly in the lumen of the large intestine where it is present in the free and bound form. The SMVT is also reported to be present in colonocytes, emphasizing the importance of the contribution of bacterially synthesized biotin in humans, and providing a mechanism for absorption in the large intestine.

The liver plays an important role in biotin homeostasis and is responsible for the majority of biotin metabolism and utilization. Although the liver contains more biotin than other tissues, unlike the other water-soluble vitamins, the liver has limited capacity to store biotin[30] and relies on circulating and extracellular biotin for its needs. Circulating biotin is extracted from the hepatic portal circulation and transported across the hepatocyte basolateral membrane by the SMVT. As with the intestinal uptake of biotin, transport is inhibited by desthiobiotin but not biocytin. Free biotin is then transported into mitochondria where it biotinylates the apocarboxylases. This intracellular transport is a nonmediated, simple diffusion process.

In human plasma, the majority (81%) of biotin circulates in the free form with the remainder being either covalently (12%) or reversibly (7%) bound. Circulating biotin is filtered by the renal glomeruli and then salvaged by reabsorption. Reabsorption occurs in the renal proximal tubule epithelial cells and is mediated by the SMVT. Exit from the renal epithelium is via a sodium-independent mechanism.

Transport of biotin across the blood-brain barrier is carrier mediated but unlike the other sodium-dependent carrier-mediated transport mechanisms, it is inhibited by pantothenic acid.

There are two main routes of biotin catabolism, β-oxidation and sulfur oxidation.[31, 32] β-Oxidation of the valeric acid side chain results primarily in the formation of bisnorbiotin and tetranorbiotin, with a smaller number of β-dehydro-, β-hydroxy-, and β-keto-intermediates also being produced. These intermediates undergo spontaneous decarboxylation to bisnorbiotin methylketone and tetranorbiotin methylketone while microorganisms cleave the heterocyclic ring of tetranorbiotin, degrading it further. Catabolism of biotin to biotin sulfoxide occurs predominantly in the liver. In the second catabolic pathway, oxidation of the sulfur atom in the heterocyclic ring results in the formation of biotin-L-sulfoxide, biotin-d-sulfoxide, and biotin sulfone and takes place mainly in the mitochondria. A third and minor route of catabolism occurs when biotin is catabolized by a combination of sulfur oxidation and β-oxidation, forming compounds such as bisnorbiotin sulfone.

Biotin and its metabolites are excreted in the urine and feces with adults typically excreting approximately 100 nmol of biotin and metabolites daily. The primary urinary metabolite is biotin itself, accounting for approximately 50% of total excretion, with the remainder being comprised of bisnorbiotin (13%–23%), biotin-D,L-sulfoxide (5%–13%), bisnorbiotin methyl ketone (3%–9%), biotin sulfone (1%–3%). Biliary excretion of biotin and catabolites is reported to be minimal.

The role of biotinidase is to cleave biotin from biocytin, thus enabling the biotin pool to be recycled and available for use as a cofactor with the biotin-dependent carboxylases. Low activity of biotinidase results in an inability to recycle biotin from biocytin and the excretion of abnormal amounts of biocytin in the urine. This progressively depletes the bodies stores, eventually resulting in biotin deficiency; the clinical and biochemical features of biotinidase deficiency are similar to those of biotin deficiency.

Clinical Application of Measurement
Biotin Insufficiency

The symptoms of nutritional biotin deficiency include dry skin, seborrheic dermatitis, fungal infections, skin rashes, brittle hair, alopecia, and hyperglycemia. If left untreated, neurological symptoms may develop: depression, altered mental status, myalgia, hyperesthesia, and paresthesia. The inherited defects of biotin metabolism give rise to similar clinical signs and symptoms. Biotin deficiency can be simply treated with a pharmacological dose of biotin which will ameliorate the clinical symptoms.

Nutritional deficiency of biotin is rare and case reports in the literature almost exclusively relate to inborn errors of biotin metabolism and transport, long-term use of parenteral nutrition,[33, 34] chronic anticonvulsant therapy,[35–37] or excessive ingestion of egg whites. Suboptimal biotin status has also been reported in pregnancy, alcoholism, smoking, inflammatory bowel disease, children with severe burns, and in patients with seborrheic dermatitis and Leiner's disease. Adherence to a vegetarian diet is not thought to affect biotin status.

Avidin

Avidin is a glycoprotein found in egg white which binds biotin very tightly and specifically and is resistant to pancreatic enzymes. Thus dietary avidin binds to dietary biotin, inhibiting its absorption and if exposure is prolonged, deficient states can result.[38]

Long-term anticonvulsant therapy

Long-term anticonvulsant therapy is associated with an increased requirement for biotin. The mechanism by which this occurs is not fully understood but competitive inhibition of intestinal biotin uptake, accelerated biotin catabolism, and impaired renal reabsorption have all been reported as contributory factors; primidone and carbamazepine have been found to inhibit the intestinal uptake of biotin,[39] likewise phenytoin, phenobarbital, and carbamazepine have been reported to displace biotin from biotinidase and this is postulated to affect the plasma transport, renal handling, or cellular uptake of biotin.

Biotin catabolism is accelerated during pregnancy. Increased urinary 3HIVA has been demonstrated in both early and late pregnancy whereas the excretion of biotin itself is only increased in late pregnancy. This is hypothesized to be a result of accelerated biotransformation of biotin to inactive metabolites and may explain why marginal biotin status is common during normal pregnancy. Furthermore, an active transport mechanism of maternal biotin appears to exist in favor of the fetus.

Inborn errors of biotin metabolism

The first case of an inherited defect in biotin metabolism was described in 1971.[40] The infant presented with metabolic acidosis, vomiting and seizures, and excessive excretion of urinary 3-methylcrotonylglycine. Fibroblast studies demonstrated reduced activity of each of the biotin-dependent carboxylases. Supplementation with a pharmacological dose of biotin resulted in both a clinical and biochemical improvement.

Since then, two separate defects in the cycle of biotin utilization have been described: HLCS deficiency and biotinidase deficiency.[41] Both disorders are autosomal recessive and have incidences of 1 in 87,000 and 1 in 60,000 live births, respectively.[42,43] Deficiency of either enzyme results in multiple carboxylase deficiency (MCD), as does nutritional deficiency of biotin. Multiple carboxylase deficiency is characterized biochemically by abnormal excretion of specific organic acids: 3HIVA and 3-methylcrotonylglycine accumulate due to deficiency of MCC, 3-hydroxypropionic acid, propionyl glycine, and 2-methylcitric acid accumulate due to deficiency of PCC and lactic aciduria reflects deficiency of PC. If left untreated, severe life-threatening illness will result.

Holocarboxylase synthetase deficiency typically presents in the neonatal period or early infancy but there have been an increasing number of reports of late onset HLCS (2–21 months). There is reduced formation of the holocarboxylase enzymes at physiological levels of biotin, due either to an increased Km of HLCS for biotin, or a reduced Vmax; the binding of biotin to the apocarboxylases is impaired. The carboxylases play a vital role in gluconeogenesis, fatty acid synthesis, and the catabolic pathways of several amino acids hence their deficiency results in significant metabolic derangements which can be life threatening. Biotinidase deficiency results in the inability to liberate and recycle endogenous biotin from biocytin and to utilize protein-bound biotin from the diet. Progressive development of biotin deficiency occurs, eventually resulting in MCD.

The initial presentation of HLCS is often precipitated by a catabolic event or period of increased dietary protein intake. Clinical features include metabolic acidosis, feeding difficulties, hypotonia, impaired consciousness, ataxia, seizures, and progressive encephalopathy and if untreated, may result in death. Conversely, the onset of isolated biotinidase deficiency tends to be insidious, occurring slightly later than HLCS deficiency, typically presenting in late infancy/early childhood dependent upon the amount of free biotin in the diet. Patients are categorized into two groups: those with <10% residual biotinidase activity are classed as profoundly deficient, whereas those with between 10 and 30% residual activity are said to be partially deficient. Clinical signs and symptoms reflect those of HLCS deficiency and include alopecia, skin rash, immune deficiency, developmental delay, hypotonia, metabolic acidosis, and progressive neurological symptoms; clinically, biotinidase deficiency may be indistinguishable from mild HLCS deficiency.

Diagnosis of MCD is made by urine organic acid analysis, where a characteristic pattern of abnormal metabolites will be present. Analysis of acylcarnitines in plasma or bloodspots will show increased 3-hydroxyisovalerylcarnitine. Hyperammonemia may also be present. The definitive diagnosis must then be made by enzyme studies to determine the activity of biotinidase in plasma and HLCS in cultured skin fibroblasts. Mutation analysis of the biotinidase gene is also available.[44]

Irrespective of which enzyme(s) is deficient, MCD can be effectively treated with pharmacological doses of oral biotin. Patients usually respond well but prompt initiation of treatment is required to prevent irreversible neurological damage. In biotinidase deficiency, daily doses of 5–10 mg oral biotin are usually effective whereas HLCS deficiency may require a higher dose, between 40 and 100 mg daily. The optimal dose is determined on an individual basis and therapy is lifelong. A small number of patients with HLCS fail to respond to biotin therapy or only partially respond.

A third inherited disorder of biotin metabolism, Biotin Responsive Basal Ganglia Disease (BRBGD) was first described in 1988.[45, 46] This neurometabolic disorder is autosomal recessive and usually presents during childhood although four distinct phenotypes have now been reported. Clinical signs and symptoms include subacute encephalopathy: confusion, convulsions, dystonia, dysarthria.[47] Prompt treatment with pharmacological doses of biotin and thiamine (Chapter 6) will resolve the symptoms in a few days; however, if treatment is delayed, residual paraparesis, mild mental retardation, or dystonia can result and if left untreated, it may be fatal. Detection of BRBGD requires a high index of clinical suspicion. Magnetic Resonance Imaging findings may assist the diagnosis.

There is also a single reported case of an inherited defect in biotin transport.[48] At presentation the proband exhibited the clinical and biochemical signs of MCD and responded to biotin supplementation; however, fibroblast and leucocyte studies confirmed normal activity of each of the four biotin-dependent carboxylases, excluding both MCD and an isolated deficiency of any one of these enzymes. The SMVT gene was sequenced and its function investigated by determining pantothenic acid transport rates. No abnormalities were identified but biochemical dependency on biotin persisted, indicating intracellular biotin deficiency. Investigation of biotin uptake by PBMCs found the uptake rate in the proband to be 10% of controls, with the parents demonstrating uptake rates consistent with heterozygous state for an autosomal recessive defect. This led to the hypothesis that the defect was caused by impaired biotin transport, but the genetic defect has yet to be elucidated.

Biotin Toxicity

There is currently no evidence to indicate that biotin is toxic.

Biotin Supplementation

Biotin is becoming an increasingly common "over the counter" vitamin supplement. It is marketed as being beneficial for strengthening nails and hair, controlling blood glucose metabolism, and easing peripheral neuropathy. Biotin supplements tend to contain between 500 and 1000 μg per tablet with some mega dose supplements containing upwards of 10 mg. Regular intake of such supplements increases the circulating concentration of biotin in the blood. Irrespective of the promised health benefits, an unfortunate consequence of high-dose biotin supplementation is the potential for it to interfere with a number of other common blood tests. Immunoassay is a well-established technique used routinely in clinical laboratories to provide analysis of a wide range of compounds including hormones, drugs, cardiac markers, tumor markers, and infectious diseases such as HIV and hepatitis C.

Many commercially available immunoassays are based on the interaction between biotin and streptavidin. In sandwich immunoassays, if a large excess of biotin is present in the serum due to supplementation, it will compete with the biotin-labeled antibody in the assay for streptavidin-binding sites and result in false negative results. Conversely, in competitive immunoassays, the excess biotin will result in false positive results for the analyte in question. There have now been numerous case reports of biotin supplementation leading to assay interference, providing evidence that most immunoassays are affected, irrespective of the manufacturer[49–52] and in November 2017 a Safety Alert was issued by the US Food & Drug Administration.[53]

Assay interference is not uncommon in clinical immunoassay; results must always be interpreted in light of clinical information and queried when they are at odds to the clinical picture. However, several cases have been described in which biotin supplementation has interfered with several tests at the same time, creating a biochemical picture of disease supported by two or more independent tests.[49] Likewise, while gross interference may be easily identifiable, particularly if at odds with the clinical presentation, identification of the more subtle and moderate biotin-induced changes will pose a far greater challenge.

Analytical Methods

More than 20 different markers of biotin status have been described and these are typically classified into direct and indirect methods.

Direct Methods for Measurement of Biotin in Biological Fluids

The measurement of biotin in blood, serum/plasma, and urine relied for many years on microbiological assays, based on the biotin requirement for growth of yeast, bacteria, and algae. Microorganisms such as *Saccharomyces cerevisiae*, *Lactobacillus plantarum*, *Escherichia coli C162*, *Lactobacillus casei*, *Kloeckera brevis*, *Amphidinium carterae*, and *Ochromonas danica* depend on biotin for growth, and all have been used for biotin assays. The principle of such an assay is to isolate an organism into a biotin-depleted medium and then add the test substance. The rate of growth is measured and is directly proportional to the concentration of biotin in the test substance. Growth of cultures is assessed by optical density of liquid cultures, colony size around sample-soaked filter paper discs, incorporation into protein of radioactively labeled amino acids, or $^{14}CO_2$ production from labeled substrate. One of the difficulties is the supply of biotin-depleted growth medium. Most such assays are insufficiently sensitive for clinical purposes and have been used predominantly for the assessment of food biotin content.[54, 55] Exceptions are the assays based on *O. danica*,[56] *L. plantarum*,[57] and *K. brevis*.[58] The *O. danica* assay offered the advantage that the organism could be grown using fully synthetic medium and offered sufficient sensitivity for the measurement of biotin in clinical studies.[56, 59, 60] This assay was used in studies of normal biotin concentrations in plasma and urine, and to assess biotin availability following oral loading.[61] Not all microbiological assays clearly distinguish between free and covalently protein-bound biotin since some of the organisms themselves express biotinidase activity, and the measured levels therefore reflect a variable population of biotin species.[62] Interestingly, *L. plantarum* does not express biotinidase, and this has been used to measure the relative amounts of biocytin and biotin by differential assay before and after sample treatment with biotinidase.[63] The lack of sensitivity and specificity in microbiological assays resulted in wide variation in published concentrations in human biological fluids.

Direct measurement of biotin, biotin-containing peptides, and biotin metabolites became possible with the development of competitive binding radioassays based on avidin and ^{3}H or ^{125}I labeled biotin.[64-67] These were initially limited in sensitivity by the available specific activity of radioactive biotin. Higher specific activity radioactive biotin, assays based on biotin-linked enzymes as the tracer molecule for avidin-binding assays, and use of labeled avidin in competitive binding formats have increased sensitivity to the levels required for clinically applicable methods.[68-71] Commercial ELISA format assays using streptavidin and biotin linked enzyme as tracer are available and practical.[72] It is to be noted that the sample preparation for assays using biotin-linked tracers may need to include inactivation of biotinidase activity in the sample by a denaturation step to prevent degradation of the tracer. Avidin-binding assays may cross-react with biocytin, and the main metabolites of biotin—bisnorbiotin and biotin sulfoxide—to a variable degree and therefore unless these species are separated before assay their results should be regarded as relating to "total avidin binding substances" (TABS) rather than a truly quantitative measure of biotin itself. Estimates of the relationship between biotin and TABS suggest that about 50% of TABS in plasma/serum is free biotin.[73, 74] In addition, biotinidase itself acts as a high affinity biotin-binding protein in human plasma,[22] reducing the renal clearance of biotin in normal but not biotinidase-deficient individuals,[63] and necessitating denaturation if total biotin concentrations are to be measured. Microbiological and avidin-binding assays have been the subject of a systematic review.[75] This review also included methods used in the determination of biotin in foodstuffs and supplements but of insufficient sensitivity for the assessment of biotin status in patients. These included spectrophotometry, gas liquid chromatography, and other physicochemical methods.

The lack of specificity in both microbiological and avidin-binding assays for biotin led to the development and use of a chromatographic step to separate the predominant forms of biotin before quantitation using available assays (microbiological or avidin binding). Reversed-phase methods on C18 columns using gradient elution[76] were incorporated as a sample preparation step.[67, 69, 77] These allowed the separation and individual quantitation of biotin, biocytin, bisnorbiotin, and biotin sulfoxide. More recently, direct analysis of biotin using stable isotope dilution LC-MS/MS has been described.[78] This technology offers high specificity and sensitivity. Liquid chromatography-tandem mass spectrometry should allow traceable and harmonized assay of plasma/serum biotin; however, there will still be the possibility of variation introduced by the sample preparation employed, since incomplete denaturation of biotin-binding proteins might lead to lower recovery and underestimation of true free biotin levels. If a measure of total biotin, including biocytin and other covalently bound biotin species is required, hydrolysis using peptidase and biotinidase treatment will still be necessary. Plasma and serum concentrations appear to be interchangeable.[72] Urinary excretion of biotin is high in biotinidase deficiency and biotin supplemented states but does not reliably distinguish between biotin deficiency and biotin-sufficient states. Measurement of plasma/serum concentrations of biotin allows better discrimination[79] and allows the investigation of biotin anomalies in specific disease states[80–83] but until the use of LC-MS/MS methods becomes widespread are not necessarily comparable between centers because of issues of specificity. It is likely that markers of end organ effect may be more reliable in the assessment of biotin status in patients.[84] Published plasma and serum biotin reference ranges are presented in Table 10.2.

Given the limitations of the existing methods for biotin measurement, it is important that any reference intervals are locally determined and specific to the assay in question. The disparate reference ranges reflect the limitations of these assays.

Indirect Methods: Functional Markers for the Assessment of Biotin Status

Measurement of biotinidase activity

When biotinylated proteins, e.g., carboxylase enzymes and histones, are broken down, biocytin and small biotin-containing peptides are released. These are the natural substrates for biotinidase. The enzyme is present in pancreatic fluid, intestinal secretions, brush border membranes, and intestinal flora for the liberation of exogenous biotin, but endogenous recycling of biocytin mainly depends on biotinidase synthesized in the liver. Much of it is secreted into the bloodstream as extracellular biotinidase. The presence of the enzyme activity in human serum was first demonstrated in the early 1950s[85] and its importance in biotin metabolism in humans was recognized with the description of the inherited deficiency resulting in late onset MCD.[86, 87] Biotinidase deficiency results in an increase of the daily requirement for exogenous biotin by approximately three to four orders of magnitude, from a few micrograms per day to tens of milligrams per day.[46, 88] Early recognition of biotinidase deficiency and treatment with high-dose biotin prevents most of the sequelae of late onset MCD, leading to the inclusion of biotinidase in many newborn screening programs.[89]

The most widely used quantitative assays for biotinidase activity have used the synthetic substrate N-(D-biotinyl) 4-amidobenzoate (BPABA) as substrate. The method using diazotization of the released 4-aminobenzoic acid and reaction with a naphthol compound to give a colored product, absorbing strongly at 546 nm, based on that of Knappe,[90] has been used clinically for plasma since 1983.[91] A fully detailed description of this method is given in Wolf et al.[88] The authors of this chapter and others have scaled down the method in their laboratories and adapted it for use on automated spectrophotometric analyzers. Variations of the same chemistry are

Table 10.2 A Summary of Published Plasma and Serum Biotin Reference Ranges

Group	N	Range	Method	Reference
Children (0–4 years)	188	0.047–0.22 µg/L	Streptavidin	72
Adults (21–50 years)	25	0.084–0.21 µg/L	ELISA	81
Adults	11	0.081 (0.05–0.11) µg/L	LCMSMS	73
Adults	15	244 ± 61 pmol/L	Avidin binding	62
Adults	600	200–700 ± 16 ng/L	*Ochromonas danica* microbiological	
Adult men	15	260 ± 91 pg/mL	Radiometric/ microbiological	58
Adult women	23	224 ± 94 pg/mL		
Adults	112	1.63 ± 0.49 nmol/L	*Lactobacillus plantarum* microbiological	35

Biotin molecular weight 244.31; pg/mL × 4.098 = pmol/L; pmol/L × 0.244 = pg/mL.

also widely used for newborn screening in a semiquantitative format for use on dried blood spots.[92] A disadvantage of the technique is that the Bratton and Marshall reaction used to quantitate the 4-aminobenzoic acid product is nonspecific and also reacts with other compounds, for example, sulfonamide drugs[93] and has insufficient sensitivity for studies in tissues with low abundance of biotinidase.[88] The same substrate can be used radioactively labeled to enhance sensitivity (e.g., to measure activities in lower abundance tissues such as cultured cells[94]). The release of PABA from the BPABA substrate can also be assayed with high specificity and sensitivity using HPLC with UV detection[95] or electrospray isotope-dilution LC-MS/MS.[96] A semiquantitative version of the LC-MS/MS method, for use on dried blood spots for newborn screening has also been described.[97]

An alternative substrate is the fluorogenic derivative of biotin, biotinyl-6-aminoquinoline, first described in 1984.[98] This has also been widely used for quantitative analysis in serum, plasma, and isolated cells, as well as for newborn screening using dried blood spots.[99] Using LC-MS/MS as the measurement technique opens the possibility of simply measuring biotin release from biocytin directly, either quantitatively using isotope dilution methodology, or semiquantitatively for newborn screening, alone or as part of an LC-MS/MS panel.[100]

Measurement of 3-hydroxyisovaleric acid

Although 3HIVA can be measured in blood, plasma, serum, amniotic fluid, and CSF, urine is widely considered the most suitable biological matrix for analysis; it contains very little protein which simplifies sample preparation and along with other organic acids it tends to be concentrated in the urine by the kidneys.[101, 102] Furthermore, evidence indicates that it is a useful marker of biotin status in a number of clinical scenarios, including pregnancy.[103]

Traditionally, urinary 3HIVA has been measured by stable isotope dilution GCMS. This is a well-established analytical technique and a comprehensive overview of this methodology can be found elsewhere.[104, 105] In brief, following addition of stable isotope internal standard, e.g., $[^2H_8]$- 3HIVA or $[^2H_6]$- 3HIVA, urine samples are acidified prior to liquid-liquid extraction. Sequential extractions with ethylacetate are recommended to ensure maximum recovery of the acid. The solvent extract is dried under nitrogen at ambient temperature; it is important that no heat is used during this stage to prevent loss of the acid. The solvent extract is reconstituted in pyridine and N,O,-bis-(trimethylsilyl)trifluoroacetamide (BSTFA) containing 1% trimethylchlorosilane (TMCS) and heated at 75°C for 30 min to convert it to the trimethylsilyl ether derivative (di TMS). This latter step is necessary to convert the free acid to a thermally stable, volatile compound, suitable for analysis by GC. The derivative is separated in a capillary gas chromatography column containing a nonpolar stationary phase and detected by electron impact mass spectrometry in selective ion mode. Quantitation is based on comparison of the ratio of 3HIVA to stable isotope internal standard in the sample, to that in the calibration standard which is prepared from pure compound.

The limitations of this method include the laborious sample preparation procedure, the potential to lose 3HIVA during the analytical process,[105] and the lack of a traceable certified reference material for 3HIVA. More importantly, while the method is reported as being able to discriminate individuals with biotin deficiency from healthy controls, false negative results are not uncommon and biotin sufficiency cannot be distinguished from biotin supplementation.

It should also be noted that the majority of laboratories which offer this test are actually providing it as part of a comprehensive organic acid profile; hence methods are unlikely to be tailored to the specific quantitation of 3HIVA. As such, the authors would recommend quantitation of urinary 3HIVA by a dedicated method such as that described by Horvarth et al.[103] which utilizes UPLC tandem mass spectrometry and has the advantages of superior sensitivity and specificity. In brief, urine samples are diluted fourfold and stable isotope internal standard added. The sample is injected onto a Waters HSS T3 (2.1×100 mm, $1.8\,\mu$m) column held at $55\,°$C and the 3HIVA is separated by gradient elution using mobile phases of 0.01% formic acid and methanol. The MS/MS operates in negative ionization mode and 3HIVA and $[^2H_8]$-3HIVA are acquired by multiple reaction monitoring of $117.0 > 59.1$ and $125.0 > 61.0$, respectively. The method is precise and sensitive, correlates with GCMS, and is capable of accurately quantitating 3HIVA across the physiological range of concentrations observed in healthy humans. The lack of a traceable reference material is still an issue.

Measurement of 3-hydroxyisovalerylcarnitine

3-Hydroxyisovalerylcarnitine (3HIVAc) is formed by conjugation of 3-hydroxyisovalerylCoA to carnitine. Analysis of 3HIVAc has been described by several techniques, including HPLC and GC-MS[106]; however, it is almost exclusively measured by LC-MS/MS, reflecting the superior sensitivity, rapid throughput, and small sample size associated with this methodology. Traditionally, analysis of 3HIVAc by LC-MS/MS necessitated a butyl ester derivatization step but as instrument sensitivity has improved, the precise and accurate quantitation of 3HIVAc is now easily achieved without a derivatization step.

Following addition of stable isotope internal standard, 3HIVAc is extracted from plasma or serum by protein precipitation with either methanol or acidified acetonitrile solution. If a butyl ester derivatization is required, the sample supernatant is evaporated and the residue derivatized with butanolic HCl, converting the 3HIVAc to its butyl ester. For underivitized analysis, the supernatant is analyzed directly.

The sample is injected onto an Agilent ZORBAX Eclipse Plus-C18 column (2.1×100 mm, $1.8\,\mu$m) with gradient elution using 0.01% formic acid and acetonitrile mobile phases. The MS/MS operates in positive ionization mode and 3HIVAc is acquired by multiple reaction monitoring: $262.0 > 85.1$ or $318 > 85.1$ for underivitized and butylated species, respectively. For full details of this method the reader is referred to Blau et al.[104]

Analysis of plasma 3HIVAc by LC-MS/MS has proved to be a sensitive marker of biotin depletion and, in principle at least, has a rapid analysis time. Another advantage of LC-MS/MS analysis is that additional acylcarnitine species can be detected simultaneously, enabling the other biotin-dependent carboxylases to be investigated at the same time. Multiple CoA carboxylase converts 3-methylcrotonyl-CoA to 3-methyglutaconyl-CoA. If activity of MCC is reduced, 3-methylcrotonyl-CoA accumulates, is shunted to 3-hydroxyisovaleryl-CoA, transesterified to 3-hydroxyisovalerylcarnitine, and subsequently excreted in the urine. Increases and decreases in 3-methyglutaconyl-CoA are reflected in an analogous fashion by urinary 3-methylglutarylcarnitine (3MGc). Likewise, PCC converts propionyl-CoA to methylmalonyl-CoA. If activity of PCC is reduced, the accumulating propionyl-CoA is transesterified to propionylcarnitine (Pc). Any changes in methylmalonyl-CoA are reflected in urinary concentration of methlmalonylcarnitine (MMc). ACC has two isoforms, both of which convert acetyl-CoA to malonyl-CoA; the related acylcarnitines being acetylcarnitine (Ac) and malonylcarnitine (Mc).

Acylcarnitine substrate to product ratios specific to biotin-dependent carboxylases have been measured in urine by LC-MS/MS and used for the assessment of biotin status.[107] Urine samples have essentially been analyzed as described before, except that the protein precipitation step is replaced with a 50-fold dilution of the sample. The evidence reported by Bogusiewicz et al. indicates that Ac:Mc, Pc:MMc, and 3HIAc:3MGc ratios are likely to be useful markers of marginal biotin deficiency with the ratio of urinary Pc:MMc having the same sensitivity of urinary 3HIVAc in distinguishing subjects with marginal biotin insufficiency. The advantage of this approach is that the use of the substrate ratios negates the need for stable isotope internal standards and formal quantitation of each carnitine species: peak area ratios of the compounds of interest being used instead.

Standardization
Biotin

The only available biotin standard reference material (NIST 3280) is intended primarily to validate analytical methods for the determination of biotin in dietary supplement tablets. As such, the material is provided in the form of whole multivitamin tablets (1.5 g) which limits its utility. It is certified by quantitative NMR and the associated uncertainty is stated on the certificate.

While there is no internationally recognized reference method for serum biotin measurement, isotope dilution LC-MS/MS methods are available; however, the lack of a suitable standard reference material remains an issue for clinical laboratories. The situation is compounded further by the absence of standardization for any of the indirect methods of biotin measurement.

There is no external quality assessment (EQA) scheme for the determination of serum biotin. However, several proficiency testing schemes exist for the determination of biotin in cereal-based products, animal feed, multivitamin powders, and infant formula. While it should be noted that these schemes distribute samples which typically have biotin concentrations in the region 1–10 ppm, several orders of magnitude higher than that seen in serum, review of results from a 2016/17 scheme highlights the analytical limitations of biotin methods. Two samples of multivitamin powder were distributed to 13 participants and analyzed in duplicate. The target standard deviation (SD) for the scheme was based on the Horwitz model and was 17.4% at 1.97 ppm and 27.5% at 0.18 ppm. The target range for interlaboratory reproducibility was achieved by 77% of laboratories. Five different methodologies were used for analysis and the intralaboratory coefficient of variation (CV) was 18.9%.

Indirect Markers

There are no traceable reference materials available for biotinidase, 3HIVc or 3HIVA although in principle suitable reference methods do exist (isotope dilution LC-MS/MS). The situation is exacerbated by the lack of an EQA scheme for the first two of these indirect markers of biotin status.

A single EQA scheme exists for urinary 3HIVA, organized by the European Research Network for evaluation and improvement of screening, Diagnosis and treatment of Inherited disorders of Metabolism (ERNDIM); the Quantitative Urine Organic Acid scheme includes 3HIVA and in 2017, 86 laboratories participated. The average intralaboratory CV was 22% and the interlaboratory CV was 51%, highlighting the difficulties associated with measuring this analyte: lack of a certified reference material, availability/choice of stable isotope internal standard, and variation in sample preparation. The utility of the scheme is also limited by the small number of samples distributed annually ($n=8$).

Best Practice

The measurement of biotin in the investigation of biotin excess states has little clinical utility because currently there is no evidence that high concentrations are toxic. However, it is worth noting that one potential application of biotin measurement in states of excess may be to confirm it is present at the supraphysiological concentrations associated with assay interference and a high risk of misdiagnosis when analytes are measured by immunoassay. While simply asking the patient if they are taking a biotin supplement may provide adequate insight in some cases, because the concentration of biotin above which interference occurs varies with both the immunoassay and the analyte in question, there may be occasions when there is merit in confirming a high circulating concentration of serum biotin. However, given the limited number of laboratories which measure serum biotin and the likelihood of obtaining a result quickly, Piketty et al.[108] have described a different approach to confirming biotin interference: excess biotin in patient samples is adsorbed on to streptavidin-coated magnetic microparticles prior to reanalysis. Alternatively, the patient should be asked to refrain from ingesting biotin for 48 h prior to repeating the blood test.[109] A more in-depth investigation of the relationship between serum biotin levels and *in vitro* immunoassay interference by Grimsey et al. concluded that for biotin supplements in the region of 300 x ADI, levels would fall below the *in vitro* interference threshold of $\geq 30 \mu g/L$ after only 8 h.[110]

In practice, the nature of the test and the clinical question posed will determine the urgency with which an accurate result is required; a cardiac marker such as troponin requested following admission to Emergency Departments with acute chest pain is likely to be dealt with differently to a thyroid hormone test requested by a Family Physician on a patient who complains of being tired all the time. When discussion with the patient does not elucidate the necessary information, local assay availability is likely to be the deciding factor.

As states of dietary biotin insufficiency are very rare, the recommended first-line investigation for suspected biotin insufficiency is measurement of serum biotinidase activity to exclude a diagnosis of biotinidase deficiency. If biotinidase activity is decreased, the measurement of a serum acylcarnitine profile is recommended. The advantage of an acylcarnitine profile rather than the isolated measurement of 3HIVAc is the ability to simultaneously measure free carnitine and other carnitine species necessary for the differential diagnosis of biotinidase and MCC deficiency (free carnitine is necessary to exclude a carnitine-depleted state which could otherwise give rise to a false negative result). Measurement of urine organic acids can also provide this information and in practice the two tests are usually considered to be complementary rather than alternative. On balance, the authors would favor acylcarnitine analysis because it is a more robust assay, less prone to analytical variability and, in principle at least, available in real time. Currently, standardization remains an issue for both of these tests.

Conclusion

Laboratory methods that have adequate sensitivity and specificity to reliably diagnose biotin deficiency and monitor biotin levels in patients receiving supplements are necessary. There are presently a limited number of reliable markers of biotin status and current evidence indicates that no single marker can reliably distinguish moderately deficient biotin from sufficient and supplemented. Measurement of 3HIVA and 3HIVc, the functional markers of biotin status, is currently the recommended approach.

It is also important to note that much of the evidence in the literature is based on results obtained with assays which lacked sensitivity and were prone to interference; the inherent limitations of these assays may be reflected in the conclusions which have been drawn thus far. The authors believe there may be merit in revisiting some of the earlier work now that superior analytical techniques are widely available. In principle, ULPC-MS/MS would confer the advantages of specificity and sensitivity, with high throughput capability.

References

1. Wildiers E. *La Cellule* 1901;**18**:313.
2. Bateman W. Egg protein. *J Biol Chem* 1916;**26**:263.
3. Boas M. The effect of desiccation upon nutritive properties of egg white. *Biochem J* 1927;**21**:712–24.

4. Allison F, Hoover S, Burk D. A respiration coenzyme. *Science* 1933;**78**:217.

5. György P. The curative factor (vitamin H) for egg white injury, with particular reference to its presence in different foodstuffs and in yeast. *J Biol Chem* 1939;**131**:733–44.

6. Vigneaud V, Hofmann K, Melville D, György P. Isolation of biotin (vitamin H) from liver. *J Biol Chem* 1941;**140**:643–51.

7. György P, Rose CS, Eakin RE, Snell EE, Williams RJ. Egg-white injury as the result of nonabsorption or inactivation of biotin. *Science* 1941;**93**:477–8.

8. Combs G, editor. *The Vitamins: fundamental aspects in nutrition and health.* San Diego, CA: Academic Press; 1992. Biotin.

9. Bonjour J. Biotin in human nutrition. *Ann NY Acad Sci* 1985;**447**.

10. Wrong O, Edmonds C, Chadwich V. Vitamins. In: Wiley NY, editor. *The large intestine: its role in mammalian nutrition and homeostasis.* Wiley; 1981. p. 157–66.

11. Streit W, Entcheva P. Biotin in microbes, the genes involved in its synthesis, its biochemical role and perspective for biotechnology production. *Appl Microbiol Biotechnol* 2003;**61**:21–31.

12. Wolf B. Disorders of biotin metabolism. In: Scriver C, Beaudet A, Aly W, Valle D, Childs B, Kinzler K, et al., editors. *The metabolic and molecular basis of inherited disease.* New York: McGraw Hill Medical Publishing division; 2001. p. 3935–62.

13. Said H, Ortiz A, McCloud E, Dyer D, Moyer M, Rubin S. Biotin uptake by the human colonic epithelial cells NCM460: a carrier mediated process shared with pantothenic acid. *Am J Physiol* 1998;**44**:C1365–71.

14. Bender D. Optimum nutrition: thiamin, biotin and pantothenate. *Proc Nutr Soc* 1999;**58**:427–43.

15. National Institutes of Health, Office of Dietary Supplements. *Biotin. Factsheet for health professionals.* Available at https://ods.od.nih.gov/factsheets/Biotin-HealthProfessional/#en1; 2018. (Accessed April 2, 2018).

16. European Commission. *EU directive 2008/100/EC.* Available at http://eur-lex.europa.eu/eli/dir/2008/100/oj; 2014. (Accessed April 2, 2018).

17. Institute of Medicine. Food and nutrition board. In: *Dietary reference intakes: thiamin, riboflavin, niacin, vitamin B6, folate, vitamin B12, pantothenic acid, biotin and choline.* Washngton, DC: The National Academies Press; 1998.

18. Donald ZJ. Bioavailability of biotin given orally to humans in pharmacologic dose. *Am J Clin Nutr* 1999;**69**:504–8.

19. McMahon R. Biotin in metabolism and human biology. *Annu Rev Nutr* 2002;**22**:221–39.

20. Mock DM. Biotin: physiology, dietary sources and requirements. In: Caballero B, Allen L, Prentice A, editors. *Encyclopedia of human nutrition.* 2nd ed. London: Academic; 2004.

21. Sweetman L, Nhyan W. Inheritable biotin-treatable disorders and associated phenomena. *Ann Rev Nutr* 1986;**6**:314–43.

22. Chauhan J, Dakshinamurti K. Role of human serum biotinidase as biotin-binding protein. *Biochem J* 1988;**256**:256–70.

23. Zempleni J, Subhashinee S, Wijeratne K, Hassan Y. Biotin. *BioFactors* 2009;**35**:36–46.

24. Manthey K, Griffin J, Zempleni J. Biotin supply affects expression of biotin transporters, biotinylation of carboxylases, and metabolism of interleukin-2 in jurkat cells. *J Nutr* 2002;**132**:887–92.

25. Dakshinamurti K, Chalifour L, Bhullar R. Requirement for biotin and the function of biotin in cells in culture. In: Dakshinamurti K, Bhagavan HN, editors. *Biotin.* New York: New York Academy of Science; 1985. p. 38–55.

26. Zempleni J. Uptake, localization, and noncarboxylase roles of biotin. *Annu Rev Nutr* 2005;**25**:175–96.
27. Ballard T, Wolff J, Griffin J, Stanley J, Calcar S, Zempleni J. Biotinidase catalyzes debiotinylation of histones. *Eur J Nutr* 2002;**41**:78–84.
28. Daberkow R, White B, Cederberg R, Griffin J, Zempleni J. Monocarboxylate transporter 1 mediates biotin uptake in human peripheral blood mononuclear cells. *J Nutr* 2003;**133**:2703–6.
29. Wolf B, Heard G, Mc Voy J, Raetz H. Biotinidase deficiency: the possible role of biotinidase in the processing of dietary protein-bound biotin. *J Inherited Metab Dis* 1984;**7**:121–2.
30. Danford D, Munro H. The liver in relation to the B-vitamins. In: Arias J, Pepper H, Schachter D, Shafritz D, editors. *The liver: biology and pathobiology*. New York: Raven Press; 1982. p. 367–84.
31. McCormick D, Wright L. The metabolism of biotin and analogues. In: Florkin M, Stotz E, editors. *Metabolism of vitamins and trace elements*. Amsterdam: Elsevier; 1971. p. 81–110.
32. Zempleni J, McCormick D, Mock D. Identification of biotin sulfone, bisnorbiotin methyl ketone, and tetranorbiotin−/−sulf-oxide in human urine. *Am J Clin Nutr* 1997;**65**:508–11.
33. Forbes G, Forbes A. Micronutrient status in patients receiving home parenteral nutrition. *Nutrition* 1997;**13**:941–4.
34. Mock D, DeLorimer A, Liebman W. Biotin deficiency: an unusual complication of parenteral alimentation. *N Engl J Med* 1981;**304**:820–3.
35. Krause K, Bonjour J, Berlit P, Kochen W. Biotin status of epileptics. *Ann NY Acad of Sci* 1985;**447**:297–313.
36. Mock DM, Mock NI, Nelson RP, Lombard KA. Disturbances in biotin metabolism in children undergoing long-term anticonvulsant therapy. *J Pediatr Gastroenterol Nutr* 1998;**26**:245–50.
37. Said H. Cell and molecular aspects of human intestinal biotin absorption. *J Nutr* 2009;**139**:158–62.
38. Peters J. A separation of the direct toxic effects of dietary raw egg white powder from its action in producing biotin deficiency. *Br J Nutr* 1967;**21**:801.
39. Zempleni J, Mock D. Biotin biochemistry and human requirements. *J Nutr Biochem* 1999;**10**:128–38.
40. Gompertz D, Draffan G, Watts J, Hull D. Biotin-responsive 3-methylcrotonylglycinuria. *Lancet 2* 1971;**2**:22–4.
41. Wolf B. Disorders of biotin metabolism. In: Scriver C, Beaudet A, Sly W, Valle D, editors. *The metabolic and molecular bases of inherited disease*. 7th ed. New York: McGraw-Hill; 1995. p. 3151–77.
42. Holocarboxylase synthetase deficiency. *Genetics home reference [internet]*. Bethesda (MD): National Library of Medicine (US), The Library; 2018. Available at https://ghr.nlm.nih.gov/condition/holocarboxylase-synthetase-deficiency; (Accessed April 8, 2018).
43. Biotinidase Deficiency. *Genetics home reference [internet]*. Bethesda (MD): National Library of Medicine (US), The Library; 2018. Available at https://ghr.nlm.nih.gov/condition/biotinidase-deficiency; (Accessed April 8, 2018).
44. Wolf BJ, Jensen K, Huner G, Demirkol M, Baykal T, Divry P. Seventeen novel mutations that cause profound biotinidase deficiency. *Mol Genet Metab* 2002;**77**:108–11.
45. Ozand P, Gascon G, Al EM. Biotin-responsive basal ganglia disease: a novel entity. *Brain* 1998;**121**:1267–79.

46. Wolf B. Biotinidase deficiency: "if you have to have an inherited metabolic disease, this is the one to have". *Genet Med* 2012;**14**:565–75.
47. Aljabri M, Kamal N, Arif M, AlQaedi A, Santali E. A case report of biotin–thiamine-responsive basal ganglia disease in a Saudi child: is extended genetic family study recommended? *Medicine* 2016;**95**:4819.
48. Mardach R, Zempleni J, Wolf B, Cannon M. Biotin dependency due to a defect in biotin transport. *J Clin Invest* 2002;**109**:1617–23.
49. Kummer S, Hermsen D, Distelmaier F. Biotin treatment mimicking graves' disease. *N Engl J Med* 2016;**375**:704–6.
50. Piketty M, Polak M, Flechtner I, Gonzales-Briceño L, Souberbielle J. False biochemical diagnosis of hyperthyroidism in streptavidin-biotin-based immunoassays: the problem of biotin intake and related interferences. *Clin Chem Lab Med* 2016;**55**:780–8.
51. Barbesino G. Misdiagnosis of graves' disease with apparent severe hyperthyroidism in a patient taking biotin megadoses. *Thyroid* 2016;**26**:860–3.
52. Elston M, Sehgal S, Du Toit S, Yarndley T, Conaglen J. Facticious graves' disease due to biotin immunoassay interference—a case and review of the literature. *J Clin Endocrinol Metab* 2016;**101**:3251–5.
53. The FDA. *Warns that biotin may interfere with lab tests: FDA safety communication.* 2017. Available at: https://www.fda.gov/MedicalDevices/Safety/AlertsandNotices/ucm586505.htm. (Accessed March 26th, 2018).
54. Carlucci A. *Amphidinium carterae* assay for biotin. *Methods Enzymol* 1970;**18**:379–83.
55. Tanaka M, Izumi Y, Yamada H. Biotin assay using lyophilized and glycerol-suspended cultures. *J Microbiol Methods* 1987;**6**:237–45.
56. Baker H, Frank O, Matovitch V, Pasher I, Aaronson S, Hutner S, et al. A new assay method for biotin in blood, serum, urine, and tissues. *Anal Biochem* 1962;**3**:31–9.
57. Wright L, Skeggs H. Determination of biotin with *Lactobacillus arabinosus*. *Proc Soc Experi Biol Med* 1944;**56**:95–8.
58. Guilarte T. Measurement of biotin levels in human plasma using a radiometric-microbiological assay. *Nutr Rep Int* 1985;**31**.
59. Nisenson A, Sherwin L. Normal serum biotin levels in infants and adults: a modified assay method. *J Pediatr* 1966;**69**:134–6.
60. Mock D, Baswell D, Baker H, Holman R, Sweetman L. Biotin deficiency complicating parenteral alimentation: diagnosis, metabolic repercussions, and treatment. *J Pediatr* 1985;**106**:762–9.
61. Clevidence B, Marshall M, Canary J. Biotin levels in plasma and urine of healthy adults consuming physiological doses of biotin. *Nutr Res* 1988;**8**:1109–18.
62. Baker H. Assessment of biotin status: clinical implications. *Ann N Y Acad Sci* 1985;**447**:129–32.
63. Baumgartner E, Suormala T, Wick H, Bausch J, Bonjour J. Biotinidase deficiency associated with renal loss of biocytin and biotin. *Ann NY Acad Sci* 1985;**447**:272–87.
64. Dashinamurti K, Chauhan J. Regulation of bitoin enzymes. *Ann Rev Nutr* 1988;**8**:211–33.
65. Horsburgh T, Gompertz D. A protein-binding assay for measurement of biotin in physiological fluids. *Clinica Chimica Acta* 1978;**82**:215–23.
66. Sanghvi R, Lemons R, Baker H, Thoene J. A simple method for determination of plasma and urinary biotin. *Clin Chim Acta* 1982;**124**:85–90.
67. Mock D, DuBois D. A sequential, solid-phase assay for biotin in physiologic fluids that correlates with expected biotin status. *Anal Biochem* 1986;**153**:272–8.

68. Hansen S, Holm J. Quantification of biotin in serum by competition with solid-phase biotin for binding to peroxidase-avidin conjugate. *Clin Chem* 1989;**35**:1721–2.

69. Mock D. Determinations of biotin in biological fluids. *Methods Enzymol* 1997;**279**:262–75.

70. Zempleni J, Mock D. Advanced analysis of biotin metabolites in body fluids allows a more accurate measurement of biotin bioavailability and metabolism in humans. *J Nutr* 1999;**129**:494S–7S.

71. Harthé C, Claustrat B. A sensitive and practical competitive radioassay for plasma biotin. *Ann Clin Biochem* 2003;**40**:259–63.

72. Wakabayashi K, Kodama H, Ogawa E, Sato Y, Motoyama K, Suzuki M. Serum biotin in Japanese children: enzyme-linked immunosorbent assay measurement. *Pediatr Int* 2016;**58**:872–6.

73. Mock D, Lankford G, Mock N. Biotin accounts for only half of the total avidin-binding substances in human serum. *J Nutr* 1995;**125**:941–6.

74. Bogusiewicz A, Stratton S, Ellison D, Mock D. Biotin accounts for less than half of all biotin and biotin metabolites in the cerebrospinal fluid of children. *Am J Clin Nutr* 2008;**88**:1291–6.

75. Livaniou E, Costopoulou D, Vassiliadou I, Leondiadis L, Nyalala J, Ithakissios D, et al. Analytical techniques for determining biotin. *J Chromatogr A* 2000;**881**:331–43.

76. Chastain J, Bowers-Komro D, McCormick D. High-performance liquid chromatography of biotin and analogues. *J Chromatogr* 1985;**330**:153–8.

77. Suormala T, Regula Baumgartner E, Bausch J, Holick W, Wick H. Quantitative determination of biocytin in urine of patients with biotinidase deficiency using high-performance liquid chromatography (HPLC). *Clinica Chimica Acta* 1988;**177**:253–69.

78. Yagi S, Nishizawa M, Ando I, Oguma S, Sato E, Imai Y. A simple and rapid ultra-high-performance liquid chromatography–tandem mass spectrometry method to determine plasma biotin in hemodialysis patients. *Biomed Chromatogr* 2016;**30**:1285–90.

79. Mock D. Biotin status: which are valid indicators and how do we know? *J Nutr* 1999;**129**:498S–503S.

80. Nagamine T, Saito S, Yamada S, Kaneko M, Uehara M, Takezawa L. Clinical evaluation of serum biotin levels and biotinidase activities in patients with various liver diseases. *Nihon Shokakibyo Gakkai Zasshi* 1990;**87**:1168–74.

81. Fujiwara M, Ando I, Yagi S, Nishizawa M, Oguma S, Satoh K. Plasma levels of biotin metabolites are elevated in hemodialysis patients with cramps. *Tohoku J Exp Med* 2016;**239**:263–7.

82. Livaniou E, Evangelatos G, Ithakissios D, Yatzidis H, Koutsicos D. Serum biotin levels in patients undergoing chronic hemodialysis. *Nephron* 1987;**46**:331–2.

83. Trüeb R. Serum biotin levels in women complaining of hair loss. *Int J Trichology* 2016;**8**:73–7.

84. Eng W, Giraud D, Schlegel V, Wang D, Lee B, Zempleni J. Identification and assessment of markers of biotin status in healthy adults. *Brit J Nutr* 2013;**110**:321–32.

85. Wright L, Cresson E, Driscoll C. Bioautography of biotin and certain related compounds. *Proc Soc Exp Biol Med* 1954;**86**:480–3.

86. Wolf B, Grier R, Parker W, Goodman S, Allen R. Deficient biotinidase activity in late-onset multiple carboxylase deficiency. *N Engl J Med* 1983;**308**:161.

87. Gaudry M, Munnich A, Saudubray J, Ogier H, Mitchell G, Marsac C. Deficient liver biotinidase activity in multiple carboxylase deficiency. *Lancet* 1983;**2**(8346):397.

88. Wolf B, Hymes J, Heard G. Biotinidase. In: Wilchek M, Bayer E, editors. *Methods in enzymology*. 184. Academic Press; 1990. p. 103–11.

89. Landau Y, Waisbren S, Chan L, Levy H. Long-term outcome of expanded newborn screening at Boston children's hospital: benefits and challenges in defining true disease. *J Inherit Metab Dis* 2017;**40**:209–18.

90. Knappe J, Bruemmer W, Biederbick K. Purification and properties of biotinidase from swine kidney and *Lactobacillus casei*. *Biochem Z* 1963;**338**:599–613.

91. Wolf B, Grier RE, Allen RJ, Goodman SI, Kien CL. Biotinidase deficiency: the enzymatic defect in late-onset multiple carboxylase deficiency. *Clinica Chimica Acta* 1983;**131**:273–81.

92. Heard G, Secor McVoy J, Wolf B. A screening method for biotinidase deficiency in newborns. *Clin Chem* 1984;**30**:125–7.

93. Bratton A, Marshall E, Babbitt D, Hendrickson A. A new coupling component for sulfanilamide determination. *J Biol Chem* 1939;**128**:537–50.

94. Wolf B, Secor MVJ. Sensitive radioassay for biotinidase activity: deficient activity in tissues of serum biotinidase-deficient individuals. *Clin Chim Acta* 1983;**135**:275281.

95. Hayakawa K, Oizumi J. Human serum biotinidase is a thiol-type enzyme. *J Biochem* 1988;**103**:773–7.

96. Turner C, Dalton N. Abstracts of the IX international congress on inborn errors of metabolism. Brisbane, Australia, 2–6 September 2003. *J Inherit Metab Dis* 2003;**26**(Suppl 2):1–237.

97. Sankaralingam A, Turner C, Dalton R. Screening for biotinidase deficiency using electrospray tandem mass spectrometry (TMS). *J Inherit Metab Dis* 1999;**22**(Suppl 1):146.

98. Wastell H, Dale G, Bartlett K. A sensitive fluorimetric rate assay for biotinidase using a new derivative of biotin, biotinyl-6-aminoquinoline. *Anal Biochem* 1984;**140**:69–73.

99. Fingerhut R, Dame T, Olgemöller B. Determination of EDTA in dried blood samples by tandem mass spectrometry avoids serious errors in newborn screening. *Eur J Pediatr* 2009;**168**:553–8.

100. Dalton R, Turner C. Simultaneous high throughput MS/MS measurement of dried blood spot proteins, enzyme activities, and metabolites for diagnosis of IEM and haemoglobinopathies. *Clinica Chimica Acta* 2010;**411**:896–7.

101. Jakob C, Sweetman L, Nyhan W, Packman S. Stable isotope dilution analysis of 3-hydroxyisovaleric acid in amniotic fluid: contribution to the prenatal diagnosis of inherited disorders of leucine catabolism. *J Inherit Met Dis* 1984;**7**:15–20.

102. Chalmers R, Lawson A, editors. *Organic acids in man: analytical chemistry, biochemistry and diagnosis of organic acidurias*. London: Chapman and Hall; 1982.

103. Horvath T, Matthews N, Stratton S, Mock D, Boysen G. Measurement of 3-hydroxyisovaleric acid in urine from marginally biotin-deficient humans by UPLC-MS/MS. *Anal Bioanal Chem* 2011;**401**:2805–10.

104. Blau N, Diran M, Gibson K, editors. *Laboratory guide to the methods in biochemical genetics*. Heidelberg: Springer-Verlag; 2008.

105. Mock D, Jackson H, Lankford G, Mock N, Weintraub S. Quantitation of urinary 3-hydroxyisovaleric acid using deuterated 3-hydroxyisovaleric acid as internal standard. *Biomed Environ Mass Spectrom* 1989;**18**:652–6.

106. Costa C, Struys E, Bootsma A. Quantitative analysis of plasma acylcarnitines using gas chromatography chemical ionization mass fragmentography. *J Lipid Res* 1997;**38**:173–82.

107. Bogusiewicz A, Horvath T, Stratton S, Mock D, Boysen G. Measurement of acylcarnitine substrate to product ratios specific to biotin-dependent carboxylases offers a combination of indicators of biotin status in humans. *J Nutr* 2012;**142**:1621–5.

108. Piketty M, Prie D, Sedel F, Bernard D, Hercend C, Chanson P. High-dose biotin therapy leading to false biochemical endocrine profiles: validation of a simple method to overcome biotin interference. *Clin Chem Lab Med* 2017;**55**:817–25.
109. Chun K. Biotin interference in diagnostic tests. *Clin Chem* 2017;**63**:619–20.
110. Grimsey P, Frey N, Bendig G, Zitzler J, Lorenz O, Kasapic D, et al. Population pharmacokinetics of exogenous biotin and the relationship between biotin serum levels and in vitro immunoassay interference. *Int J Pharmacokinet* 2017;**2**(4):247–56.

Methods for assessment of folate (Vitamin B₉) 11

Agata Sobczyńska-Malefora,

Nutristasis Unit, Viapath, St. Thomas' Hospital, London, United Kingdom
Faculty of Life Sciences and Medicine, King's College London, London, United Kingdom

Chapter Outline

Laboratory Assessment of Vitamin Status. https://doi.org/10.1016/B978-0-12-813050-6.00011-5

Historical Overview

The double ringed structure of folate (pteroylglutamic acid) was first isolated from the wing pigment of Brimstone and Cabbage White butterflies, hence the term pteridine which is derived from the Greek word *pteron* (wing).[1] Purified folate was isolated in 1941 from four tons of spinach and named after the Latin word *folium* (leaf).[2] A few years later folate was also isolated from liver extracts and *Corynebacterium* species cultures as a factor that supported *Lactobacillus rhamnosus* (previously known as *Lactobacillus casei*) growth.[3–5] The isolation of pteroylglutamic acid led to a series of studies in humans, monkeys, chicks, and microorganisms. In the early 1930s, the British physician Lucy Wills was the first to recognize macrocytic anemia in poor pregnant textile workers in India.[6] This anemia was very similar to pernicious anemia that had previously been described in elderly people and prompted her to call it "pernicious anemia of pregnancy". However, this anemia regressed when the pregnancy was over, while in the elderly it led to death. Wills continued her research using rats and monkeys by feeding them diets that were deficient in proteins and vegetables, and subsequently curing them with yeast extracts.

Soon, it became recognized that the administration of folate not only improved anemia, but also enhanced the growth of existing tumors; thus folate metabolism was considered a promising target for anticancer drug design, leading to the development of folate antagonists, such as methotrexate.[7, 8] More than half a century after the isolation of folate, its importance in the prevention of spina bifida was also demonstrated.[9–11] Periconceptional folic acid supplementation began and was followed by recommendations that all women of a reproductive age should consume 400 µg of folic acid daily from supplements or fortified foods.[12] Mandatory fortification of flour began in many countries, including the United States in 1998 and Canada in 2000.[12, 13]

More recently, a meta-analysis published in 1995 from 27 studies involving >4000 patients showed an association between folates and vascular disease.[14] The understanding of the importance of folates in the prevention and treatment of a wide range of disorders has resulted in an increased interest in folate disease-related research.[15]

Structure and Function

The structures of some of the most important natural folates, and the synthetic compound folic acid, are shown in Fig. 11.1. The structure consists of a pteridine ring which is linked through a methylene bridge to p-aminobenzoic acid and then to

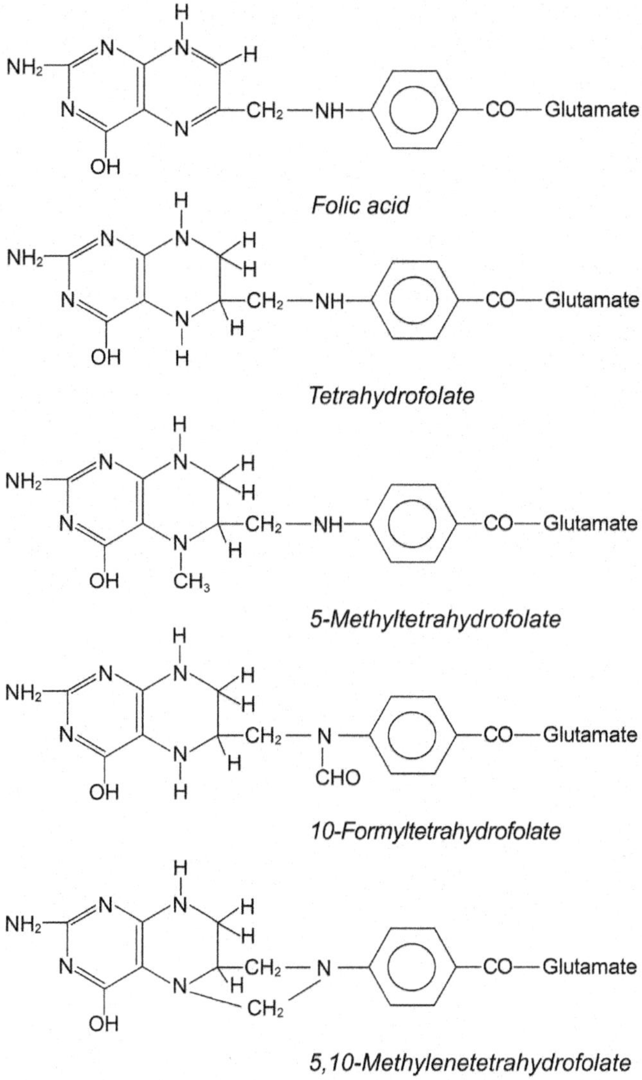

FIG. 11.1

The chemical structure of folic acid and some natural folates.

L-glutamic acid residues. Folates differ from each other in the state of the oxidation of the pteridine ring, the number of glutamic acid residues that are linked one to another via γ-glutamyl linkages to form an oligo-γ-glutamyl chain, and also in the nature of one of the carbon substituents at the N5 and N10 positions. The pteridine ring of folic acid is not reduced; hence this form is resistant to chemical oxidation. To be enzymatically active, the pteridine ring of folic acid must be reduced to dihydrofolate (DHF) by the action of a dihydrofolate reductase (DHFR). In cells, this enzyme reduces DHF to tetrahydrofolate (THF). In the reduced form, cellular folates function conjugated to a polyglutamate chain. There are a mixture of unsubstituted polyglutamyl tetrahydrofolates and various substituted one-carbon forms of THF such as 5-methylterahydrofolate (5-MTHF) or 5-formylTHF.[16] The reduced forms of folates, particularly the unsubstituted dihydro- and tetrahydro-forms, are chemically unstable. The substituted forms are susceptible to oxidative chemical rearrangements and as a consequence may lose their activity. The main source of one carbon units comes from the C3 of the amino acid serine, which is synthesized in three enzymatic steps by the glycolytic pathway.[17] The carbon unit is transferred to THF in a reaction catalyzed by serine hydroxymethyltransferase, generating 5,10-methyleneTHF. In the cytoplasm, 5,10-methyleneTHF can transfer its methylene group to deoxy-uridine monophosphate, to synthesize deoxythymidine monophosphate which is then utilized for DNA biosynthesis and repair (Fig. 11.2). 5,10-MethyleneTHF can also be oxidized to 5,10-methenylTHF, which in turn is converted to 10-formylTHF

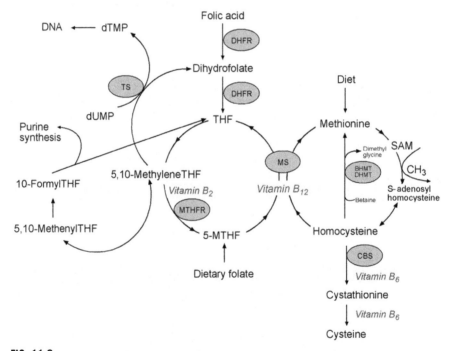

FIG. 11.2

Folate and homocysteine metabolism. THF, tetrahydrofolate; 5-MTHF, 5-methyltetrahydrofolate; dUMP, deoxyuridine monophosphate; dTMP, deoxythymidine monophosphate; TS, thymidylate synthase; MTHFR, methylene tetrahydrofolate reductase; MS, methionine synthase; CBS, cystathionine beta-synthase; SAM, S-adenosyl methionine; BHMT, betaine-homocysteine methyltransferase; DHMT, dimethylthetin-homocysteine methyltransferase.

(Fig. 11.2). In rapidly growing cells, 10-formylTHF is required for the synthesis of purines and hence for DNA formation. Pyrimidine biosynthesis generates DHF which is reduced to THFs, while purine biosynthetic reactions generate THF (Fig. 11.2). Methylenetetrahydrofolate reductase (MTHFR) and vitamin B_2, which serves as a cofactor, are required for the conversion of 5,10-methyleneTHF to 5-MTHF. This reaction is irreversible as opposed to other reactions in the folate cycle. Following 5-MTHF synthesis, the remaining 5,10-methyleneTHF is used for the synthesis of thymidylate and purines.

Absorption, Cellular Uptake, and Catabolism

The naturally occurring folates found in foods are a mixture of reduced folate polyglutamates. Before absorption of dietary folate can take place, polyglutamated forms of folates must be hydrolyzed in the gut lumen to monoglutamate forms by folylpolyglutamate carboxypeptidase. Monoglutamates are then transported to the mucosal cells of the jejunum by a saturable active energy-dependent carrier-mediated process at physiological concentrations and by passive diffusion at higher concentrations.[1] Absorbed folate monoglutamates are converted to 5-MTHF during the transit through the intestinal mucosa before they reach the hepatic portal vein. 5-MTHF enters the plasma circulation via the liver and in this form is taken up by cells via reduced folate carrier (RFC) or through receptor-mediated endocytosis (proton-coupled folate transporter (PCFT)). To be retained by the cell, folate must be converted into polyglutamates. 5-MTHF and synthetic folic acid are poor substrates for the conversion to polyglutamates.[18] Therefore for 5-MTHF plasma folate to be conjugated, it must have its methyl group removed once it enters the cell. Demethylation can only be performed by the vitamin B_{12}-dependent methionine synthase (MS).[19] If the concentration of 5-MTHF polyglutamate in a cell is low, as in folate deficiency or after cell division has used the intracellular folate, the utilization and retention of the plasma 5-MTHF monoglutamate is high. Folic acid is first reduced to DHF via DHFR and then to THF. Folic acid has poor affinity for RFC, hence its transport to the tissue occurs mainly via PCFT.[18] Not all tissues express PCFT.

One product of folate catabolism is p-aminobenzoylglutamate, which is excreted as p-acetamidobenzoylglutamate. The urinary excretion of this catabolite in healthy females is $200 \pm 10\,\mu g/day$, rising in pregnancy to $600\,\mu g/day$.[17] Urinary excretion of intact folates is very low, reflecting high renal retention.

Distribution and Tissue Concentration

Folates of bacterial, plant, and animal origin are predominantly in the form of folate polyglutamates. The composition of folates differs between tissues, and between extracellular and intracellular locations. The polyglutamate forms are found mainly in cells. In *L. rhamnosus*, almost all folate is in polyglutamate forms. In orange juice, 40%–45% of folates are present as 5-methyltetrahydro derivatives of pentaglutamate, 10%–15% as tetraglutamate and diglutamate, and 30%–40% as monoglutamate. Human red cells contain folate pentaglutamate and some tetra- and hexaglutamates. The proportion of polyglutamates is in the range of 60%–90%. In monkeys and sheep, hexaglutamate appears to predominate with lesser amounts of penta- and heptaglutamates.[1] The distribution of folate coenzymes also differs for cellular organelles. The cytosol and mitochondria of liver cells contain high concentrations of folate, while nuclei contain low concentrations. Moreover, the distribution of folate-dependent enzymes is linked to their function. Thus cytoplasmic folate coenzymes participate in the biosynthesis of purines and thymidylate and in methyl group biogenesis, while mitochondrial folate coenzymes participate in protein synthesis.[12, 19]

In human plasma, folate is typically present at 3–20 µg/L and in red cells at 150–600 µg/L. Cerebrospinal fluid (CSF) contains around 15–35 µg/L of folate, whereas ~40 µg/L is typical in saliva. White blood cells contain 60–123 µg/L of folate. There is 4–11 µg of folate per gram of wet tissue in a normal liver.[1]

The distribution of various folate coenzymes is also dependent on genotype. Fazili et al. showed that the distribution of folates differed by MTHFR variant.[20] They found that in individuals with the wild-type genotype (cytosine/cytosine (C/C)), the major folate was 5-MTHF (90% of total folate) with a smaller proportion of 5-formylTHF (10%). Individuals with the (cytosine/thymine (CT)) genotype displayed the same folate pattern. However, in individuals with the T/T genotype, significant amounts of 5,10-methenylTHF and THF were also detected. They found that 30% of folates in T/T individuals are made up of formylated folates.

Sources and Stability

The main natural sources of dietary folates are green leafy vegetables, fruits, and dairy products (Table 11.1).[21]

Diets containing adequate amounts of fresh vegetables are good sources of folates. Folates are unstable and their activity is lost over several days or weeks. Therefore appropriate storage, processing, and preparation of food may help protect from a loss of activity. Conversely, folic acid, a synthetic form of folate, is highly stable for months or even years.

Enterohepatic recirculation of folate is high with as much as 100 µg of folate a day being reexcreted by bile. Deficiency of vitamin B$_{12}$, vitamin C, iron, and zinc can reduce the efficiency of the utilization of this process.[22] Folate utilization is also affected by certain drugs. It is also known that microflora in the large intestine produce large quantities of folate. It has been shown that physiological doses of natural folate are absorbed across the intact large intestine in humans.[23] $^{13}C_5$-Glutamyl-5-formylTHF was infused during a colonoscopy procedure in six healthy adults and>3 weeks later, the same compound was administered through an intravenous injection. The rate of folate absorption from the large intestine was determined by the appearance of labeled 5-MTHF in plasma. The rate of folate absorption infused into the cecum was 0.7 nmol/h and 2.8 nmol/h from the intravenous injection.[23]

Table 11.1 Examples of Foods High in Folate

Food Type	Folate Content (µg) Per 100 g Raw Edible Matter
Broccoli (sprouting)	90
Cabbage	44
Spinach	150
Broad beans	145
Lettuce	55
Cauliflower	66
Beetroot	150
Carrots (old)	12
Potato	35
Orange	31
White flour	31
Brown flour	51

Requirements and Recommendations in Humans

Following on from the historical experiment by Victor Herbert which demonstrated that a folate-deficient diet leads to the development of early megaloblastic changes after 133 days,[24] folate requirements were defined as the amount of folate needed daily to produce a hematological response in folate-deficient megaloblastic anemia.[1] This value was set at between 50 and 200 µg of folate daily. However, the bioavailability of folates from natural foods is estimated to be around 50%[25, 26] and with advances in folate science, recommendations for folate intake have changed over the years. Today, folate recommendations and the units used to express folate intake differ between countries. In the United Kingdom, the reference nutrient intake (RNI) is used to define the amount of a nutrient that is considered sufficient to meet the requirements of 97.5% of the population. For infants and children, the RNIs per day are 0–12 months, 50 µg; 1–3 years, 70 µg; 4–6 years, 100 µg; 7–10 year, 150 µg; for children 11 years and above and adults the RNI for folate is 200 µg/d.[27] For all women who could become pregnant, the UK government recommends to take 400 µg/d of folic acid prior to conception and up until the twelfth week of pregnancy. Women with a history of a previous neural tube defect (NTD)-affected pregnancy and those who are diabetic are advised to take 5 mg/d of folic acid also prior to conception and until the twelfth week of pregnancy.

In the United States, Canada, Europe, and many other countries worldwide, 300–400 µg/d of folate is recommended for adults. For pregnancy and lactation these vary between 400 and 600 µg/d. A tolerable Upper Intake Level (UL), which represents the highest level of a nutrient that is likely to pose no risk of adverse health effects to almost all individuals in the general population, was set at 1 mg/d of folic acid for adults.[28] The UL was based on the risk of progression of neurological symptoms in vitamin B$_{12}$ deficient patients.

Causes and Clinical Signs of Deficiency

The most common cause of folate deficiency is an inadequate dietary intake of folate. This may be due to selective dietary habits, social circumstances, eating disorders, alcoholism, or parenteral nutrition insufficient in folate supplements. Coeliac disease, tropical sprue, Crohn's disease, and chronic pancreatitis all lead to malabsorption of nutrients and are another major causative factor for folate deficiency. Increased folate turnover such as is seen in pregnancy, breastfeeding, skin conditions, and hematological disorders may also result in folate deficiency. Furthermore, a number of folate antagonist drugs, e.g., aminopterin, methotrexate (amethopterin), pyrimethamine, trimethoprim, and triamterene will produce folate deficiency by inhibiting folate enzymes. Anticonvulsants, antituberculosis drugs, and oral contraceptives may also contribute to folate deficiency.[29] There are also many inborn errors of folate metabolism, e.g., deficiencies in DHFR or MTHFR enzymes,[30] leading to severe folate deficiency.

An acute folate deficiency can occur after administration of folate antagonists. It may cause diarrhea and nausea, ulceration of the mouth and pharynx, hair loss, and dermatitis. In chronic folate deficiency, common symptoms include a loss of energy and tiredness. A sore mouth and tongue may occur. In children, growth and puberty may be delayed. Chronic folate deficiency will lead to megaloblastic anemia (refer to Chapter 12). Neurological damage may also occur as a result of folate deficiency.[31] Hereditary folate deficiency often leads to severe anemia and mental retardation. Folate deficiency is also the most common cause of elevated plasma homocysteine.

Homocysteine

Historical overview

Homocysteine was first described by Butz and du Vigneaud in 1932.[32] They synthesized it by treating methionine with sulfuric acid. The amino acid which they obtained was similar in structure to cysteine, thus they named it homocysteine. In 1933, the first case of homocystinuria was published.[33] In this case an eight-year-old boy being treated at the Massachusetts General Hospital had mental impairment, dislocated lenses, and skeletal abnormalities. He died with symptoms of a stroke in 1932. The postmortem revealed a thrombosis of the carotid artery and a cerebral infarct. The carotid arteries were narrowed by atherosclerotic plaques such as those seen in the elderly. Throughout the 1930s and up to the early 1960s, the identification of the metabolites, enzymes, and reactions involved in the metabolism of homocysteine and methionine was heavily investigated.[34]

In 1962, homocysteine was identified in the urine of some mentally impaired children.[35, 36] Two years later, genetic defects of cystathionine β-synthase (CBS) causing homocystinuria and grossly elevated plasma concentrations of homocysteine were identified.[37] Patients with this CBS defect were found to suffer from premature arteriosclerosis and frequent thromboembolisms. More than50% of these patients had recurrent cardiovascular events and about 25% died of such events before the age of 30 years.

In 1969 McCully first proposed that homocysteine causes atherosclerosis.[38] His hypothesis was based on the finding of atherosclerotic plaques in autopsies of young people with homocystinuria, including the first homocystinuria case published which was restudied by McCully.[33, 38, 39] As the accumulation of homocysteine was caused by different enzyme abnormalities in these patients, McCully concluded that homocysteine causes arteriosclerosis plaques due to a direct effect of the amino acid on the cells and tissues of the arteries. In 1976, a child with the third major cause of homocystinuria—MTHFR deficiency—was found to have similar arteriosclerotic plaques throughout the body which independently confirmed his earlier conclusions.[40] The occurrence of premature atherothrombotic events in individuals with homocystinuria prompted the hypothesis that more modest elevations of homocysteine may be relevant to cardiovascular disease in the general population.[38]

Homocysteine chemistry

Homocysteine is a sulfur-containing amino acid that is closely related to methionine and cysteine (Fig. 11.3).

Homocysteine is not present in naturally occurring proteins, but is an important metabolic branch point in the pathway from methionine to cysteine (Fig. 11.2). Thus homocysteine found in organisms is formed during the metabolism of the essential amino acid methionine in the methylation cycle. In cells, homocysteine is normally present in its reduced form. In plasma, only about 1% of homocysteine exists in the free reduced form. Reduced homocysteine has a highly reactive free thiol group, which participates in redox reactions and is susceptible to auto-oxidation at a physiological pH, thereby forming a disulfide bond between two homocysteine molecules (homocystine) or a mixed disulfide with cysteine (Fig. 11.3). Alternatively, reduced homocysteine can form disulfide bonds with proteins. About 70% of plasma homocysteine is bound to albumin.[41] Three homocysteine fractions can therefore be measured: reduced homocysteine (1% of the total), oxidized homocysteine (homocystine, homocysteine disulfide-cysteine) which represents about 10%–20% of the total homocysteine, and a protein-bound fraction.[42] Because of the instability, high variability and low concentrations of the free homocysteine fraction, total plasma homocysteine (tHcy) is measured and used in clinical practice. Furthermore, tHcy correlates well with free homocysteine.[43]

Homocysteine metabolism

Demethylation of dietary methionine during essential methylation reactions produces homocysteine. Approximately 50% of homocysteine is remethylated via the remethylation cycle to generate methionine (Fig. 11.2). In every cell, methionine is partitioned between protein synthesis and the formation of S-adenosylmethionine (SAM). The formation of SAM is mediated by adenosine triphosphate (ATP) and

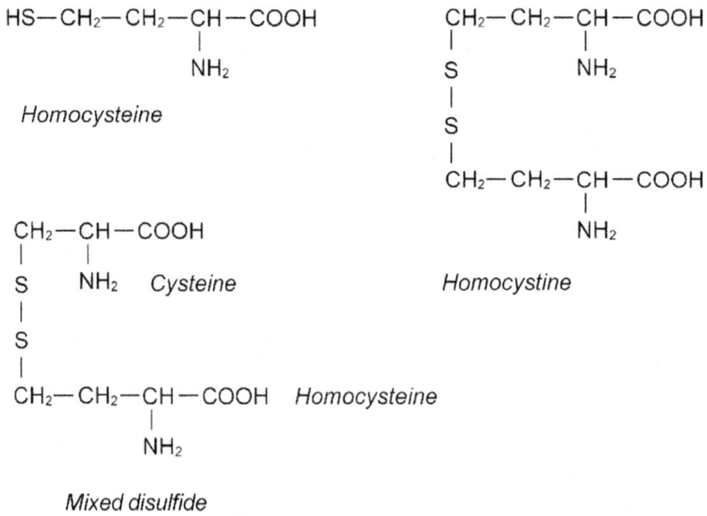

HS—CH$_2$—CH$_2$—CH—COOH
 |
 NH$_2$

Homocysteine

CH$_2$—CH$_2$—CH—COOH
 | |
 S   NH$_2$
 |
 S
 |
CH$_2$—CH$_2$—CH—COOH
 |
 NH$_2$

Homocystine

CH$_2$—CH—COOH
 | |
 S   NH$_2$ *Cysteine*
 |
 S
 |
CH$_2$—CH$_2$—CH—COOH   *Homocysteine*
 |
 NH$_2$

Mixed disulfide

FIG. 11.3

The chemical structures of homocysteine, homocystine, and mixed disulfide.

other enzymes such as methionine adenosine transferase. SAM is the principal biological methyl group donor in humans and is required for many methylation reactions (e.g., amino acid residues in proteins, DNA, RNA, phospholipids, and small proteins). There have been over 115 different cellular methyltransferase reactions identified in mammals in which SAM serves as a methyl donor.[11] S-adenosylmethionine has a positively charged sulfonium ion and carbon atoms that are adjacent to the sulfur atom and have electrophilic properties. The principle of these methyltransferases reactions is that the positively charged sulfonium provides the adjoining electron-poor methyl group which then attacks electron-rich nucleophiles (e.g., DNA, RNA).[44] Methyltransferases transfer methyl groups in this cycle to many acceptors such as proteins (myelin), phospholipids, or DNA (Fig. 11.2).

A product of all methylation reactions is S-adenosylhomocysteine (SAH), which is then hydrolyzed by S-adenosyl homocysteine hydrolase to homocysteine and adenosine in a reversible reaction with a thermodynamic equilibrium that favors SAH synthesis rather than hydrolysis. In particular, SAH is a potent competitor to SAM at different binding sites and can therefore inhibit methylation. When homocysteine increases, SAH also increases. Decreased remethylation of homocysteine to methionine and SAM may impair the methylation reactions. If the methionine balance is negative, and in the presence of low concentrations of SAM, homocysteine is primarily directed toward the remethylation pathway to form methionine by the vitamin B$_{12}$-dependent MS-mediated reaction. During this vitamin B$_{12}$-dependent cycle, a methyl group from 5-MTHF is donated to homocysteine to reform methionine. An additional minor remethylation pathway present in the liver utilizes betaine as a methyl donor to form dimethylglycine in a reaction catalyzed by betaine-homocysteine methyltransferases (BHMT) and dimethylthetin-homocysteine methyltransferase (DHMT) (Fig. 11.2).[42]

An alternative route of homocysteine utilization is catabolism via the transsulfuration pathway, where homocysteine is condensed with serine to form cystathionine in a vitamin B_6-dependent reaction (Fig. 11.2).[45] This reaction is irreversible and a deficiency of CBS results in the abnormal accumulation of homocysteine. Cystathionine is then hydrolyzed to cysteine and α-ketobutyrate by γ-cystathionase. The transsulfuration pathway has a very limited tissue distribution—it is restricted to the liver, kidney, intestine, and pancreas.

Hyperhomocysteinemia determinants

Nutritional deficiencies and/or inborn errors in homocysteine metabolism can lead to hyperhomocysteinemia. Severe hyperhomocysteinemia (defined as tHcy >100 µmol/L) is usually caused by rare inborn errors of metabolism. Deletions in the CBS enzyme are the most common genetic cause of severe hyperhomocysteinemia (1 per 300,000 live births). It is an autosomal recessive disorder characterized by severely elevated concentrations of homocysteine and methionine in blood and urine.[46] Around 1% of the general population are heterozygous for CBS deficiency, and this may be associated with mild elevations of plasma homocysteine.[47] A more frequent polymorphism, affecting 5%–15% of the population, is a thermolabile variant of MTHFR which arises due to a C-T substitution in a gene (C677T) encoding this enzyme. Individuals with this polymorphism have a MTHFR activity of about 50% of the wild type[22] and about 25% higher tHcy concentrations than individuals with the CC variant. The frequency of the C677T polymorphism of MTHFR varies between racial and ethnic groups. A low prevalence of the T/T genotype has been reported for African-Americans, while in Caucasians there are geographical variations with the highest prevalence being reported for South Europeans.[48, 49] The MS 2756A to G variant has a minor impact on homocysteine.[50] The dietary intake of folate, vitamin B_{12} and B_6 are the main nutritional determinants of tHcy.[51–53] Besides vitamin deficiencies and genetic defects, there are many other factors leading to an increased homocysteine concentration (Table 11.2). Therefore an elevated homocysteine concentration is a frequent abnormality with multiple possible etiologies, rather than being caused by a single factor.

Clinical Application of Measurement

Serum folate, red cell folate (RCF), and tHcy are the most commonly used laboratory markers for the assessment of folate status. Serum folate reflects recent dietary intake and is the earliest marker suggestive of suboptimal folate status. RCF is an indicator of long-term folate status (the preceding 120 days) and represents folate accumulated during red cell synthesis. Plasma homocysteine is a functional marker of folate status. Folate deficiency is the most common reason for elevated tHcy. Others reasons for elevated tHcy concentrations include deficiencies of

Table 11.2 Determinants of Elevated Plasma Homocysteine

Genetic factors: defects and polymorphisms in CBS, MS and MTHFR genes

Lifestyle: smoking, high alcohol and coffee consumption intake, suboptimal nutrition leading to vitamin B$_2$, B$_6$, B$_9$, B$_{12}$ deficiencies, sedentary lifestyle, stress

Age: plasma tHcy increases throughout the life. The higher concentrations seen in the elderly may be a consequence of an increased prevalence of intestinal malabsorption, reduced enzymatic activity, reduced kidney function, and other physiological age-related changes

Gender: males have higher homocysteine levels than females

Hormonal changes: tHcy gradually increases after the menopause

Drugs: nitrous oxide, methotrexate, anticonvulsants, oral antidiabetic drugs, lipid-lowering drugs, sex hormones

Clinical conditions affecting homocysteine metabolism: gastrointestinal disturbances, renal failure, diabetes mellitus, HIV/AIDS, hypothyroidism

vitamins B$_{12}$, B$_6$, B$_2$, and renal impairment. Elevated homocysteine has been also implicated in many other disease states.

Other laboratory markers of folate status with potential clinical use include unmetabolized folic acid (UFA), urinary folate and folic acid, serum and urinary para-aminobenzoylglutamate (pABG) and para-acetamidobenzoylglutamate (apABG), DNA methylation, uracil misincorporation into DNA and micronuclei, plasma and urinary formiminoglutamate, and aminoimidazole carboxamide riboside.[3, 18] The availability of these markers is currently limited and clinical utility has not been fully explored. However, a test that requires a 24 h urine collection will not likely be used widely in a diagnostic setting because of the inherent inter- and intraindividual variability, relative lack of sensitivity, and cumbersome sample collection, even though unique insights may be gained when used in combination with serum and RCF measurements.[18] The presence of unmetabolized serum folic acid has been linked with various adverse health effects including carcinogenesis, the masking of vitamin B$_{12}$ deficiency, childhood obesity, and both type 2 and gestational diabetes.[54] The measurement of this marker may be clinically useful, especially in certain population groups such as pregnant women or in the elderly. Global DNA methylation correlates well with serum folate and is potentially a sensitive tool for the identification of folate depletion.[55, 56] Studies have shown that, for example, a diet containing 50–120 µg of folate per day for 7–9 weeks was sufficient to induce genomic DNA hypomethylation of blood mononuclear cells, and that folate repletion over several weeks caused hypomethylation to return to normal.[57, 58] Based on their study, Jacobs et al. suggested that "normal" folate concentrations (>3 µg/L) are borderline at best and are probably too low to maintain DNA methylation, since tHcy begins to rise recognizably before the concentration of plasma folate falls into a true deficient state.[57] Moreover, rises in plasma tHcy appear earlier than in DNA hypomethylation.[55]

The clinical utility of the most commonly used markers of folate status is briefly described as follows.

Serum/Plasma Folate

Serum folate is the most common first-line test used for folate assessment (some types of plasma are also suitable). This marker reflects recent dietary intake and is the earliest marker suggestive of suboptimal folate status. Folate concentrations increase following folate or folic acid ingestion for up to 2 h and then decline rapidly. It has been suggested that this rapid clearance indicates that fasting folate concentrations represent reduced folates released by tissues.[18] However, in a typical diagnostic setting, nonfasting samples are usually received for analysis. Therefore it is recommended that serum folate results are interpreted in light of clinical symptoms and other results available at the time of presentation such as full blood count (FBC), as well as previous folate results if available. In most patients, serum results reflect folate status well and correlate well with RCF and tHcy. A decline in concentration over time is also worth addressing since the risk of NTD such as spina bifida,[59] certain types of cancers,[60] Alzheimer's disease, and cardiovascular disease[61–63] increases in folate insufficiency.

High serum folate warrants vitamin B_{12} status assessment in view of the potential for the masking of vitamin B_{12} deficiency. In vitamin B_{12}-deficient cells, it is possible that folate becomes metabolically trapped as 5-MTHF and cannot be converted to THF, and is therefore not available for purine and pyrimidine synthesis. This is because in the absence of MS activity (the enzyme is vitamin B_{12} dependent), unusable 5-MTHF accumulates, producing a pseudo folate-deficient state.[64] The methyl-trap hypothesis suggests that cells, in an attempt to resynthesize methionine and SAM, respond inappropriately by increasing the production of 5-MTHF. The methyl-trap explains why vitamin B_{12} deficiency produces an apparently identical megaloblastic anemia to that seen in folate deficiency, as in both cases, the cells would be deficient in folate cofactors required for DNA and RNA synthesis. It also explains the clinical observation that treating a patient with megaloblastic anemia, caused by vitamin B_{12} deficiency, with a pharmacological dose of folic acid, allows the cells of the bone marrow to start to divide again. It is thought that the newly replaced folate would carry out several such cycles before being trapped as 5-MTHF. Patients with severe B_{12} deficiency often have a low RCF because impaired methionine synthesis results in the accumulation of 5-MTHF monoglutamate, which diffuses out of cells resulting in a corresponding high serum folate.[65]

RCF

The analysis of folate in red cells is considered to be a strong indicator of folate adequacy because it reflects intracellular status and is not influenced by recent or transient changes in dietary folate intake. The concentration of folate in red cells is much greater than it is in serum and is fixed at the time of erythropoiesis in the bone marrow, thus providing an integrated average index representative of a retrospective four-month period.[66] This test is preferred for the assessment of folate status as low values strongly suggest folate deficiency. However, the analysis of RCF is challenging because of the chemical instability of folates, the presence of different polyglutamate chain lengths attached to a folate molecule, and the inefficient extraction of folate from cells. This leads to significant differences in folate concentration estimation between methodologies and laboratories. As a result, laboratories need to establish their own reference intervals in order to report clinically useful values.

Individual Folate Forms in Serum/Plasma and Red Cells

The clinical utility of measuring individual folate forms, and UFA, has been explored in many studies. In particular, the utility of 5-MTHF in relation to polymorphisms in the MTHFR gene which leads to a redistribution of folate forms in red cells, and possibly other tissues, has been investigated.[20, 67, 68] Individuals with the MTHFR 677 TT genotype are predisposed to lower tissue stores of 5-MTHF as a consequence of impaired folate binding and a diminished capacity to convert 5,10-methyleneTHF to 5-MTHF. These individuals also tend to have a greater proportion of folates in the 5,10-methyleneTHF, THF, and formyl folate forms when compared with those with the Cytosine - Thymine (CT) and Cytosine - Cytosine (CC) genotype.[20, 69] Poor 5-MTHF availability leads to an increase in tHcy which has been associated with many diseases and health complications including cardiovascular disease.[40, 70] There is also evidence to suggest that 5-MTHF deficiency may be a cardiovascular risk factor independent of homocysteine.[71, 72]

Measurements of other forms of biologically active folate are currently poorly utilized in the diagnostic setting. The methods capable of their measurements require costly instrumentation and a high degree of technical expertise, and are currently being utilized mainly in research laboratories studying one carbon metabolism.

The analysis of folic acid has gained interest in view of findings of UFA in fasting serum samples from people consuming fortified foods and folic acid supplements,[73] and the correlation with some adverse health effects.[54, 74] High serum concentrations of folic acid have been linked to increased cancer risk and progression within certain patient groups. For example, folic acid was found to reduce the natural killer cell response to cancer cells.[75] There is also a suggestion that UFA may cause insulin resistance in children, interact with epilepsy medications, mask vitamin B$_{12}$ deficiency, and may be hepatotoxic at high concentrations.[54, 76–78] Because of the poor availability of folic acid measurement, the test has not been utilized in diagnostic laboratories. The test has been used in the National Health and Nutrition Examination Survey (NHANES) population study.[79–81] Its clinical usefulness remains to be established. It is predicted that folic acid estimations would be of value in countries where folic acid fortification of wheat flour produced in industrial mills is mandatory, for pregnant women, children, the elderly, and people with vitamin B$_{12}$ deficiency or insulin resistance.

5-MTHF in Cerebrospinal Fluid (CSF)

Measurements of 5-MTHF in CSF have been utilized for the diagnosis and monitoring of cerebral folate deficiency (CFD). CFD is a neurological condition associated with low concentrations of 5-MTHF in the CSF with normal 5-MTHF (and also normal serum folate) in the peripheral circulation.[82] In CFD, the age of onset for symptoms (e.g., deceleration of head growth, irritability, and insomnia) is 4–6 months. After 6 months there is a delay in neurodevelopmental milestones including psychomotor retardation and hypotonia. After 6 years of age, visual disorders may also develop. Some children also develop spasticity, speech difficulties, and epilepsy.[83] There have been a number of suggested causes for CFD including decreased transport across the blood-brain barrier. The suggested mechanisms for the decreased transport across the blood-brain barrier include blocking of folate receptor 1 by autoantibodies and mutations in the FOLR1 gene.[84] Treatment of CFD with folinic acid, also known as leucovorin (5-formylTHF), over a long period of time can significantly improve clinical symptoms in patients with CFD.[82, 85] Low 5-MTHF in the CSF alone should be interpreted with caution because there are a number of other conditions that can cause secondary CSF folate deficiency. Conditions causing secondary CSF folate deficiency include: Rett syndrome (from chronic use of anticonvulsant drugs and antifolate drugs), Aicardi-Goutières syndrome, Kearns-Sayre syndrome, inherited neurometabolic disorders including dihydropteridine reductase deficiency, 5,10-MTHF reductase deficiency, and mitochondrial disorders including white matter lesions.[82, 83, 86]

Total Plasma Homocysteine

tHcy is a very sensitive marker of folate status decline and correlates well with serum folate. Since the 1990s, evidence has accumulated that elevated concentrations of tHcy are associated with cardiovascular disease,[40, 70, 87] cognitive impairment,[88, 89] neuropsychiatric disorders,[90] cancer,[91, 92] diabetic complications,[93, 94] and pregnancy outcome.[95] Hyperhomocysteinemia is a well-established indicator and risk determinant for thromboembolic disease.[40, 96–98] Patients with untreated hyperhomocysteinemia are at high risk of premature occlusive vascular disease. Thromboembolism, strokes, and coronary occlusions are the major causes of morbidity and early mortality in these patients.[99]

In 1976, Wilcken and Wilcken reported that the concentration of homocysteine-cysteine mixed disulfide was higher in patients with coronary heart disease than in age- and sex-matched controls. This work had led to clinical studies investigating the associations between plasma homocysteine concentration and cardiovascular disease. In 1991, Clarke et al. showed abnormally high homocysteine concentrations after methionine loading in 30%–40% of patients with cerebrovascular, peripheral, or coronary artery diseases.[100]

Important associations of homocysteine concentrations to cardiovascular risk and myocardial infarctions in particular have been shown from population/epidemiological studies. One example of such evidence came from a five-year prospective study involving 14,916 male physicians, which estimated a threefold increased risk of myocardial infarction in those with homocysteine concentrations above the 95th percentile.[101] Other population-based studies have also suggested that homocysteine is an independent risk factor for coronary heart disease. A case-control study among

21,826 subjects, aged 12–61 years, from Norway showed that 123 subjects who later developed coronary heart disease had concentrations of homocysteine higher than in controls. In this study, the relative risk for a 4 µmol/L increase in serum homocysteine was 1.41 (95% confidence interval (CI):1.16–1.71).[102] A similar nested case-control study design was conducted in the United Kingdom and examined the association between tHcy and stroke within the British Regional Heart Study cohort.[103] Between 1978 and 1980, serum was saved from 5661 apparently healthy men, aged 40–59 years, randomly selected from general UK practices. During the follow-up to 1991, there were 141 incident cases of stroke among men with no history of stroke at the initial screening. Homocysteine concentrations were significantly higher in these cases than in controls, and there was a graded increase in the relative risk of stroke in the second, third, and fourth quartiles of the tHcy distribution (odds ratios 1.3, 1.9, 2.8; trend $P=0.005$) relative to the first quartile. The authors concluded that tHcy is a strong and independent risk factor for stroke.[103]

Elevated homocysteine concentrations have also been linked with mortality.[104] A study by Taylor et al. demonstrated that elevated plasma homocysteine (>14 µmol/L) concentrations were significantly associated with overall mortality from cardiovascular disease, and with the progression of coronary heart disease in patients with symptomatic cerebrovascular or occlusive disease.[105]

However, the placebo-controlled intervention studies which lowered homocysteine with B vitamins in various groups of patients failed to demonstrate a protective effect against future cardiovascular events.[106, 107] Many reasons have been suggested for this lack of positive response, e.g., insufficient power, recruitment of participants from folate fortified areas, participants with normal tHcy at baseline and short duration of the studies.[108] Therefore more research is required in this area. On the contrary, the intervention studies in people with stroke and neurological conditions have demonstrated the benefits of lowering homocysteine, especially in people with elevated concentrations at baseline.[89]

Analytical Methods
Brief Historical Overview of Folate Method Development

Microbiological assays were the first methods used for folate status determination.[3] They used either *L. rhamnosus* or *Streptococcus faecalis* and were considered as gold standard methods since they measured multiple folate forms that had vitamin activity.[109, 110] Although these methods have been largely superseded today by faster and less laborious protein-binding assays, recent improvements to microbiological assays which include automation, development of a new *Lactobacillus* strain, and cryopreservation of the inoculum, have led to a renewed interest in this methodology.[18]

The identification of folate-binding protein (FBP) in cow's milk in 1967[111] led quickly to the development of competitive protein-binding assays for serum folate concentrations. One of the earliest methods developed in the early 1970s was a radioisotopic assay by Waxman et al. that used ³H 5-MTHF, took around 4 h to perform, and had comparable results to the microbiological assay that utilized *L. rhamnosus*.[112] For many years, the Quantaphase folate assays by Biorad Laboratories, Hercules, CA were the most popular tests for serum and RCF analysis. The test was initially introduced in the mid-1970s as Quanta-Count and was calibrated to *L. rhamnosus*. In the mid-1980s, Quantaphase I, calibrated to Quanta-Count, went live. The accuracy of this latter test was challenged by Levine[113] and as a result, Quantaphase II was introduced in 1991. In brief, samples were combined with folate (^{125}I) in a solution containing dithiothreitol and cyanide. The mixture was boiled to inactivate endogenous binding proteins. The reduced folate and its analogs were stabilized with dithiothreitol during heating. The mixture was then cooled and combined with immobilized FBP. The reaction mixture was incubated for 1 h, during which the endogenous and labeled folate competed for the limited number of binding sites based on their relative concentrations. After incubation, the mixture was centrifuged and decanted and the radioactivity of a pellet formed at the bottom of the tube was counted using a gamma counter. The concentration of folates was determined from the standard curves containing folic acid. The Quantaphase tests also measured vitamin B_{12}. Although the results were accurate for vitamin B_{12}, underrecovery of certain folate forms, especially 5-MTHF (61%), was reported for Quantaphase II[114]. These tests were discontinued in 2007. Radioassays for folate analysis are no longer available. In addition to being reagent/kit specific, they posed a health hazard to the user from radioisotopes, as well as presenting an environmental disposal problem.

The FBP-based assays used today are available on several commercial platforms. They offer high throughput analysis with quick turnaround times at a relatively low cost but FBP assays are unable to differentiate the different folate forms. Chromatographic-based methods began to be explored by research laboratories with an interest in folate metabolism. However, methods using High Performance Liquid Chromatography (HPLC) coupled with UV detection had insufficient sensitivity to detect the different folate forms present in biological samples at endogenous concentrations.[115, 116] Methods using electrochemical[117] or fluorescence detection[118, 119] were more sensitive, but hampered by a lack of appropriate internal standards. More sensitivity was achieved with early Gas Chromatography (GC) and Liquid Chromatography-Mass Spectrometry (LC-MS) methods, although they were laborious and required derivatization of the folates.[120, 121] With the arrival of LC-MS/MS in the late 1990s came methods capable of detecting individual folate species.

Microbiological Folate Assays

Folate-dependent bacteria such as *L. rhamnosus* or *S. faecalis* are used in microbiological assays. The bacteria grow proportionally to the amount of folate present in a sample. Following on average 2 days of incubation at 37°C, the turbidity of the inoculated medium is measured and the folate concentration calculated. The main feature of *L. rhamnosus* is its equal response to all folate monoglutamates. *L. rhamnosus* also responds well to di- and triglutamates, therefore incomplete hydrolysis of folylpolyglutamates to monoglutamates is not an issue in this assay.[3] Folylpolyglutamates in red cells are normally hydrolyzed to monoglutamates not only for microbiological assays but in all other folate assays. *L. rhamnosus* can also grow in the complete absence of folate if other products of one carbon metabolism are present, e.g., thymidine, purines, methionine, serine, glycine, and panthothenate.[3] In human plasma, all these compounds are present with the exception of thymidine; because of this, *L. rhamnosus* media for folate assays must contain amino acids, purines, and pantothenate. In these conditions, the growth of *L. rhamnosus* in response to folate is due to the *L. rhamnosus* folate requirement for thymidylate synthesis.[3]

S. faecalis, like *L. rhamnosus,* responds well to folate mono- and diglutamates, but this organism cannot use 5-MTHF for growth. This feature of *S. faecalis* is utilized to measure the amount of nonmethylated folates. Using both assays, the amount of 5-MTHF is calculated by subtracting *S. faecalis* from the *L. rhamnosus* value.

Similarly to *S. faecalis, Pediococcus cerevisiae* (previously known as *Leuconostoc citrovorum*) does not respond to 5-MTHF, but this organism also does not grow on folic acid, DHF, or 10-formyl folic acid (oxidation product of reduced folates).[3] Therefore differences in folate concentrations achieved using *S. faecalis* and *P. cerevisiae* have been used to calculate folic acid content. However, the presence of DHF and 10-formyl folic acid reduces the accuracy of such analysis. In addition to the above, all these compounds are present in low quantities and a comparison of growth responses would require high precision and accuracy that would be hard to achieve with this methodology.

The *L. rhamnosus* assay has been successfully used until today. The method does not require expensive equipment; it offers high sensitivity and uses small sample volume. It is the only method that can utilize dried blood spot samples. The introduction of a 96-well microplate in the late 1980s simplified the assay and shortened the analysis time significantly.[122, 123]

However, in comparison with FBP assays, the technique is relatively laborious, high throughput analyses are not always possible and duplicate analyses are recommended due to low precision. Because the growth medium is very rich, aseptic conditions need to be maintained.[3] The development of folate-depleted cryoprotective preparations of *L. rhamnosus* enabled reproducible growth, stability of the organism for many months at −18°C, and helped with the standardization of the assay.[124] The presence of antibiotics in blood samples may also interfere with the assay. This issue was partially circumvented by the introduction of the chloramphenicol-resistant strain of *L. rhamnosus.*[125]

Detailed information about key factors which need to be considered when optimizing microbiological folate assays are described elsewhere.[18] A typical method which uses the chloramphenicol-resistant strain of *L. rhamnosus* maintained by cryopreservation and microtiter plates was described thoroughly by Molloy and Scott.[126] In brief, blood collected into EDTA vacutainers is used. Plasma is separated by centrifugation for the plasma folate assay (serum samples can also be used) and whole blood is mixed with freshly made 1% ascorbic acid (ratio 1:10) for the RCF assay, then the mixtures are left for 30 min to allow the conversion of folate polyglutamates to monoglutamates. Folic acid is used to prepare a standard curve.

The medium is prepared by dissolving folic acid broth powder in ultrapure water, adding chloramphenicol, Tween 80 and boiling the mixture until the folic acid is fully dissolved. Following this, and once the medium had cooled, ascorbic acid and cryopreserved *L. rhamnosus* are added. If an EDTA sample is used, the addition of manganese sulfate is recommended to prevent the inhibition of growth due to residual EDTA. Samples, QC, and standards are diluted with 0.5% of sodium ascorbate solution. For each diluted sample, duplicate aliquots of 100 and 50 μL are transferred to four separate wells of a 96-well microtiter plate, resulting in an overall use of eight wells for every specimen analyzed. A separate place is used for a calibration curve. An inoculated folate medium is then added to all wells using a multichannel pipette. Plates are then sealed and incubated at 37°C for ~42 h. After the 42 h have passed, the plates are inverted and mixed carefully to produce an even cell suspension. Sealers are removed and the plates are read at 590 nm in a microtiter plate reader.

FBP Assays

FBP-based assays for folate are available as kits for use on several commercial platforms and are the most popular methods currently used in clinical laboratories. They offer rapid automated analysis for a large number of samples, are not dependent on highly skilled laboratory staff, and the assay reagents are low cost. The principles of these tests are the competition between a labeled folate standard and endogenous folates for a folate binder (or a noncompetitive binding where excess of FBP is added to a sample) followed by a labeled folate conjugate. Because different forms of folates have various affinities for folate binders, efforts have been made to optimize assay conditions that maximize specificity of the test. For example, folic acid has a higher affinity for folate binders than 5-MTHF; however, at a pH 9.3, the affinity of both compounds to FBP is equal.[18] Because folic acid is more stable than 5-MTHF, it is often used as calibrant in this type of assay. However, 5-MTHF is the predominant folate form, both in plasma and red cells. Therefore differences in affinity for FBP will result in lower results if assay conditions are not optimized.

As with microbiological assays, FBP assays require the conversion of polyglutamates to monoglutamates. Incomplete conversion may lead to higher results as folylpolyglutamates have a higher affinity for FBP. The conditions for deconjugation vary not only with FBP, but also microbiological and chromatographic assays, which is one of the factors contributing to the difference in RCF folate results across all methods.

Similarly to the microbiological assay, the linear concentration range for FBP-based assays is narrow at ~1.5–20 μg/L. The upper limit is too low, especially when used in countries where mandatory folic acid supplementation of food has been implemented, and for patients taking folic acid supplements. Based on the author's work (performed in the United Kingdom where fortification of flour is not mandated), it is estimated ~4% of all samples measured for serum folate in a hospital setting will exceed the concentration of 20 µg/L.[127] Diluting the samples does not offer a solution, as FBP assays are known to be strongly affected by matrix effects, and unless samples are diluted with human albumin or folate free serum/plasma, the accuracy of folate results will be diminished. Diluting all samples >20 µg/L analyzed in a routine diagnostic setting would incur an additional cost to the analysis.

Current UK NEQAS Haematinics reports show Roche Cobas, Abbott Architect, Siemens Centaur, and Beckman Dxl as the most commonly used commercial platforms for measuring serum folate and Abbott Architect, Beckman Dxl, and Siemens Centaur as the most commonly used platforms for measuring RCF. Other platforms include Beckman Access, Siemens Immulite 2000, Siemens Dimension/Vista, and Ortho Vitros. The Abbott Architect and Siemens use a chemiluminescence microparticle technology (CMIA) for the tests. This technique is very popular as it offers excellent sensitivity with detection limits as low as attomole (10^{-18}) or zeptomole (10^{-21}) for some other tests.[128] Chemiluminescence refers to the emission of light generated from a chemical reaction; in this assay it is oxidation after the addition of hydrogen peroxide or sodium hydroxide. In Abbott's method, following the release of folates from endogenous FBP in a two-step reaction, FBP coated with paramagnetic microparticles is added. Folate in the patient's sample binds to FBP coated with microparticles. After the washing step, the acridinium-labeled (chemiluminescent compound) conjugate is added, which binds to unoccupied sites on the FBP-coated microparticles. Hydrogen peroxide or sodium hydroxide is then added to induce chemiluminescence, measured as relative light units (RLUs). An inverse relation exists between the amount of folate in the sample and the RLUs detected by the system. The same technology with the acridinium label is applied on the Siemens Centaur.

The Beckman's assay is a chemiluminescence immunoassay (CIA). In this method, milk FBP with folate coupled with alkaline phosphatase attached in the preceding reaction binds to paramagnetic microparticles coated with goat antimouse IgG/antifolate-binding protein MAb complexes. The addition of dioxetane phosphate generates chemiluminescence.

The Roche Elecsys (Cobas) utilizes electrochemiluminescence. Following the release of folate from its binding proteins, ruthenium-labeled FBP is added. A folate complex is formed, the amount of which depends upon the folate concentration in the sample. Streptavidin-coated microparticles are added, as well as folate labeled with biotin. The unbound sites of the ruthenium-labeled FBP become occupied, and a ruthenium-labeled FBP-folate biotin complex is formed, which is bound to the solid phase via the interaction of biotin and streptavidin. The magnetic particles are captured onto the surface of the electrode and unbound substances are removed. A voltage is applied to the electrode which induces chemiluminescent emission, measured by a photomultiplier.[65]

All methods for serum folate are robust, with a coefficient of variation of (CV) <10% between laboratories using the same technology and overall CV of ~10% for all methods combined.

High Performance Liquid Chromatography and Gas Chromatography Folate Assays

HPLC- and GC-based assays make the quantification of individual folate forms including folate polyglutamates possible. Many of these methods were developed in the 1980s and 1990s. Although they are still used today for the quantification of certain folate forms in blood or plasma, 5-MTHF in CSF and folate forms in foods or other tissues, they have now largely been superseded by the more sensitive and quicker LC-MS/MS-based assays. Many HPLC methods were designed to measure folates in their monoglutamate form; they use reversed-phase columns and a mobile phase with low pH, typically a phosphate buffer with an organic component such as acetonitrile to facilitate separation.

Red cell lysate preparation

For whole blood and RCF analysis, lysates are prepared with hypotonic diluents (often 1:10 dilution with ascorbic acid) in the absence of hemoglobin denaturation. Ascorbic acid also protects folate from oxidation. Other methods use mercaptoethanol or dithiothreitol as preservatives.[121]

Enzymatic deconjugation of erythrocyte polyglutamyl folates to monoglutamates

The enzymatic deconjugation of erythrocyte polyglutamyl folates to monoglutamates follows cell lysate preparation. This is usually accomplished by using an endogenous folylpolyglutamate hydrolase. An optimum pH, temperature, and deconjugation time need to be established prior to the implementation of the method. The highest yield of folate monoglutamates is achieved for whole-blood lysates prepared at pH 4.0 and 37°C for 4 h.[129] Similar results are achieved for hemolysates prepared at pH 4.7 and 37°C for 3 h.[129]

Extract purification procedures

Following deconjugation, the sample extracts require extensive cleaning and the addition of internal standards. Reversed-phase cartridges are most often used for solid-phase extraction (SPE) purification but affinity columns have also been used.[117]

Calibration

The preparation of calibration stock solutions requires great care because of the poor stability of folates. The concentration of assay calibrants is determined spectrophotometrically, and stocks are further diluted in solutions that allow long-term storage. The author's own work, and that of others, show that a high concentration stock solution for 5-MTHF, one of the most stable forms of natural folates, is stable for at least 10 years when kept at −80°C.[68, 81]

The method described below, shown as an example, can be used to quantify 5-MTHF in plasma, red cells, and CSF.[68] It uses ascorbic acid as the antioxidant, SPE to purify the sample, and native fluorescence for the detection of 5-MTHF. The principles of the analytical procedure are based on the method described by Pfeiffer et al.[121] The effects of pH, composition of the mobile phase, and sample solvents on the ionization of 5-MTHF have been investigated to optimize chromatographic conditions. A pH <3 is used to suppress the ionization of the two carboxyl groups in the glutamate part of folate. Under these conditions, 5-MTHF elutes at around 10.2 min. Plasma samples are prepared by centrifugation. Whole-blood lysates are prepared by the addition of 100 μL of whole blood to 1000 μL of 1% ascorbic acid. CSF is collected into universal plain tubes. 4-Aminoacetophenone is used as an internal standard[130] and Bond Elut C$_{18}$ (100 mg, 1 mL reservoir) cartridges (Varian Inc.) are utilized for SPE with an elution strategy based on that of Pfeiffer et al[121] with some modifications. Sample components are separated using an ACE C$_{18}$, 3 μm column (125×4.6 mm) supplied by Hichrom, UK, with a mobile phase composition of 0.033 mol/L potassium phosphate (pH 2.3) buffer-acetonitrile-methanol (89: 6.6: 4.4, by volume) at a flow rate of 0.34 mL/min. The fluorescence detector wavelength settings are: excitation 290 nm and emission 365 nm.

A primary stock solution of 5-MTHF was prepared by dissolving ~5 mg of 5-MTHF powder in 10 mL of 20 mM potassium phosphate buffer pH 7.2 with 1 g/L cysteine. An aliquot (1 mL) of this stock solution was removed to determine the concentration by UV spectrophotometry and 90 mg of ascorbic acid powder was immediately added to the remaining stock solution. The absorbance of 5-MTHF was measured at 290 nm using the molar extinction coefficient of 32,000 L mol^{-1} cm^{-1} after 1 in 40 and 1 in 33.3 dilution with 20 mmol/L potassium phosphate buffer pH 7.2 containing 1 g/L cysteine.

The primary stock solution (1018.70 μmol/L) was diluted 1 in 5 in 10 g/L ascorbic acid and this stock solution was aliquoted and stored at −70°C (secondary stock-203.74 μmol/L). One aliquot of the secondary stock solution was diluted further with 1 g/L ascorbic acid to the concentration of 20.37 μmol/L (9374 μg/L) (tertiary stock). This stock was used on the day of analysis to prepare a calibration curve of a minimum of four points.

The accuracy of the 5-MTHF calibration standard was checked against the SRM 1955 (National Institute of Standards & Technology, US) which included three reference samples with certified 5-MTHF concentrations ±uncertainty of 4.26±0.25, 9.73±0.24, and 37.1±1.4 nmol/L.[131] The RCF concentration was calculated according to the formula:

RCF 5-MTHF = {whole blood 5-MTHF—[plasma 5-MTHF (1-hematocrit)]}/ hematocrit.[72, 132]

Although the method is robust with intra- and interassay CV below 10% and good recoveries for spiked samples, other natural folates or UFA cannot be quantified with this technique because of a lack of sensitivity or fluorescence activity, respectively. Each sample run takes ~40 min, making the method unsuitable for processing a large number of samples. However, the method is well suited for measuring 5-MTHF in CSF.

A combined HPLC and microbiological assay was developed by Sweeney et al. to measure UFA in serum.[133] The method was able to detect UFA in serum after oral doses of 200 μg of folic acid. Although the authors claimed this method to be inexpensive and simple, it included a laborious sample preparation: first the HPLC separation of folic acid, followed by SPE, and then the microbiological assay.

HPLC methods with electrochemical detection were also successfully applied for measuring folates in biological fluids[134] and tissues.[117] Similarly, GC methods were developed but due to lengthy sample preparation and derivatization procedures, they did not gain much popularity. Some methods included an additional HPLC step instead of SPE, followed by GC analysis.[135]

Liquid Chromatography-Mass Spectrometry Folate Assays

Liquid chromatography-mass spectrometry folate measurements require high cost equipment and a high level of expertise to develop the method. As with the HPLC and GC methods, a manual sample cleanup is needed if automated sample handlers with a SPE extraction option are not available. The preparation of calibration stocks and storage conditions are the same as for HPLC methods. The advantages of LC-MS/MS over HPLC and GC methods include the availability of stable isotope-labeled internal standards for most common folate forms. The methods are more sensitive and many of these techniques can detect all of the most biologically active folate vitamins present in whole blood (5-MTHF, THF, 5,10-methylene-THF, methenyl-THF/10-formyl-THF, and formyl-THF).

The development of LC-MS methods was initiated in the late 1990s. A few years later tandem LC-MS/MS followed. The improvements to this methodology continue today and are targeted mainly at postlysis chemical interconversions of folate forms[136] and susceptibility to oxidation.[137] The initial LC-MS methods were able to measure 5-MTHF only in human plasma.[138] The separation and detection of other folate forms was achieved for standards, but they lacked the sensitivity to detect these minor folate fractions in human samples.[139] They used electrospray ionization (ESI) either in negative or positive ion modes and often time-consuming FBP affinity columns to purify samples.

An LC-MS/MS method for 5-MTHF and folic acid was presented by Kok et al.[140] The main application of this method was to study the metabolism of folic acid following the administration of labeled folic acid. The method used 2 mL of plasma. The development of the first isotope-dilution LC-MS/MS method for the simultaneous determination of other folate vitamins besides 5-MTHF in human serum offered a breakthrough in the assessment of various folate forms in biological fluids.[121] An automated 96-well plate was soon introduced to this method to increase sample throughput and minimize time spent on sample preparation.[20] The method required 275 μL of serum and was also utilized for whole blood with an automated SPE that could process 96 samples in 5 h. It offered an excellent linearity in the range 0.22–220 nmol/L (0.1–96.4 μg/L). Using this method, various concentrations of folate forms according to the MTHFR genotype were found in whole-blood samples. Most notably, individuals with the homozygous genotype had significantly less 5-MTHF and more nonmethylated folates than wild or heterozygous types. The method served as a reference method for the assessment of serum folate assays by UK NEQAS Haematinics.[141]

Another fast and automated method was developed, which could process 192 samples in 24 h and only used 60 μL of sample.[142] The SPE step in this method was substituted with protein precipitation and evaporation. Furthermore, this method, in addition to individual folate forms, also measured 4-α-hydroxy-5-methyltetrahydrofolate (hmTHF) as well as folate catabolites pABG and apABG. The clinical utility of folate catabolite measurements in serum has been hitherto poorly described. hmTHF is an intermediate product of 5-MTHF oxidation. In the absence of reducing agents, the compound undergoes structural changes to form a pyrazino-s-triazine derivative also known as MeFox.[137] It is currently unknown if MeFox is only formed in vitro or also in vivo. MeFox and 5-formylTHF are isobaric compounds forming the same mass to charge during ionization, hence they coelute with each other, making analysis of 5-formylTHF imprecise. The LC-MS/MS methods for various folate vitamins need to take this into account and separate these compounds, even though the contribution of 5-formylTHF to the total folate pool is small. For example, in a large serum sample set, 5-formylTHF was detected in 15% of all samples with a concentration in some of >1 nmol/L.[137] The authors also found that MeFox was present in 99% of samples from the large set of specimens. The median concentration of MeFox in this study was 1.35 nmol/L.[137] The LC-MS/MS method developed by this group was further improved to include less sample volume (150 μL), and the cleanup procedure was modified to increase sample throughput.[81]

Recently, a method was published which addresses folate interconversions by denaturing enzyme activity in the lysates through brief heat treatment.[136] A derivatization technique using sodium cyanoborodeuteride reduction to convert methylene-THF to deuterated 5-MTHF was also introduced in this method. This derivatization enables accurate measurements of THF and methylene-THF, which were indistinguishable due to interconversions in most of the methods described before. Using this method, the authors showed that the total folate pool was relatively evenly spread between THF, methylene-THF, and the combined pool of methenyl-THF and 10-formylTHF, with 5-formylTHF and 5-MTHF being lower in abundance. The method has so far been used in cultured cells and is expected to be used to study and gain a quantitative understanding of one carbon metabolism.

Analytical Methods for Homocysteine Assessment

The analysis of homocysteine for diagnostic purposes began in the early 1960s when the first patients with homocystinuria were described. The large amount of homocysteine in urine and plasma excreted by these patients was determined by simple tests or by amino acid analysis.[143, 144] A growing understanding of the links between homocysteine and folate and vitamin B$_{12}$ metabolism led to the development of methods able to measure mild homocysteine elevations. A variety of methodologies exist today for the measurement of tHcy. Immunoassays and LC-MS/MS are the most popular techniques. The HPLC with fluorescence detection methods which were commonly used during the 1990s are still used in many laboratories today.

HPLC coupled with ion-exchange chromatography is mainly used in metabolic laboratories performing a full amino acid screen. Radioenzymatic assays are now obsolete, although enzymatic assays are still in use. Gas chromatography and CE-based assays are rare and have been superseded by LC-MS/MS methods.

In the European Research Network for evaluation and improvement of screening for Diagnosis and treatment of Inherited disorders of Metabolism (ERNDIM) proficiency testing scheme out of 88 participants who report homocysteine results, 47% report results using LC-MS/MS methods, 19% use immunoassays, 11% use HPLC, 6% use enzymatic methodology while the remaining 6% use other methods (2017 data).

Enzymatic assays

The main advantage of enzymatic assays over other techniques is that they do not require specialized instruments and use simple chemistry analyzers or a microplate reader for colorimetric assays. They use a very small sample volume of ~5 μL, have a higher linear range than immunoassays, and some methods are suitable for the analysis of plasma derived from a finger prick.[145] Radioenzymatic assays were the first methods used for the assessment of homocysteine in plasma. Dithioerythritol was used to reduce protein-bound homocysteines to free homocysteine. Homocysteine was converted to SAH in the presence of [^{14}C] adenosine and SAH hydrolase.[146, 147] Radioactive SAH was then quantified by HPLC, thin-layer chromatography or paper chromatography, and scintillation counting.

More recently, various commercial enzymatic assays have been developed. Tan et al. developed a 96-well microtiter plate assay which uses a highly specific homocysteine α,γ-lyase which converts homocysteine to α-ketobutyrate, ammonia, and H_2S.[148] In the second step, H_2S combines with *N,N*-dibutylphenylene diamine to form a highly fluorescent product. In a method by Matsuyama et al. homocysteine methyltransferase is used, which generates L- and D-methionine. In the second step, D-amino oxidase and N-ethylmaleimide are used to oxidize the redox indicator and generate the colored product.[149] Other enzymatic assays developed for commercial use are listed elsewhere.[18]

Gas chromatography mass spectrometry (GC-MS) assays

Stabler et al. developed one of the first GC-MS methods for homocysteine[150] which was further modified and simplified later.[151, 152] The major advantages of these methods include high accuracy and precision. With this technique, methylmalonic acid and amino acids such as methionine can be concomitantly measured.[153] Most methods use internal standards; dithiothreitol or 2-mercaptoethanol are the reducing agents and an ion-exchange step is used to separate amino acids. For derivatization, sialyl (gives an intense ion and prolongs the lifetime of the column) or alkyl esters (2–3 min reaction at room temperature) are used.[145] However, the cumbersome sample preparation and scarce availability of GC-MS instruments did not popularize this technology for plasma homocysteine, even though the methods can be used in high throughput settings.

HPLC assays

Assays based on HPLC were the most common for the determination of plasma homocysteine during the 1990s. Apart from homocysteine, they can simultaneously measure other thiols and depending on the detection method as well methionine concentration. The analytical performance is excellent with intra- and interassays CVs being <5%. As with other methods, they require the reduction of disulfide bonds and protein-bound homocysteine to free homocysteine. This is usually achieved with the use of dithioerythritol, sodium or potassium borohydride, and 2-mercaptoethanol. The extent of sample preparation varies according to the method of detection. The fluorescence detection method is the most widely used by far, however ion exchange and electrochemical, and UV detections are also used.

HPLC methods can perform analysis at a much reduced cost when compared with immunoassays because of lower ongoing reagent costs. HPLC also compares favorably to LC-MS/MS because of favorable instrument costs, but HPLC is labor intensive and can process a limited number of samples within one analytical run. HPLC methods with fluorometric detection require derivatization. Monobromobimane and halogenosulfonylbenzofurazans (SBD-F and ABD-F) are most frequently used. The advantage of the former is that it reacts rapidly with thiols at room temperature thus allowing for automation, but the compound itself is fluorogenic, hence the impurities and products of hydrolysis fluoresce which may lead to chromatographic interferences.[145] The issue is overcome by the use of gradient elution. The same is not necessary if SBD-F or ABD-F (not fluorescent) is used as they form very stable products with thiols, but these agents only react with thiols at high temperatures: 50°C and 60°C, respectively, at a tightly controlled pH, hence automation may be problematic if suitable sample pretreatment instruments are not available. However, SBD-thiol adducts are light sensitive and require protection against it to ensure reliable results.[154]

A typical sample preparation method with the use of monobromobimane comprises of the following steps: reduction of homocysteine with sodium borohydride, followed by the addition of sodium hydroxide, dimethyl sulfoxide, EDTA, and octanol. After 3 min are allowed for reduction to take place, monobromobimane is added in the presence of an ethylmorpholine buffer.[155] The monobromobimane contains a reactive functional group (allylic bromide) that reacts with thiol groups of homocysteine (not fluorescent), forming a stable and highly fluorescent product. The derivatization is terminated by the addition of glacial acetic acid and the samples are injected onto an HPLC system with fluorescence detection (excitation 365 nm, emission 475 nm). A formate buffer and acetonitrile are the components of the mobile phase and a gradient elution is used. An ACE 5 C16 (100×4.6 mm) column can be used for sample separation.[156] The method can be automated.

Ion-exchange chromatography is often used in laboratories performing a full amino acids screen. Methionine is part of these measurements and helps to distinguish between inherited homocystinurias. For example, a Biochrom 30 amino acid analyzer can be used. Following the precipitation of proteins with sulfosalicylic acid and their subsequent removal, the clear supernatant containing a mixture of amino acids is placed into the analyzer. Sulfosalicylic acid protects homocysteine from re-oxidation. Cation-exchange chromatography coupled with postcolumn derivatization using ninhydrin leads to a colorimetric detection at 440 and 570 nm. A column containing cation-exchange resin (polystyrene) is sulfonated to provide a negative electrical charge and reacts with divinylbenzene. Lithium citrate buffers of varying pH and ionic strengths are then pumped through the column to separate the individual amino acids. The column temperature is accurately controlled. The eluent is mixed with the ninhydrin reagent and is moved to the reaction coil where it forms colored compounds. The amount of colored compound is directly proportional to the concentrations of amino acids present.

The first HPLC techniques with electrochemical detection methods for tHcy were developed in the 1970s.[157] They do not require derivatization and are very sensitive and specific. The detection of nonderivatized thiols can occur using dual mercury gold amalgam or platinum electrodes for amperometric detectors. Porous carbon electrodes have also been used for coulometric detectors, but this material is less selective than a mercury/gold amalgam electrode.[145]

HPLC with UV detection is the method of choice for laboratories that do not have a fluorescence detector. They require postcolumn derivatization with, for example, 4,4'-dithiodipyridine and subsequent detection at 324 nm.

Immunoassays

Commercially available immunoassays for tHcy provide fast and high throughput analysis and are the methods of choice in laboratories with a high sample turnover or laboratories that do not have LC-MS/MS analyzers. They were developed in the 1990s as a result of increased clinical interest in homocysteine and its links, most notably with cardiovascular disease. The main drawback of these methods is the relatively high cost of reagents that include patented antibodies. They also have a limited dynamic range, set at 50 μmol/L. However, unlike HPLC and LC-MS/MS, the assays are highly automated and require little technical expertise, therefore the overall cost of analyses may be comparable to other methods. Following reduction with, e.g., dithiothreitol, free homocysteine is enzymatically converted to S-adenosyl-L-homocysteine (SAH) with the recombinant enzyme SAH hydrolase and excess adenosine. Excess adenosine drives the conversion of homocysteine to SAH by the recombinant SAH hydrolase, and not the physiological SAH hydrolase. The SAH then competes with labeled compounds for the monoclonal antibody. Depending on the detection, the methods are subdivided into fluorescence polarization immunoassay (FPIA), CIA, and enzyme-linked immunoassay (EIA).

Examples of FPIA include homocysteine analysis on IMx and AxSYM by Abbott Diagnostics. The methods are no longer available, since these analyzers have been superseded by the family of Architect instruments. In the AxSYM method, following the conversion of homocysteine to SAH, S-adenosyl-L-cysteine fluorescein tracer was added which competed for the sites on the mouse monoclonal antibody molecule (anti-SAH). The intensity of a polarized fluorescent light was measured by the FPIA optical assembly. The assay was calibrated using a 6-point procedure and utilized a four-parameter logistic curve fit method to generate a standard curve. The accuracy of the AxSym assay was compared to the earlier developed IMx homocysteine assay (correlation coefficient was 0.985).

The initial steps of the chemiluminescence method available on the Architect (Abbott Diagnostics) are the same as for the AxSym method. Here, the SAH competes with acridinium-labeled S-adenosyl cysteine for a particle-bound anti-SAH monoclonal antibody. Following a wash stage and magnetic separation, hydrogen and sodium peroxide are added to induce oxidation, resulting in chemiluminescence measured as RLUs. The accuracy of the results was compared to the AxSym homocysteine assay (correlation coefficient was 0.98).

In the method on Immulite by DPC, an anti-SAH antibody is labeled alkaline phosphatase.

Axis Biochemicals used the principles of the EIA in their homocysteine assay. SAH in the sample competes with SAH bound to the microtiter plates for binding sites on an anti-SAH antibody. The unbound anti-SAH antibody is removed and a secondary rabbit antimouse antibody labeled with horseradish peroxidase is added.[145] Sulfuric acid is added to stop the reaction. A substrate, tetramethylbenzidine, is added and peroxidase activity is measured spectrophotometrically. The absorbance is inversely correlated to the homocysteine concentration.

Immunological assays underwent intensive scrutiny when they were first developed. Many studies compared their performance with gold standard methods at that time such as GC-MS and HPLC. In 2000, Nexo et al. demonstrated good performance for FPIA on IMx and EIA methods. Both methods used the calibrators manufactured by AXIS.[158] The bias for FPIA was −2% to 3% and 2% to 4% for EIA. A study by La'ulu et al. which included ADVIA Centaur and Immulite (Siemens Healthcare Diagnostics), Architect $i2000_{SR}$ and AxSYM by Abbott, Catch enzymatic assay (Equal Diagnostics) on Modular P analyzer by Roche, and Diazyme enzymatic assay by Diazyme Laboratories showed a good correlation of all methods with HPLC with correlation coefficients of 0.95–0.99.[159] Bland-Altman plots demonstrated a percentage bias of between −29.3% (Immulite) and 7.2% (Centaur). Based on the criteria set for this study, the authors concluded that with the exception of the Immulite, all other homocysteine methods included in this work gave comparable results and could be used interchangeably.

Capillary electrophoresis

Capillary electrophoresis methods offer good separation and a shorter analysis time compared to HPLC, they use a small analyte volume, little solvents, can be automated, and measure other thiols. The appearance of much simpler LC-MS/MS homocysteine methods, however, has diminished the development of new CE methods, but has not halted them completely. A recently developed CE method coupled with liquid-liquid extraction for total cysteine and homocysteine[160] is the best example of this. The method is rapid and selective. The analytes are derivatized with 1,1'-thiocarbonyldiimidazole and samples are then purified by chloroform-ACN extraction. Electrophoretic separation is performed using 0.1 M phosphate with various additives. The total analysis time is less than 9 min and the limit of detection is 0.2 μmol/L for tHcy. For a list of other homocysteine CE methods, the reader is referred to a review by Ducros et al.[145]

Liquid chromatography mass spectrometry assays

A number of LC-MS/MS methods have been developed to date.[161, 162] They offer excellent sensitivity and specificity in respect to all other homocysteine methods. They normally do not require a derivatization step, can be used on blood spot samples, and many methods allow for the simultaneous determination of other metabolites related to one carbon metabolism, vitamin B_{12}, and homocysteine metabolism, e.g., folate, methionine, methylmalonic acid, SAM or SAH. Apart from the high initial cost of LC-MS/MS system, the reagent cost is relatively low and high throughput analyses are possible.

Calibrants for LC-MS/MS should be prepared in the same matrices as the samples to avoid matrix effects. An isotopically labeled internal standard is added to both samples and calibrants. A reducing agent, most often dithiothreitol, is then added to break the bonds between homocysteine dimers, protein, and cysteine bonds. The mixture is left to stand at room temperature for 15 min which was found to be sufficient for the complete release of free homocysteine.[163] Precipitation of proteins follows with, for example, acetonitrile, trichloroacetic acid at various concentrations, trifluoroacetic acid, ascorbic acid, or mixtures of trifluoroacetic acid with acetonitrile. Samples are then centrifuged and supernatants are used directly for LC-MS/MS analysis. Ultracentrifugation has also been used for protein removal and found to be superior over acetonitrile precipitation in enhancing signal sensitivity.[163, 164] Before entering the mass spectrometer, samples are most commonly volatized either by ESI or atmospheric pressure chemical ionization (APCI). In ESI, a high voltage is applied, leading to the formation of charged droplets. Droplets then move to the inlet of the MS and generate ions during evaporation. In APCI, the analyte is sprayed by a heated nebulizer probe, producing a mist of fine droplets that are converted into gas by the combination effects of heat and gas flow. In the ionization region under atmospheric pressure, molecules are ionized at corona discharge (electric discharge). Sample ions then pass through a small orifice into the MS. One limitation of all LC-MS/MS methods is matrix effects which can lead to ion suppression or enhancement due to coeluting compounds with the same retention time as the analyte of interest, leading to an increase in background noise and reduction of sensitivity. Salts and detergents present in biological samples often cause ion suppression. They can alter the efficiency of the droplet formation or evaporation, hence changing the amount of charged ions in the gas phase in the source. APCI methods are believed to be able to withstand ion suppression better than ESI.[163] Enhanced cleanup procedures, using the same matrix for calibrators and alterations to chromatographic conditions can minimize matrix effects.

Sample Types and Sample Stability
Total Plasma Serum and Plasma Folate

A fasting sample (at least for 3h) is preferred for serum and plasma folate analysis since the concentration of folate elevates significantly within one to 2h after foods that are high in folate contents or supplements. Samples should be collected in either serum tubes or lithium heparin tubes. EDTA tubes have also been used but the recent findings suggest that 5-MTHF degrades in those tubes at an estimated rate of 1.9% per hour, while MeFox increases at a rate of 25.7% per hour.[18, 165] MeFox can be measured using LC-MS/MS methods.

Serum or plasma should be separated as soon as possible to avoid folate degradation and hemolysis, which can also influence results. Lipemia does not interfere with many assays. Our investigations showed that lipemia did not affect folate concentrations analyzed using the Abbott Architect chemiluminescence method. Twenty-five highly lipemic samples were analyzed both with lipids and following lipid removal, using Cataclear. There was no statistical difference in folate concentration between the two sets of samples; the folate concentration range was 2.3–16.5 µg/L, the mean concentration difference between all samples with lipids and without was 0.34 µg/L (unpublished data).

It is recommended that samples should be light protected. However, our work and that of others suggest that folates may not be as light sensitive as previously thought. Only a 1.7% of decrease in folate concentration was found in light-exposed samples compared to 1% from those stored in dark conditions ($N=25$, stored for 7 days).[166]

If testing is not performed immediately (1–2 days) after specimen collection, samples may be stored for up to 7 days at 2–8°C or 30 days at −20°C. Samples are stable for many years if they are kept at −70°C.

Total Red Cell/Whole-Blood Folate

Whole-blood EDTA nonfasting samples are used for RCF/whole-blood analysis. The hematocrit needs to be determined prior to sample pretreatment. Lysates are usually prepared with ascorbic acid within 1–2 days from blood collection. Ideally, samples should not be frozen before lysate preparation. Based on the UK NEQAS Haematinics report, the storage conditions of samples prior to RCF analysis vary between laboratories. It was reported that 48% of laboratories store whole blood at 4°C and prepare lysates just before analysis, 26% prepare lysates upon receiving the samples and freeze them at −20°C until analysis, 13% freeze the whole blood at −20°C and prepare lysates before analysis, and the remaining 13% use other methods[167].

The concentration of lysing reagents and duration of lysis step also varies between methodologies. Hemolysates with ascorbic acid are stable for several weeks at −20°C and for several years at −70°C[18]. At least three cycles of freezing and thawing do not affect folate concentration in hemolysates. The freezing and thawing of undiluted whole-blood samples is not permitted, as folate degrades significantly in these conditions. Immediate testing is recommended following the first thawing cycle.

Although most manufacturers recommend protecting whole-blood samples from light and processing within 2 days of blood collection, we have demonstrated that if these conditions are not maintained, the whole-blood specimens may still be used for the RCF analyses.[168] In brief, 11 whole-blood EDTA samples were divided in half. One part was kept at 2–8°C, and the other at room temperature in dark conditions until analysis. Lysates were prepared within 2 days of blood collection (baseline sample) and for up to 7 days for both sets of samples. A paired samples t-test showed that there was no difference in RCF concentration between individual samples kept at 2–8°C and those at room temperature ($p = 0.724$ for samples processed on day one; $p = 0.335$ for samples processed 8 days after blood collection). Repeated ANOVA measures demonstrated no difference in RCF concentrations for samples lysed at different time intervals: $p = 0.923$ at 2–8°C and $p = 0.2$ at room temperature. The storage conditions did not affect RCF concentrations ($p = 0.218$).

It was found that whole-blood EDTA samples were stable for RCF analysis for up to 7 days if kept at 2–8°C or room temperature.

Cerebrospinal Fluid 5-Methyltetrahydrofolate Analysis

The CSF should be collected into universal tubes or other standard tubes and immediately stored, ideally at −70°C or −80°C until analysis to prevent any degradation of 5-MTHF. The sample should also be protected from light. Stability studies in the author's laboratory of samples kept at 4°C demonstrated 25%, 32%, and 43% of 5-MTHF degradation after 2, 3, and 4 weeks, respectively. The freeze and thaw stability was checked by subjecting samples to three freeze-thaw cycles and comparing the results with an aliquot kept at −80°C. The results following three freeze-thaw cycles were only 5% lower, suggesting no significant effect on 5-MTHF concentration.

Total Plasma/Serum Homocysteine

Postprandial and orthostatic variations are the preanalytical factors that need to be taken into account in plasma homocysteine analysis. To exclude the influence of recent dietary intake, it is preferred that blood samples should be collected after an overnight fast. Factors affecting albumin concentrations will alter homocysteine because homocysteine is protein bound, and venepuncture should not be performed after venous stasis or following the subject resting in a supine position. Serum-separating tube (SST), EDTA, or lithium heparin tubes are often used for tHcy analysis. For example, a Biochrom 30 amino acid analyzer uses lithium heparin samples. The Diazyme assay does not use EDTA samples. Analysis can also be performed on citrated plasma, in which case the result should be multiplied by 1.1 to take into account the volume of citrate in the tube. Preanalytical sample handling is of critical importance since a delayed removal of red blood cells results in artificially high tHcy concentrations because of homocysteine release from the blood cells; tHcy increases by 1 μmol/L in 3h if the sample is kept at room temperature.[18] Plasma should be separated from red cells ideally within 60 min and kept on ice prior to separation.

Alternatively, samples can be collected into special homocysteine tubes containing a preservative, for example, Kabevette Vacuum HCY 837V 3.5 (Kabe Labortechnik GmbH). According to our validation data, the Kabevette tubes prolong the stability of unspun samples for up to 72 h (Fig. 11.4).[169] These investigations were carried out on blood samples collected into Kabevette and SST tubes from 14 healthy volunteers. Baseline samples were separated from red cells within 30 min of collection. Other Kabevette samples were separated after storage at an ambient temperature in dark conditions for 6, 36, and 72 h. The delay in separation did not affect tHcy results.

Reference Ranges
Serum and RCF

Serum concentrations <3 µg/L (6.8 nmol/L) usually indicate folate insufficiency. Traditionally a value of 140 µg/L (317 nmol/L) for RCF is considered as a cutoff for folate adequacy.[28] However, significant differences in folate measurements exist between different methodologies. FBP methods in particular differ in the detection of various folate forms; hence this will result in concentration variations as discussed earlier. Because of methodology bias, it is recommended that each laboratory establishes its own reference ranges. The bias is especially pronounced for RCF assays. In addition to method differences, there is no agreement on the preparation and storage of red cell lysates. Although many laboratories use ascorbic acid to prepare lysates, the concentration and pH of the ascorbic acid solutions vary. Some laboratories have reported storing samples at 4°C, while others freeze them before analysis. There are also differences in the incubation times used to deconjugate polyglutamate to monoglutamate species of folate. All these variables contribute to the efficiency of folate extraction from red cells. This lack of harmonization has prevented the inter-laboratory adoption of reference range data. Based on the UK NEQAS Haematinics Assays survey of 2007/08, 40% of laboratories participating in the scheme use their own established reference range for RCF, 30% use data supplied by the manufacturers with the kits purchased, 22% utilize historical data, and the remaining 8% adopt ranges from other laboratories which use the same instrumentation or employ ranges from published literature.[167] Table 11.3 presents the diversity of the lower and upper limits of the reference ranges for serum and RCF assays used, both within the same and different methodologies, along with the manufacturer's cut-offs.

Multimarker Status Evaluation

An approach to assess the adequacy of folate which combines both direct and functional indicators of folate status was proposed by Selhub et al.[170] Irrespective of age and gender, and after adjustment for vitamin B₁₂ and creatinine, the relationship between folate and homocysteine was found to be biphasic: at low folate concentrations, homocysteine concentrations increase as folate concentration falls, but at higher folate concentrations, homocysteine remains unchanged. Therefore two-phase

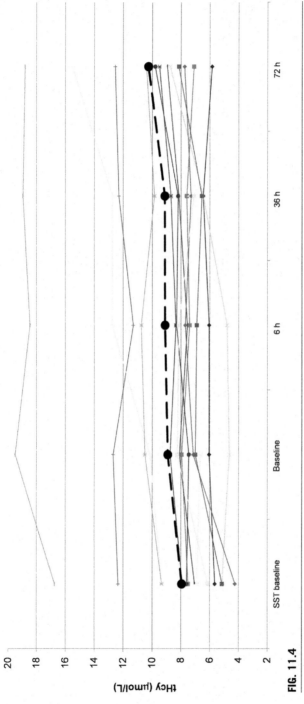

FIG. 11.4

Comparison of serum homocysteine (SST tubes) with homocysteine results from Kabevette tubes (baseline and separated after 6, 36, and 72 h from blood collection), $N = 14$. The *dashed line* represents the mean concentration of all samples.

Table 11.3 RCF and Serum Folate Ranges of Lower and Upper Reference Ranges Used by Laboratories and Manufacturer Cutoffs for Deficiency of RCF and Serum Folate (UK NEQAS Haematinics 2008)

	RCF			Serum Folate		
Method	**Lower Limit Range (µg/L)**	**Upper Limit Range (µg/L)**	**Manufacturer Deficiency Cutoffs (µg/L) (Year)**	**Lower Limit Range (µg/L)**	**Upper Limit Range (µg/L)**	**Manufacturer Deficiency Cutoffs (µg/L) (Year)**
Beckman Access	135–240	750–900	<164 (2003) <237 (2005)	1.9–3	8.5–20.0	<3.0 (2003) <5.21 (2007)
Beckman DXI	160–237	600–1200	<164 (2003) <237 (2005)	1.8–3.4	12.4–20.0	<3.0 (2003) <5.21 (2007)
Architect	60–272	337–1948	<160 (2003)	2–5	12.3–45.3	<2.3 (2003) <4.82 (2007)
Centaur	75–300	300–1000	<280 (2001)	1.1–6.1	12.0–26.0	<3.37 (2001)
Immulite				2.8–3.3	13.5–39.0	<3.0 (2006)
Roche E170	192	775		2.0–4.0	9.0–20.0	<2.0 (2001) <3.8 (2004)

regression models were constructed to estimate the serum and RCF concentration at which the functional indicators achieved a minimum or optimum level. According to this model, the value at which an optimum homocysteine concentration is achieved is approximately 4.4 µg/L (10 nmol/L) for serum folate and 150 µg/L (340 nmol/L) for RCF. However, the authors stated that this model does have a caveat, since a similar model for adequacy of plasma B_{12} which was based on two functional indicators (tHcy and methylmalonic acid (MMA)) provided different minimum cutoff points. In addition, the value for RCF of 150 µg/L cannot be applied to expecting mothers because this concentration is not adequate enough for the prevention of NTDs.[170]

Total Plasma Homocysteine

The most commonly used upper limit for normal tHcy concentration is 15 µmol/L.[143] Other cutoffs in use include 10 µmol/L,[14] 12 µmol/L,[171] 14 µmol/L,[172] 16 µmol/L,[173] and 18 µmol/L,[69] It is important to remember that the circulatory concentration of tHcy increases with age and renal function. The higher tHcy concentrations seen in the elderly may be a consequence of an increased prevalence of intestinal malabsorption, reduced enzymatic activity, reduced kidney function, and other physiological age-related changes. In addition, males have higher tHcy concentrations than females and homocysteine is lower in pregnancy.[174] Therefore all these factors need to be taken into account when interpreting tHcy results. In spite of these differences, most laboratories use only one cutoff value to distinguish normal and elevated tHcy concentrations. Furthermore, one epidemiological study has demonstrated that the risk for vascular disease starts to increase with homocysteine concentrations of between 10.5 and 11.7 µmol/L.[101] The cutoff values for tHcy are usually calculated as the 2.5th–97.5th percentile interval from the "normal" population without disease. However, studies suggest that a large proportion of the "normal" population (~40%) does not consume enough folate to keep the concentration of homocysteine low, hence establishing appropriate reference values for tHcy using this approach is questionable.[174] Varied exclusion criteria can be implemented to overcome the inclusion of individuals with potentially suboptimal folate status when establishing tHcy reference ranges. As an example, Rasmussen et al. found in their study that subjects <30 years should have tHcy <8.1 µmo/L, women between 30 and 59 years <7.9 µmol/L, men (30–59 years) <11.2 µmol/L, and all >60 years <11.9 µmol/L.[174] Likewise, Rossi et al recommends the use of a plasma homocysteine reference range based only on the upper quartile of folate values: 6.5–11.9 µmol/L for women and 8.3–13.7 µmol/L for men.[175]

External Quality Assurance (EQA), Standardization, and Reference Materials

External Quality Assurance or Proficiency Testing schemes primarily assist laboratories in assessing the accuracy of their results and precision of measurements. Precision is also monitored internally via QC materials used with each analytical run, as well as serial measurements of patient samples. Achieving accuracy for all folate assays has been particularly challenging since no formal standardization program exists for folate measurements, thus good performance in an EQA scheme does not guarantee the accuracy of measurements. Moreover, many EQA schemes assess performance using means for the method rather than the overall mean. Such an approach makes it difficult to identify methods with poor performance. In addition, EQA schemes may occasionally provide modified samples, e.g., diluted specimens of animal origin, with various additives which may not be suitable for all methods. Therefore laboratories must seek additional means for accuracy assessment. Comparing results with a higher order reference procedure and analyzing standard reference material can often provide more reassurance in terms of accuracy rather than a satisfactory EQA performance.

Isotope dilution LC-MS/MS methods are accepted as reference methods for the quantification of serum folate.[141] Using this method as reference, UK NEQAS Haematinics assessed the performance of all other methods. The method with lowest bias to the reference method was found to be traceable to the WHO 03/178 standard.[141]

All homocysteine assays demonstrate good performance and several EQA schemes are available, e.g., ERNDIM.

External Quality Assurance schemes for serum folate report an intra-method coefficient of variation of 5%–12%, with overall CV for all methods being <10%. However, significant variability exists for RCF, with intramethod CV between 12 and 32%, and overall CV ~40% (UK NEQAS Haematinics, November 2017).

Serum-based international reference materials have been available for some time but they only provide certified concentrations for 5-MTHF. The National Institute of Standards and Technology (NIST) has developed SRM 1955 and SRM 1950. SRM 1955 became available in 2006; three levels (3 × 1 mL) of serum are provided. This SRM gives certified values for 5-MTHF and homocysteine and reference values for folic acid, 5-formylTHF, and total folate. SRM 1950 uses human frozen plasma, has one level only and like SRM 1955, gives certified values for 5-MTHF and homocysteine, and a reference value for folic acid. The UK-based National Institute for Biological Standards and Control and WHO international Laboratory for Biological Standards have developed 03/178. One level of a freeze-dried specimen is provided and reference values for 5-MTHF, 5-formylTHF, and total folate are given. The same institutions have developed a whole-blood folate international standard (95/528); 1 level of freeze-dried human whole-blood hemolysate with consensus values for total folate.

Acknowledgments

Dr Dominic J. Harrington is greatly acknowledged for his ongoing support of my work. Many thanks go to Renata Gorska of Nutristasis Unit for her artistic skills in drawing the figures and to Kaiya Chowdhary of Nutristasis Unit for conducting

RCF stability study. Thanks also go to Matthew Critcher, Rachel Hann, and Oluwakemi Ajanlekoko of Nutristasis Unit (placement students from the University of Nottingham) for their work with Kabevette tubes and lipemic samples. Special thanks go to Arianna Malefora for her proofreading.

References

1. Chanarin I. The folates. In: Barker BM, Bender DA, editors. *Vitamins in medicine.* William Heinemann Medical Books Ltd; 1980. p. 247–314.
2. Mitchell HK, Snell EE, Williams RJ. The concentration of "folic acid". *J Am Chem Soc* 1941;**63**:2284.
3. Shane B. Folate status assessment history: implications for measurement of biomarkers in NHANES. *Am J Clin Nutr* 2011;**94**:337S–42S.
4. Angier RB, Boothe JH, Hutchings BL, Mowat JH, Semb J, Stokstad EL, et al. Synthesis of a compound identical with the *L. casei* factor isolated from liver. *Science* 1945;**102**:227–8.
5. Hutchings BL, Stokstad EL, Bohonos N, Slobodkin NH. Isolation of a new *Lactobacillus casei* factor. *Science* 1944;**99**:371.
6. Wills L, Stewart A. Experimental an ae mia in monkeys, with special reference to macrocytic nutritional an ae mia. *Br J Exp Pathol* 1935;**16**:444–53.
7. Farber S, Diamond LK. Temporary remissions in acute leukemia in children produced by folic acid antagonist, 4-aminopteroyl-glutamic acid. *N Engl J Med* 1948;**238**:787–93.
8. Hoffbrand AV, Weir DG. The history of folic acid. *Br J Haematol* 2001;**113**:579–89.
9. Czeizel AE, Dudas I. Prevention of the 1St occurrence of neural-tube defects by periconceptional vitamin supplementation. *N Engl J Med* 1992;**327**:1832–5.
10. MRC Vitamin Study Res Group. Prevention of neural tube defects results of the medical research council vitamin study. *Lancet* 1991;**338**:131–7.
11. Scott JM, Weir DG, Molloy A, McPartlin J, Daly L, Kirke P. Folic acid metabolism and mechanisms of neural tube defects. *CIBA Found Symp* 1994;**181**:180–7.
12. Dietary Reference Intakes. *Folate, other B vitamins and choline.* Washington, D.C.: National Academy Press; 1998.
13. US Department of Health and Human Services Food and Drug Administration. *Food standards: amendment of the standards of identity for enriched grain product to require addition of folic acid.* 61. Fed Regist; 1996. p. 8781.
14. Boushey CJ, Beresford SA, Omenn GS, Motulsky AG. A quantitative assessment of plasma homocysteine as a risk factor for vascular disease. Probable benefits of increasing folic acid intakes. *JAMA* 1995;**274**:1049–57.
15. Lucock M. Is folic acid the ultimate functional food component for disease prevention? *BMJ* 2004;**328**:211–4.
16. FAO/WHO. Human Vitamin and Mineral Requirements. In: *Rome, report of a joint FAO/WHO expert consultation Bangkok.* Thailand: FAO; 2002.
17. Thurnham DI, Bender DA, Scott J, Halsted CH. Water-soluble vitamins. In: Garrow JS, James WPT, Ralph A, editors. *Human nutrition and dietetics.* Churchill Livingstone; 2000. p. 249–87.
18. Bailey LB, Stover PJ, McNulty H, Fenech MF, Gregory III JF, Mills JL, et al. Biomarkers of nutrition for development-folate review. *J Nutr* 2015;**145**:1636S–80S.
19. Appling DR. Compartmentation of folate-mediated one-carbon metabolism in eukaryotes. *FASEB J* 1991;**5**:2645–51.

20. Fazili Z, Pfeiffer CM. Measurement of folates in serum and conventionally prepared whole blood lysates: application of an automated 96-well plate isotope-dilution tandem mass spectrometry method. *Clin Chem* 2004;**50**:2378–81.

21. McCance RA, Widdowson EM. *The Composition of foods*. 6th ed. Food Standards Agency; 2002.

22. Finglas PM, Wright AJ, Wolfe CA, Hart DJ, Wright DM, Dainty JR. Is there more to folates than neural-tube defects? *Proc Nutr Soc* 2003;**62**:591–8.

23. Aufreiter S, Gregory III JF, Pfeiffer CM, Fazili Z, Kim YI, Marcon N, et al. Folate is absorbed across the colon of adults: evidence from cecal infusion of (13)C-labeled [6S]-5-formyltetrahydrofolic acid. *Am J Clin Nutr* 2009;**90**:116–23.

24. Herbert V. Experimental nutritional folate deficiency in man. *Trans Assoc Am Phys* 1962;**75**:307–20.

25. Gregory JF. Bioavailability of folate. *Eur J Clin Nutr* 1997;**51**:S54–9.

26. Baily LB. Dietary reference intakes for folate: the debut of dietary folate equivalents. *Nutr Rev* 1998;**56**:294–9.

27. Scientific Advisory Committee on Nutrition (SACN). *Update on folic acid*. GOV.UK; 2017.

28. Food and Nutrition Board IoM. *Dietary reference intakes for thiamin, riboflavin, niacin, vitamin B12, folate, vitamin B12, pantothenic acid, biotin and choline*. Washington, DC: National Academy Press; 1998.

29. Lambie DG, Johnson RH. Drugs and folate metabolism. *Drugs* 1985;**30**:145–55.

30. Erbe RW. Inborn-errors of folate metabolism. *N Engl J Med* 1975;**293**:753–7.

31. Scott JM. Folate vitamin-B12 interrelationships in the central-nervous-system. *Proc Nutr Soc* 1992;**51**:219–24.

32. Butz LW, DuVigneaud V. The formation of a homologue of cystine by the decomposition of methionine with sulphuric acid. *J Biol Chem* 1932;**99**:135–42.

33. Anonymous. Case Records of the Massachusetts General Hospital, Case 19471. Marked cerebral symptoms following a limp of three months' duration. *N Engl J Med* 1933;**209**:1063–6.

34. Finkelstein JD. Homocysteine: a history in progress. *Nutr Rev* 2000;**58**:193–204.

35. Bolander-Gouaille C. *Focus on homocysteine and the vitamins involved in its metabolism*. 2nd ed Springer; 2002.

36. Gerritsen T, Vaughn JG, Waisman HA. The identification of homocystine in the urine. *Biochem Biophys Res Commun* 1962;**9**:493–6.

37. Mudd SH, Skovby F, Levy HL, Pettigrew KD, Wilcken B, Pyeritz RE, et al. The natural-history of homocystinuria due to cystathionine beta-synthase deficiency. *Am J Hum Genet* 1985;**37**:1–31.

38. McCully KS. Vascular pathology of homocysteinemia: implications for the pathogenesis of arteriosclerosis. *Am J Pathol* 1969;**56**:111–28.

39. Mudd SH, Levy HL, Abeles RH. A derangement in B12 metabolism leading to homocystinemia cystathioninemia and methylmalonic aciduria. *Biochem Biophys Res Commun* 1969;**35**:121–3.

40. McCully KS. Homocysteine, vitamins, and vascular disease prevention. *Am J Clin Nutr* 2007;**86**:1563S–8S.

41. Clarke R, Stansbie D. Assessment of homocysteine as a cardiovascular risk factor in clinical practice. *Ann Clin Biochem* 2001;**38**:624–32.

42. Langman LJ, Cole DE. Homocysteine: cholesterol of the 90s? *Clin Chim Acta* 1999;**286**:63–80.

43. Moat SJ, Bonham JR, Tanner MS, Allen JC, Powers HJ. Recommended approaches for the laboratory measurement of homocysteine in the diagnosis and monitoring of patients with hyperhomocysteinaemia. *Ann Clin Biochem* 1999;**36**:372–9.

44. Brosnan JT, Brosnan ME. The sulfur-containing amino acids: an overview. *J Nutr* 2006;**136**(Suppl):1636S–40S.

45. Mudd SH, Levy HL, Skovby F. Disorders of transsulfuration. In: Scriver C, editor. *The metabolic and molecular bases of inherited disease*. New York: McGraw-Hill; 1995. p. 1279–327.

46. Finkelstein JD, Mudd SH, Irreverre F, Laster L. Homocystinuria due to cystathionine synthetase deficiency: the mode of inheritance. *Science* 1964;**146**:785–7.

47. Boers GHJ, Smals AGH, Trijbels FJM, Fowler B, Bakkeren JAJM, Schoonderwaldt HC, et al. Heterozygosity for homocystinuria in premature peripheral and cerebral occlusive arterial-disease. *N Engl J Med* 1985;**313**:709–15.

48. Franco RF, Araujo AG, Guerreiro JF, Elion J, Zago MA. Analysis of the 677 C-->T mutation of the methylenetetrahydrofolate reductase gene in different ethnic groups. *Thromb Haemost* 1998;**79**:119–21.

49. Gudnason V, Stansbie D, Scott J, Bowron A, Nicaud V, Humphries S. C677T (thermolabile alanine/valine) polymorphism in methylenetetrahydrofolate reductase (MTHFR): its frequency and impact on plasma homocysteine concentration in different European populations. EARS group. *Atherosclerosis* 1998;**136**:347–54.

50. Klerk M, Lievers KJ, Kluijtmans LA, Blom HJ, den Heijer M, Schouten EG, et al. The 2756A>G variant in the gene encoding methionine synthase: its relation with plasma homocysteine levels and risk of coronary heart disease in a Dutch case-control study. *Thromb Res* 2003;**110**:87–91.

51. Nygard O, Refsum H, Ueland PM, Vollset SE. Major lifestyle determinants of plasma total homocysteine distribution: the Hordaland homocysteine study. *Am J Clin Nutr* 1998;**67**:263–70.

52. Refsum H, Nurk E, Smith AD, Ueland PM, Gjesdal CG, Bjelland I, et al. The Hordaland homocysteine study: a community-based study of homocysteine, its determinants, and associations with disease. *J Nutr* 2006;**136**(Suppl):1731S–40S.

53. Jacques PF, Bostom AG, Wilson PW, Rich S, Rosenberg IH, Selhub J. Determinants of plasma total homocysteine concentration in the framingham offspring cohort. *Am J Clin Nutr* 2001;**73**:613–21.

54. Patel KR, Sobczynska-Malefora A. The adverse effects of an excessive folic acid intake. *Eur J Clin Nutr* 2017;**71**:159–63.

55. Mason JB. Biomarkers of nutrient exposure and status in one-carbon (methyl) metabolism. *J Nutr* 2003;**133**(Suppl):941S–7S.

56. Bednarska-Makaruk M, Graban A, Sobczynska-Malefora A, Harrington DJ, Mitchell M, Voong K, et al. Homocysteine metabolism and the associations of global DNA methylation with selected gene polymorphisms and nutritional factors in patients with dementia. *Exp Gerontol* 2016;**81**:83–91.

57. Jacob RA, Gretz DM, Taylor PC, James SJ, Pogribny IP, Miller BJ, et al. Moderate folate depletion increases plasma homocysteine and decreases lymphocyte DNA methylation in postmenopausal women. *J Nutr* 1998;**128**:1204–12.

58. Rampersaud GC, Kauwell GP, Hutson AD, Cerda JJ, Bailey LB. Genomic DNA methylation decreases in response to moderate folate depletion in elderly women. *Am J Clin Nutr* 2000;**72**:998–1003.

59. Bower C, Stanley FJ. Dietary-folate as a risk factor for neural-tube defects—evidence from a case-control study in Western Australia. *Med J Aust* 1989;**150**:613–9.
60. Sauer J, Mason JB, Choi SW. Too much folate: a risk factor for cancer and cardiovascular disease? *Curr Opin Clin Nutr Metab Care* 2009;**12**:30–6.
61. McNulty H, Scott JM. Intake and status of folate and related B-vitamins: considerations and challenges in achieving optimal status. *Br J Nutr* 2008;**99**:S48–54.
62. Quadri P, Fragiacomo C, Pezzati R, Zanda E, Forloni G, Tettamanti M, et al. Homocysteine, folate, and vitamin B-12 in mild cognitive impairment, alzheimer disease, and vascular dementia. *Am J Clin Nutr* 2004;**80**:114–22.
63. Verhaar MC, Stroes E, Rabelink TJ. Folates and cardiovascular disease. *Arterioscler Thromb Vasc Biol* 2002;**22**:6–13.
64. Scott J, Weir D. Folate/vitamin B12 inter-relationships. *Essays Biochem* 1994;63–72.
65. Harrington DJ. Investigation of megaloblastic anaemia: Cobalamin, folate and metabolite status. In: Bain ML&IB B, editor. *Dacie and lewis practical haematology*. Elsevier; 2016. p. 187–213.
66. Leeming RJ, Pollock A, Melville LJ, Hamon CG. Measurement of 5-methyltetrahydrofolic acid in man by high-performance liquid chromatography. *Metabolism* 1990;**39**:902–4.
67. Shmeleva VM, Kapustin SI, Papayan LP, Sobczynska-Malefora A, Harrington DJ, Savidge GF. Prevalence of hyperhomocysteinemia and the MTHFR C677T polymorphism in patients with arterial and venous thrombosis from North Western Russia. *Thromb Res* 2003;**111**:351–6.
68. Sobczynska-Malefora A, Harrington DJ, Voong K, Shearer MJ. Plasma and red cell reference intervals of 5-methyltetrahydrofolate of healthy adults in whom biochemical functional deficiencies of folate and vitamin B 12 had been excluded. *Adv Hematol* 2014;**2014**:465623.
69. Smulders YM, Smith DE, Kok RM, Teerlink T, Gellekink H, Vaes WH, et al. Red blood cell folate vitamer distribution in healthy subjects is determined by the methylenetetrahydrofolate reductase C677T polymorphism and by the total folate status. *J Nutr Biochem* 2007;**18**:693–9.
70. McNulty H, Pentieva K, Hoey L, Ward M. Homocysteine, B-vitamins and CVD. *Proc Nutr Soc* 2008;**67**:232–7.
71. Antoniades C, Shirodaria C, Warrick N, Cai S, de Bono J, Lee J, et al. 5-methyltetrahydrofolate rapidly improves endothelial function and decreases superoxide production in human vessels: effects on vascular tetrahydrobiopterin availability and endothelial nitric oxide synthase coupling. *Circulation* 2006;**114**:1193–201.
72. Quere I, Perneger TV, Zittoun J, Bellet H, Gris JC, Daures JP, et al. Red blood cell methylfolate and plasma homocysteine as risk factors for venous thromboembolism: a matched case-control study. *Lancet* 2002;**359**:747–52.
73. Sweeney MR, Staines A, Daly L, Traynor A, Daly S, Bailey SW, et al. Persistent circulating unmetabolised folic acid in a setting of liberal voluntary folic acid fortification. Implications for further mandatory fortification? *BMC Public Health* 2009;**9**:295.
74. Selhub J, Rosenberg IH. Excessive folic acid intake and relation to adverse health outcome. *Biochimie* 2016;**126**:71–8.
75. Troen AM, Mitchell B, Sorensen B, Wener MH, Johnston A, Wood B, et al. Unmetabolized folic acid in plasma is associated with reduced natural killer cell cytotoxicity among postmenopausal women. *J Nutr* 2006;**136**:189–94.
76. Yajnik CS, Deshpande SS, Jackson AA, Refsum H, Rao S, Fisher DJ, et al. Vitamin B12 and folate concentrations during pregnancy and insulin resistance in the offspring: the Pune maternal nutrition study. *Diabetologia* 2008;**51**:29–38.

77. Girotto F, Scott L, Avchalumov Y, Harris J, Iannattone S, Drummond-Main C, et al. High dose folic acid supplementation of rats alters synaptic transmission and seizure susceptibility in offspring. *Sci Rep* 2013;**3**:1465.

78. Christensen KE, Mikael LG, Leung KY, Levesque N, Deng L, Wu Q, et al. High folic acid consumption leads to pseudo-MTHFR deficiency, altered lipid metabolism, and liver injury in mice. *Am J Clin Nutr* 2015;**101**:646–58.

79. Yetley EA. Monitoring folate status in population-based surveys. *Biofactors* 2011;**37**:285–9.

80. Yetley EA, Pfeiffer CM, Phinney KW, Bailey RL, Blackmore S, Bock JL, et al. Biomarkers of vitamin B-12 status in NHANES: a roundtable summary. *Am J Clin Nutr* 2011;**94**:313S–21S.

81. Fazili Z, Whitehead Jr. RD, Paladugula N, Pfeiffer CM. A high-throughput LC-MS/MS method suitable for population biomonitoring measures five serum folate vitamers and one oxidation product. *Anal Bioanal Chem* 2013;**405**:4549–60.

82. Gordon N. Cerebral folate deficiency. *Dev Med Child Neurol* 2009;**51**:180–2.

83. Ramaekers VT, Blau N. Cerebral folate deficiency. *Dev Med Child Neurol* 2004;**46**:843–51.

84. Perez-Duenas B, Ormazabal A, Toma C, Torrico B, Cormand B, Serrano M, et al. Cerebral folate deficiency syndromes in childhood: clinical, analytical, and etiologic aspects. *Arch Neurol* 2011;**68**:615–21.

85. Hansen FJ, Blau N. Cerebral folate deficiency: life-changing supplementation with folinic acid. *Mol Genet Metab* 2005;**84**:371–3.

86. Verbeek MM, Blom AM, Wevers RA, Lagerwerf AJ, van de GJ, Willemsen MA. Technical and biochemical factors affecting cerebrospinal fluid 5-MTHF, biopterin and neopterin concentrations. *Mol Genet Metab* 2008;**95**:127–32.

87. Castro R, Rivera I, Blom HJ, Jakobs C, Tavares DA, I.. Homocysteine metabolism, hyperhomocysteinaemia and vascular disease: an overview. *J Inherit Metab Dis* 2006;**29**:3–20.

88. Miller JW. Homocysteine, Alzheimer's disease, and cognitive function. *Nutrition* 2000;**16**:675–7.

89. Smith AD, Refsum H. Homocysteine, B vitamins, and cognitive impairment. *Annu Rev Nutr* 2016;**36**:211–39.

90. Obeid R, McCaddon A, Herrmann W. The role of hyperhomocysteinemia and B-vitamin deficiency in neurological and psychiatric diseases. *Clin Chem Lab Med* 2007;**45**:1590–606.

91. Phelip JM, Ducros V, Faucheron JL, Flourie B, Roblin X. Association of hyperhomocysteinemia and folate deficiency with colon tumors in patients with inflammatory bowel disease. *Inflamm Bowel Dis* 2008;**14**:242–8.

92. Gatt A, Makris A, Cladd H, Burcombe RJ, Smith JM, Cooper P, et al. Hyperhomocysteinemia in women with advanced breast cancer. *Int J Lab Hematol* 2007;**29**:421–5.

93. van Guldener C, Stehouwer CD. Diabetes mellitus and hyperhomocysteinemia. *Semin Vasc Med* 2002;**2**:87–95.

94. Huijberts MS, Becker A, Stehouwer CD. Homocysteine and vascular disease in diabetes: a double hit? *Clin Chem Lab Med* 2005;**43**:993–1000.

95. Hague WM. Homocysteine and pregnancy. *Best Pract Res Clin Obstet Gynaecol* 2003;**17**:459–69.

96. den Heijer M, Rosendaal FR, Blom HJ, Gerrits WB, Bos GM. Hyperhomocysteinemia and venous thrombosis: a meta-analysis. *Thromb Haemost* 1998;**80**:874–7.

97. Makris M. Hyperhomocysteinemia and thrombosis. *Clin Lab Haematol* 2000;**22**:133–43.

98. Morris MS, Jacques PF, Rosenberg IH, Selhub J. Elevated serum methylmalonic acid concentrations are common among elderly Americans. *J Nutr* 2002;**132**:2799–803.

99. Mudd SH, Levy HL, Krauss JP. Disorders of transsulfuration. In: Childs B, Kinzler KW, editors. *The metabolic and molecular bases of inherited disease*. New York: McGraw-Hill; 2001. p. 2007–56.

100. Clarke R, Daly L, Robinson K, Naughten E, Cahalane S, Fowler B, et al. Hyperhomocysteinemia: an independent risk factor for vascular disease. *N Engl J Med* 1991;**324**:1149–55.

101. Stampfer MJ, Malinow MR, Willett WC, Newcomer LM, Upson B, Ullmann D, et al. A prospective study of plasma homocyst(e)ine and risk of myocardial infarction in US physicians. *JAMA* 1992;**268**:877–81.

102. Arnesen E, Refsum H, Bonaa KH, Ueland PM, Forde OH, Nordrehaug JE. Serum total homocysteine and coronary heart-disease. *Int J Epidemiol* 1995;**24**:704–9.

103. Perry IJ, Refsum H, Morris RW, Ebrahim SB, Ueland PM, Shaper AG. Prospective-study of serum total homocysteine concentration and risk of stroke in middle-aged British men. *Lancet* 1995;**346**:1395–8.

104. Nygard O, Nordrehaug JE, Refsum H, Ueland PM, Farstad M, Vollset SE. Plasma homocysteine levels and mortality in patients with coronary artery disease. *N Eng J Med* 1997;**337**:230–6.

105. Taylor LM, Moneta GL, Sexton GJ, Schuff RA, Porter JM. Homocysteine progression AS. Prospective blinded study of the relationship between plasma homocysteine and progression of symptomatic peripheral arterial disease. *J Vasc Surg* 1999;**29**:8–19.

106. Bonaa KH, Njolstad I, Ueland PM, Schirmer H, Tverdal A, Steigen T, et al. Homocysteine lowering and cardiovascular events after acute myocardial infarction. *N Eng J Med* 2006;**354**:1578–88.

107. Lonn E, Yusuf S, Arnold MJ, Sheridan P, Pogue J, Micks M, et al. Homocysteine lowering with folic acid and B vitamins in vascular disease. *N Eng J Med* 2006;**354**:1567–77.

108. Antoniades C, Antonopoulos AS, Tousoulis D, Marinou K, Stefanadis C. Homocysteine and coronary atherosclerosis: from folate fortification to the recent clinical trials. *Eur Heart J* 2009;**30**:6–15.

109. Baker H, Herbert V, Frank O, Pasher I, Hutner SH, Wasserman LR, et al. A microbiologic method for detecting folic acid deficiency in man. *Clin Chem* 1959;**5**:275–80.

110. Chanarin I, Elmes PC, Mollin DL. Folic-acid studies in megaloblastic anaemia due to primidone. *Br Med J* 1958;**2**:80–2.

111. Ghitis J. The folate binding in milk. *Am J Clin Nutr* 1967;**20**:1–4.

112. Waxman S, Schreiber C, Herbert V. Radioisotopic assay for measurement of serum folate levels. *Blood* 1971;**38**:219–28.

113. Levine S. Analytical inaccuracy for folic acid with a popular commercial vitamin B12/folate kit. *Clin Chem* 1993;**39**:2209–10.

114. Fazili Z, Pfeiffer CM, Zhang M. Comparison of serum folate species analyzed by LC-MS/MS with total folate measured by microbiologic assay and bio-rad radioassay. *Clin Chem* 2007;**53**:781–4.

115. Wegner C, Trotz M, Nau H. Direct determination of folate monoglutamates in plasma by high-performance liquid chromatography using an automatic precolumn-switching system as sample clean-up procedure. *J Chromatogr* 1986;**378**:55–65.

116. Hoppner K, Lampi B. Reversed phase high pressure liquid chromatography of folates in human whole blood. *Nutr Rep Int* 1983;**27**:911–9.

117. Bagley PJ, Selhub J. Analysis of folate form distribution by affinity followed by reversed-phase chromatography with electrical detection. *Clin Chem* 2000;**46**:404–11.

118. Pfeiffer CM, Gregory III JF. Enzymatic deconjugation of erythrocyte polyglutamyl folates during preparation for folate assay: investigation with reversed-phase liquid chromatography. *Clin Chem* 1996;**42**:1847–54.

119. Huang L, Zhang J, Hayakawa T, Tsuge H. Assays of methylenetetrahydrofolate reductase and methionine synthase activities by monitoring 5-methyltetrahydrofolate and tetrahydrofolate using high-performance liquid chromatography with fluorescence detection. *Anal Biochem* 2001;**299**:253–9.

120. Santhosh-Kumar CR, Kolhouse JF. Molar quantitation of folates by gas chromatography-mass spectrometry. *Methods Enzymol* 1997;**281**:26–38.

121. Pfeiffer CM, Fazili Z, McCoy L, Zhang M, Gunter EW. Determination of folate vitamers in human serum by stable-isotope-dilution tandem mass spectrometry and comparison with radioassay and microbiologic assay. *Clin Chem* 2004;**50**:423–32.

122. Newman EM, Tsai JF. Microbiological analysis of 5-formyltetrahydrofolic acid and other folates using an automatic 96-well plate reader. *Anal Biochem* 1986;**154**:509–15.

123. Horne DW, Patterson D. *Lactobacillus casei* microbiological assay of folic acid derivatives in 96-well microtiter plates. *Clin Chem* 1988;**34**:2357–9.

124. Grossowicz N, Waxman S, Schreiber C. Cryoprotected *Lactobacillus casei*: an approach to standardization of microbiological assay of folic acid in serum. *Clin Chem* 1981;**27**:745–7.

125. Davis RE, Nicol DJ, Kelly A. An automated method for the measurement of folate activity. *J Clin Pathol* 1970;**23**:47–53.

126. Molloy AM, Scott JM. Microbiological assay for serum, plasma, and red cell folate using cryopreserved, microtiter plate method. *Methods Enzymol* 1997;**281**:43–53.

127. Sobczynska-Malefora A, Critcher MS, Harrington DJ. The application of holotranscobalamin and methylmalonic acid in hospital patients and total vitamin B12 in primary care patients to assess low vitamin B12 status. *J Hematol Thromb* 2015;**1**:8.

128. Weeks I, Woodhead JS. Chemiluminescence immunoassay. In: Kemeny DM, Challacombe SJ, editors. *ELISA and other solid phase immunoassays. Theoretical and practical aspects.* Cichester, UK: John Wiley & Sons; 2017. p. 265–78.

129. Fazili Z, Pfeiffer CM, Zhang M, Jain R. Erythrocyte folate extraction and quantitative determination by liquid chromatography-tandem mass spectrometry: comparison of results with microbiologic assay. *Clin Chem* 2005;**51**:2318–25.

130. Chladek J, Sispera L, Martinkova J. High-performance liquid chromatographic assay for the determination of 5-methyltetrahydrofolate in human plasma. *J Chromatogr B Biomed Sci Appl* 2000;**744**:307–13.

131. Ihara H, Watanabe T, Hashizume N, Totani M, Kamioka K, Onda K, et al. Commutability of National Institute of Standards and Technology standard reference material 1955 homocysteine and folate in frozen human serum for total folate with automated assays. *Ann Clin Biochem* 2010;**47**:541–8.

132. Hoffbrand AV, Newcombe FA, Mollin DL. Method of assay of red cell folate activity and the value of the assay as a test for folate deficiency. *J Clin Pathol* 1966;**19**:17–28.

133. Sweeney MR, McPartlin J, Weir DG, Scott JM. Measurements of sub-nanomolar concentrations of unmetabolised folic acid in serum. *J Chromatogr B Anal Technol Biomed Life Sci* 2003;**788**:187–91.

134. Opladen T, Ramaekers VT, Heimann G, Blau N. Analysis of 5-methyltetrahydrofolate in serum of healthy children. *Mol Genet Metab* 2006;**87**:61–5.

135. Lin Y, Dueker SR, Jones AD, Clifford AJ. A parallel processing solid phase extraction protocol for the determination of whole blood folate. *Anal Biochem* 2002;**301**:14–20.

136. Chen L, Ducker GS, Lu W, Teng X, Rabinowitz JD. An LC-MS chemical derivatization method for the measurement of five different one-carbon states of cellular tetrahydrofolate. *Anal Bioanal Chem* 2017;**409**:5955–64.

137. Fazili Z, Pfeiffer CM. Accounting for an isobaric interference allows correct determination of folate vitamers in serum by isotope dilution-liquid chromatography-tandem MS. *J Nutr* 2013;**143**:108–13.

138. Nelson BC, Dalluge JJ, Margolis SA. Preliminary application of liquid chromatography-electrospray-ionization mass spectrometry to the detection of 5-methyltetrahydrofolic acid monoglutamate in human plasma. *J Chromatogr B Biomed Sci Appl* 2001;**765**:141–50.

139. Stokes P, Webb K. Analysis of some folate monoglutamates by high-performance liquid chromatography-mass spectrometry I. *J Chromatogr A* 1999;**864**:59–67.

140. Kok RM, Smith DE, Dainty JR, Van Den Akker JT, Finglas PM, Smulders YM, et al. 5-Methyltetrahydrofolic acid and folic acid measured in plasma with liquid chromatography tandem mass spectrometry: applications to folate absorption and metabolism. *Anal Biochem* 2004;**326**:129–38.

141. Blackmore S, Pfeiffer CM, Lee A, Fazili Z, Hamilton MS. Isotope dilution-LC-MS/MS reference method assessment of serum folate assay accuracy and proficiency testing consensus mean. *Clin Chem* 2011;**57**:986–94.

142. Hannisdal R, Ueland PM, Svardal A. Liquid chromatography-tandem mass spectrometry analysis of folate and folate catabolites in human serum. *Clin Chem* 2009;**55**:1147–54.

143. Ueland PM, Refsum H, Stabler SP, Malinow MR, Andersson A, Allen RH. Total homocysteine in plasma or serum: methods and clinical applications. *Clin Chem* 1993;**39**:1764–79.

144. Wannmacher CM, Wajner M, Giugliani R, Filho CS. An improved specific laboratory test for homocystinuria. *Clin Chim Acta* 1982;**125**:367–9.

145. Ducros V, Demuth K, Sauvant MP, Quillard M, Causse E, Candito M, et al. Methods for homocysteine analysis and biological relevance of the results. *J Chromatogr B Anal Technol Biomed Life Sci* 2002;**781**:207–26.

146. Kredich NM, Kendall HE, Spence Jr. FJ. A sensitive radiochemical enzyme assay for S-adenosyl-L-homocysteine and L-homocysteine. *Anal Biochem* 1981;**116**:503–10.

147. Refsum H, Helland S, Ueland PM. Radioenzymic determination of homocysteine in plasma and urine. *Clin Chem* 1985;**31**:624–8.

148. Tan Y, Tang L, Sun X, Zhang N, Han Q, Xu M, et al. Total-homocysteine enzymatic assay. *Clin Chem* 2000;**46**:1686–8.

149. Matsuyama N, Yamaguchi M, Toyosato M, Takayama M, Mizuno K. New enzymatic colorimetric assay for total homocysteine. *Clin Chem* 2001;**47**:2155–7.

150. Stabler SP, Marcell PD, Podell ER, Allen RH. Quantitation of total homocysteine, total cysteine, and methionine in normal serum and urine using capillary gas chromatography-mass spectrometry. *Anal Biochem* 1987;**162**:185–96.

151. Stabler SP, Lindenbaum J, Savage DG, Allen RH. Elevation of serum cystathionine levels in patients with cobalamin and folate deficiency. *Blood* 1993;**81**:3404–13.

152. Ducros V, Schmitt D, Pernod G, Faure H, Polack B, Favier A. Gas chromatographic-mass spectrometric determination of total homocysteine in human plasma by stable isotope dilution: method and clinical applications. *J Chromatogr B Biomed Sci Appl* 1999;**729**:333–9.

153. Windelberg A, Arseth O, Kvalheim G, Ueland PM. Automated assay for the determination of methylmalonic acid, total homocysteine, and related amino acids in human serum or plasma by means of methylchloroformate derivatization and gas chromatography-mass spectrometry. *Clin Chem* 2005;**51**:2103–9.

154. Ubbink JB. Assay methods for the measurement of total homocyst(e)ine in plasma. *Semin Thromb Hemost* 2000;**26**:233–41.

155. Fiskerstrand T, Refsum H, Kvalheim G, Ueland PM. Homocysteine and other thiols in plasma and urine: automated determination and sample stability. *Clin Chem* 1993;**39**:263–71.

156. Sobczynska-Malefora A, Harrington DJ, Rangarajan S, Kovacs JA, Shearer MJ, Savidge GF. Hyperhomocysteinemia and B-vitamin status after discontinuation of oral anticoagulation therapy in patients with a history of venous thromboembolism. *Clin Chem Lab Med* 2003;**41**:1493–7.

157. Saetre R, Rabenstein DL. Determination of cysteine in plasma and urine and homocysteine in plasma by high-pressure liquid chromatography. *Anal Biochem* 1978;**90**:684–92.

158. Nexo E, Engbaek F, Ueland PM, Westby C, O'Gorman P, Johnston C, et al. Evaluation of novel assays in clinical chemistry: quantification of plasma total homocysteine. *Clin Chem* 2000;**46**:1150–6.

159. La'ulu SL, Rawlins ML, Pfeiffer CM, Zhang M, Roberts WL. Performance characteristics of six homocysteine assays. *Am J Clin Pathol* 2008;**130**:969–75.

160. Ivanov AV, Bulgakova PO, Virus ED, Kruglova MP, Alexandrin VV, Gadieva VA, et al. Capillary electrophoresis coupled with chloroform-acetonitrile extraction for rapid and highly selective determination of cysteine and homocysteine levels in human blood plasma and urine. *Electrophoresis* 2017;**38**:2646–53.

161. Fu X, Xu YK, Chan P, Pattengale PK. Simple, fast, and simultaneous detection of plasma total homocysteine, methylmalonic acid, methionine, and 2-methylcitric acid using liquid chromatography and mass spectrometry (LC/MS/MS). *JIMD Rep* 2013;**10**:69–78.

162. Da Silva L, Collino S, Cominetti O, Martin FP, Montoliu I, Moreno SO, et al. High-throughput method for the quantitation of metabolites and co-factors from homocysteine-methionine cycle for nutritional status assessment. *Bioanalysis* 2016;**8**:1937–49.

163. Rafii M, Elango R, House JD, Courtney-Martin G, Darling P, Fisher L, et al. Measurement of homocysteine and related metabolites in human plasma and urine by liquid chromatography electrospray tandem mass spectrometry. *J Chromatogr B Anal Technol Biomed Life Sci* 2009;**877**:3282–91.

164. Hempen C, Wanschers H, van der Sluijs G. A fast liquid chromatographic tandem mass spectrometric method for the simultaneous determination of total homocysteine and methylmalonic acid. *Anal Bioanal Chem* 2008;**391**:263–70.

165. Hannisdal R, Ueland PM, Eussen SJ, Svardal A, Hustad S. Analytical recovery of folate degradation products formed in human serum and plasma at room temperature. *J Nutr* 2009;**139**:1415–8.

166. Clement NF, Kendall BS. Effect of light on vitamin B12 and folate. *Lab Med* 2017;**40**:657–9.

167. UK NEQAS Haematinics. *Report on reference range data collected from haematinics scheme participants in July 2007*. Good Hope Hospital, Haematology Department, Heart of England Foundation Trust; 2008. p. 1–16. Ref Type: Report.

168. Sobczynska-Malefora A, Chowdhary K, Harrington DJ. Red cell folate stability in whole EDTA blood samples. *Ann Clin Biochem* 2016;**53**(Suppl 1):94. Ref Type: Abstract.

169. Sobczynska-Malefora A, Critcher MS, Harrington DJ. *The suitability of Kabevette sample collection tubes for measurement of total plasma homocysteine using the Abbott Architect chemiluminescence method.* In: *10th International Conference: One carbon metabolism, vitamins B and homocysteine*; 2015. P2968.

170. Selhub J, Jacques PF, Dallal G, Choumenkovitch S, Rogers G. The use of blood concentrations of vitamins and their respective functional indicators to define folate and vitamin B12 status. *Food Nutr Bull* 2008;**29**(2 Suppl):S67–73.

171. Hvas AM, Ellegaard J, Nexo E. Vitamin B12 treatment normalizes metabolic markers but has limited clinical effect: a randomized placebo-controlled study. *Clin Chem* 2001;**47**:1396–404.

172. Selhub J, Jacques PF, Wilson PW, Rush D, Rosenberg IH. Vitamin status and intake as primary determinants of homocysteinemia in an elderly population. *JAMA* 1993;**270**:2693–8.

173. Lewerin C, Nilsson-Ehle H, Matousek M, Lindstedt G, Steen B. Reduction of plasma homocysteine and serum methylmalonate concentrations in apparently healthy elderly subjects after treatment with folic acid, vitamin B12 and vitamin B6: a randomised trial. *Eur J Clin Nutr* 2003;**57**:1426–36.

174. Rasmussen K, Moller J, Lyngbak M, Pedersen AM, Dybkjaer L. Age- and gender-specific reference intervals for total homocysteine and methylmalonic acid in plasma before and after vitamin supplementation. *Clin Chem* 1996;**42**:630–6.

175. Rossi E, Beilby JP, McQuillan BM, Hung J. Biological variability and reference intervals for total plasma homocysteine. *Ann Clin Biochem* 1999;**36**:56–61.

Methods for assessment of Vitamin B$_{12}$

12

Dominic J. Harrington

Nutristasis Unit, Viapath, St. Thomas' Hospital, London, United Kingdom
Faculty of Life Sciences and Medicine, King's College London, London, United Kingdom

Chapter Outline

Laboratory Assessment of Vitamin Status. https://doi.org/10.1016/B978-0-12-813050-6.00012-7

Structure and Function

Vitamin B$_{12}$ (cobalamin, B$_{12}$) is a corrinoid that consists of a corrin ring and a single central cobalt atom held in place by two ligands (Fig. 12.1). The lower ligand is a benzimidazole group attached to the ring through a ribose-phosphate group. The metabolic utility of vitamin B$_{12}$ in humans is conferred by an upper ligand consisting of either a methyl or 5′-deoxyadenosyl moiety.[1] De novo biosynthesis of vitamin B$_{12}$ by prokaryotes occurs through aerobic and anaerobic pathways in bacteria and archaea, respectively. To conserve energy, gram-negative bacteria salvage preformed corrinoids and transport them into the cell for vitamin B$_{12}$ assembly.[2] Seemingly vitamin B$_{12}$ synthesis did not make the transition to eukaryotes. Higher plants have no metabolic requirement for the vitamin and have developed no mechanism for its production or storage.

Man, however, has evolved with a dependency on exogenous sources of vitamin B$_{12}$ to support two essential reactions that guard against debilitating hematological and neurological disorders[3, 4] The human serum metabolome in vitamin B$_{12}$ deficiency, and the changes that occur after supplementation, have recently been characterized and revealed connections between vitamin B$_{12}$ status and serum metabolic markers of mitochondrial function, myelin integrity, oxidative stress, and peripheral nerve function, including some previously implicated in Alzheimer and Parkinson diseases.[5] One of the two essential vitamin B$_{12}$-dependent reactions takes place in the cytosol and uses methylcobalamin as a cofactor for methionine synthase during the

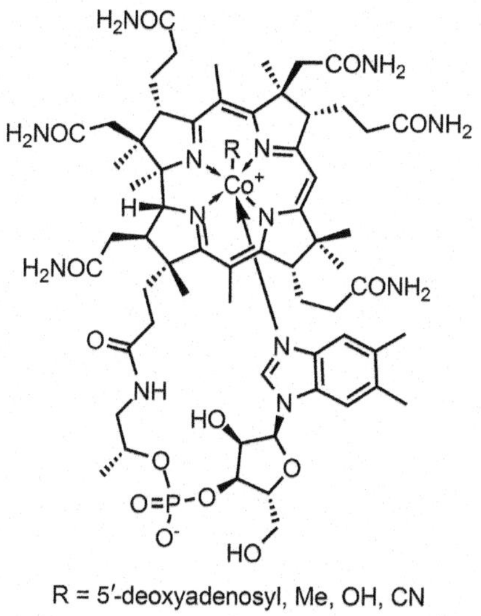

R = 5′-deoxyadenosyl, Me, OH, CN

FIG. 12.1

The basic corrin ring structure of vitamin B_{12} compounds.

remethylation of methionine from homocysteine. Methionine is subsequently converted to adenosylmethionine to supply methyl groups that are critical for the methylation of proteins, phospholipids, neurotransmitters, RNA, and DNA. The second reaction takes place in the mitochondria following the generation of adenosylcobalamin from hydroxocobalamin through the action of vitamin B_{12} coenzyme synthetase. Adenosylcobalamin is essential for the methylmalonyl-CoA mutase-mediated conversion of methylmalonyl-CoA to succinyl-CoA.[1]

Hydroxo- and cyanocobalamin are the most commonly available forms of vitamin B_{12} used to correct deficient states, and originally vitamin B_{12} was a term that referred solely to cyanocobalamin before its use was extended to denote all forms of vitamin B_{12}.[1] Interestingly, hydroxocobalamin is a well-recognized cyanide antidote—with the hydroxo ligand preferentially replaced with cyanide—and it has been proposed that vitamin B_{12} may perform a protective role through the inactivation of low levels of cyanide that are present in foods such as fruits, beans, and nuts.

Dietary Intake

Vitamin B_{12} enters the diet with food of animal origin. Daily intakes of 1 µg and 1.5 µg are recommended by the European Union and UK government, respectively. The United States recommends the higher intake of 2.4 µg.[3] With the exception of those consuming restricted diets, e.g., vegans and vegetarians, dietary intake of vitamin B_{12} comfortably exceeds metabolic requirement and leads to the accumulation of substantial body stores of ~1–5 mg. Typical Western diets provide ~4–6 µg of vitamin

B_{12} each day from which 1–5 µg is absorbed.[1] However, absorption of vitamin B_{12} from a single meal is unable to exceed 1.5–2.5 µg because of the finite capacity of ileal receptors for the vitamin. Maximum absorption efficacy is reestablished within four to 6 h.[6]

Despite the modest daily metabolic requirement, and substantial body reserves of vitamin B_{12}, deficiency is relatively common and may develop for a variety of reasons (Table 12.1). Depleted tissue stores of vitamin B_{12} are signaled biochemically through an increase in the circulatory concentration of homocysteine and methylmalonic acid

Table 12.1 The Causes of Vitamin B_{12} Deficiency

Deficit	Origin
Restricted intake	Malnutrition Reduced intake of food of animal origin Breastfed infants of vitamin B_{12} deficient mothers
Impaired gastric absorption	Atrophic gastritis with achlorhydria Gastrectomy Zollinger-Ellison syndrome
Loss or inactivity of intrinsic factor	Pernicious anemia
Pancreatic insufficiency	Insufficient trypsin to release of haptocorrin bound vitamin B_{12}
Impaired ileal absorption of vitamin B_{12}-intrinsic factor complex	Ileal resection Ileal disease, e.g., Crohn's Inflammatory bowel disease and tuberculous ileitis Tropical sprue Luminal disturbances: chronic pancreatic disease and gastrinoma Parasites: giardiasis, bacterial overgrowth, and fish tapeworm Blind loop syndrome
Congenital	Intrinsic factor receptor deficiency/defect Imerslund–Gräsbeck syndrome Congenital deficiency of intrinsic factor–"Juvenile" pernicious anemia Inborn errors of cobalamin metabolism
Transport protein defects	Haptocorrin deficiency Transcobalamin deficiency
Increased requirement	Hemolysis HIV infection Pregnancy
Acquired	Alcohol *(impedes absorption as consequence of gastritis)* Nitrous oxide *(irreversibly binds to cobalt atom in vitamin B_{12} and deactivates it)* Proton pump inhibitors *(reduce gastric acid production)* H_2 receptor antagonists *(reduce gastric acid production)* Metformin *(impedes absorption)* Colchicine *(reduces IF-B12 receptors)* Slow K *(impedes absorption)* Cholestyramine *(decreases gastric absorption)*

as the functionality of methionine synthase and methylmalonyl-CoA mutase in cells becomes impaired (Chapter 11, Fig. 11.2). Advanced deficient states present clinically with variable sequelae.[3, 4] Since no "Gold Standard" definition of what constitutes a deficient state has been agreed, estimates of the prevalence of vitamin B_{12} deficiency are dependent on the local criteria used to define a deficient state. In the 2001–2004 National Health and Nutrition Examination Survey, 1.6% of subjects over 51 years of age in the United States were defined as deficient using serum concentrations of vitamin $B_{12} < 147$ pmol/L and methylmalonic acid concentrations of $>0.27\,\mu$mol/L as diagnostic cutoffs.[7] There is a consensus view that the prevalence of deficiency increases with age. Using serum concentrations of vitamin $B_{12} < 147$ pmol/L and homocysteine concentrations of $>20\,\mu$mol/L the prevalence of vitamin B_{12} deficiency was ~5% in people 65–74 years of age, and >10% in people 75 years of age or older.[8] Deficient states are also common (~10%) in patient populations.[9]

Absorption

The efficient absorption of dietary vitamin B_{12} is dependent on several key sequential stages. A discrete error in any one of these may lead to deficiency (Fig. 12.2). Initially vitamin B_{12} must be freed from food proteins by the action of pepsin and gastric acid. Once liberated, two proteins compete in the stomach for the free vitamin: intrinsic factor made in gastric parietal cells and haptocorrins (also known as R-binders) which are produced by the salivary glands. At acidic pH, the haptocorrins have a higher affinity for vitamin B_{12} than intrinsic factor and serve to protect the vitamin from acid degradation by producing a haptocorrin-vitamin B_{12} complex. Metabolically inert dietary cobinamides (an intermediate in porphyrin and chlorophyll metabolism) are also bound.

As the contents of the stomach enter the first part of the duodenum the surrounding pH increases leading to partial digestion of the haptocorrins by proteases secreted by the pancreas which frees vitamin B_{12} permitting it to attach preferentially to intrinsic factor. It is the intrinsic factor-cobalamin complex that attaches to cubam receptors, which consist of amnionless and cubilin,[10] that is taken up by endocytosis into the ileal cell. After internalization, vitamin B_{12} is freed once again and transported into the blood, where it meets binding proteins transcobalamin and haptocorrin. Transcobalamin is synthesized in enterocytes and circulates in a predominately unsaturated form, most newly absorbed vitamin B_{12} therefore binds to transcobalamin. Transcobalamin has a rapid turnover and is responsible for the daily transport of ~4 nmol of vitamin B_{12} into cells.[11] There are receptors for holotranscobalamin on the surface of all DNA-synthesizing cells. Once bound to the target tissue, holotranscobalamin undergoes endocytosis before lysosomal degradation, releasing cobalamin for metabolic reactions. Conversely, haptocorrin is almost fully saturated with vitamin B_{12} and inactive vitamin B_{12} analogs in blood and although this protein carries the major part of the vitamin in the circulation, few sites are available to freshly absorb vitamin B_{12}. This protein attaches to cell surface receptors on the liver and other storage cells, and is cleared slowly when compared with holotranscobalamin, with a turnover of 0.1 nmol of vitamin B_{12} daily.[12]

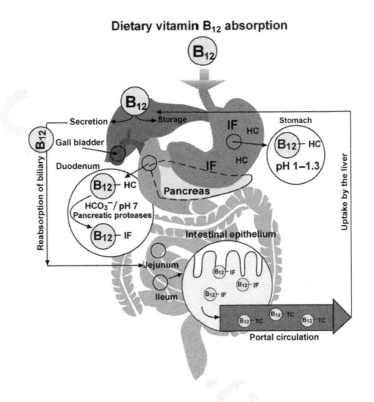

FIG. 12.2

Mechanism of dietary vitamin B_{12} absorption. HC, holohaptocorrin; IF, intrinsic factor; TC, holotranscobalamin.

With thanks to Renata Gorska from the Nutristasis Unit for figure preparation.

Cobalamin undergoes enterohepatic circulation via the liver and bile ducts with 1.4 µg excreted daily in the bile, of which 1 µg is reabsorbed in the ileum. Little, if any, metabolism of the corrinoid ring system takes place, and vitamin B_{12} is excreted as the intact cobalamin.

Clinical Application of Measurement

The prompt detection of vitamin B_{12} deficiency offers the opportunity to correct or prevent megaloblastic anemia and potentially irreversible neuropathy and neuropsychiatric changes.[3, 4, 13] Since the clinical manifestation of vitamin B_{12} deficiency is highly variable, it is key to note that patients who have developed peripheral neuropathy or subacute combined degeneration of the cord may have no discernable hematological diathesis.[14]

Severe deficiency affects all DNA-synthesizing tissues with rapidly dividing cells affected initially—bone marrow is commonly affected, followed by tissues in the epithelial cell surfaces of the gastrointestinal tract, gonads, and skin. Infertility and skin hyperpigmentation can also occur.[3]

Many patients clinically express cobalamin deficiency because of malabsorption caused by an intrinsic factor deficit, which stems from autoimmune disease—a condition known as pernicious anemia. Typical clinical features include: anemia, thrombocytopenia, neutropenia, glossitis, cardiomyopathy, jaundice, weight loss, neurological symptoms. Pernicious anemia has an incidence of ~1:1000 and is arguably unfortunately named since although pernicious anemia is a cause of megaloblastic anemia, deficiency may manifest without hematological symptoms. Pernicious anemia usually manifests in those aged 30 years and more, and affects both sexes equally. It is most frequent in northern Europeans, and relatively infrequent in oriental populations. Two types of antibody have been detected in the sera of patients with pernicious anemia: Type I block the binding of vitamin B_{12} to intrinsic factor; type II prevents the attachment of intrinsic factor or the intrinsic factor-B_{12} complex to ileal receptors. Assay methods have been reviewed.[15] Enzyme-linked immunosorbent assay (ELISA)-based assay kits that detect both type I and type II antibodies are available from a number of companies with the UK NEQAS Haematinics External Quality Assurance Scheme showing poor diagnostic agreement between these kits for a minority of patient-derived samples. In patients who are vitamin B_{12} deficient but intrinsic factor antibody negative, it remains important to establish whether the capacity to absorb the vitamin is normal.

Megaloblastic Anemia

Macrocytosis is the most common reason that vitamin B_{12} status is investigated. Mean red cell volume (MCV) cannot, however, be used diagnostically to evaluate status—the test has poor specificity and sensitivity and values >100 fL as a consequence of vitamin B_{12} insufficiency reflect an advanced deficient state with regards to hematological dysfunction only.[16] The condition is characterized by the presence of megaloblastic red cell precursors in the bone marrow and occasionally also in the blood as a consequence of impaired DNA synthesis. Megaloblasts have a chromatin pattern and increased cytoplasm as a result of asynchrony of nuclear and cytoplasmic maturation with a relatively immature nucleus for the degree of cytoplasmic hemoglobinization. The delay in nuclear maturation is seen in all lineages. A MCV of up to 130 fL may occur, with oval macrocytes, poikilocytes, and hypersegmentation of neutrophils (>5% with more than five nuclear lobes).[17] A decrease in the mean platelet volume and increased platelet anisocytosis may also be seen. As megaloblastic change advances, red cell fragments and small poikilocytes may appear leading to a fall in the MCV.[17] Example peripheral blood film and bone marrow biopsy illustrating megaloblastic features is shown in Fig. 12.3A and B, respectively.

Macrocytic red cells are also seen in myelodysplastic syndromes and myelodysplastic/myeloproliferative neoplasms. Excess alcohol consumption, hypothyroidism,

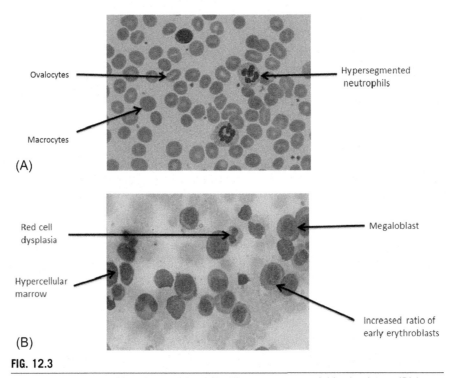

FIG. 12.3

Photomicrographs of (A) peripheral blood film illustrating a megaloblastic picture (B) bone marrow biopsy illustrating some megaloblastic features.

liver disease, aplastic anemia, and the rare inherited orotic aciduria and Lesch-Nyhan syndromes may result in a mild increase in MCV. Automated reticulocyte counts facilitate detection of increased red cell turnover and high MCV as a result of hemolysis or bleeding. Coexisting iron deficiency or thalassemia trait may mask macrocytic changes. Congenital dyserythropoietic anemias types I and III and erythroleukemia exhibit some features of megaloblastic erythropoiesis that are unrelated to vitamin B$_{12}$. Drugs interfering with DNA synthesis result in macrocytosis and megaloblastic erythropoiesis[17] (Table 12.1).

Response to Treatment

Patients frequently report a sense of improvement in their symptoms within 24 h of vitamin B$_{12}$ replacement. However, hematological improvement does not begin for several days and takes up to 2 months to fully resolve.[3] The serum concentration of homocysteine and methylmalonic acid should fall to be within the normal range within a week of treatment, failure to do so suggests an incorrect diagnosis, unless renal failure or other causes of metabolite elevation coexist. There is no advantage in measuring serum vitamin B$_{12}$ or holotranscobalamin in the days postreplacement since any increase merely reflects the pharmacological intervention and not necessarily therapeutic effectiveness.[18] Vitamin B_{12} deficiency-induced neurological deficits may not fully resolve in response to treatment in approximately 6% of affected patients. A delay in the diagnosis and treatment of over 6 months increases the likelihood that the damage will be irreversible.[18]

Analytical Methods
Historical Methods

Laborious bioassays using chicks and rats were the only methods available to isolate the uncharacterized liver constituent that would be an effective treatment for patients with pernicious anemia.[19]

Once vitamin B_{12} had been identified, physical methods based on ultraviolet and visible absorption spectra of the cobalamins were developed to facilitate the measurement of the vitamin in solutions free from interfering substances. The molar extinction coefficient is $27,500\,cm^{-1}/M$ at 361.0 nm.[20] These were complemented with chemical methods. For example, a method based on the hydrolysis of the vitamin to yield 5,6-dimethylbenzimidazole which was benzoylated to 4,5-dimethyldibenzoyl-o-phenylenediamine. This compound was treated with concentrated sulfuric acid to form 4,5-dimethyl-o-phenylenediamine, which was determined colorimetrically or fluorometrically.[21] A more specific method followed from the same authors which was dependent upon the determination of cyanide liberated from vitamin B_{12} by photolysis with monochromatic light and subsequent colorimeric determination.[22]

Microbiological Assays

Microbiological-based assays for vitamin B_{12} were the first to have utility in the clinical laboratory. They proved to be both sensitive and specific, but intrinsically time consuming. Of course, these are a functional assay dependent on utilization of the vitamin by the chosen microbe rather than a direct estimation of vitamin abundance. The first microbiological assay for the antianemic factor became available in 1949 when Hutner et al. exploited the vitamin B_{12}-dependency of *Euglena gracilis*, a chlorophyll-containing protozoon found in pond water.[23] Ross modified the assay for the determination of vitamin B_{12} (as the antianemic factor had become known) in body fluids.[24] Assays using several different strains were developed: *Lactobacillus lactis* Dorner (ATCC 8000), *Lactobacillus leichmannii* (ATCC 7830), *L. leichmannii* (ATCC 4797), and *E. gracilis* Z strain have all been used.[25] Of these *L. leichmannii* (ATCC 7830) gave the most robust assay and was used most frequently. It was a microbiological assay that was first used to assign a potency value to the British Standard for human serum vitamin B_{12},[26] and this was later reclassified as the 1st WHO International Standard (IS) (81/563). Microbiological bioassays for serum vitamin B_{12} are still used, by a small minority of laboratories, and play an important role in the evaluation of new automated methods. They are also used in population studies where they are useful in providing information on the long-term comparability of results.[27]

Radioisotope-Based Dilution Assays for Serum Vitamin B_{12}

The first radioisotope-based dilution assay for vitamin B_{12} was published in 1961.[28] This type of assay was faster to perform than the microbiological-based assays and had the advantage of not being influenced by inhibitory substances such as antibiotics which may be present in serum samples. Disadvantages included the requirement to use radioisotopes, a need for larger sample volumes, and the expensive counting equipment. Results generated by the radioisotope assays were broadly comparable with microbiological assays—although tended to be higher.[29]

Contemporary Methods

At least four distinct laboratory markers of vitamin B_{12} status are used today. Of these, it is the total abundance of vitamin B_{12} in serum that has been used habitually since the 1960s by the majority of clinical and research laboratories. The popularity of this assay reflects successful automation of serum vitamin B_{12} tests by manufacturers of clinical chemistry platforms rather than clinical utility. However, automated assays for holotranscobalamin (marketed as "active B12") have been available for the past decade and offer an alternative approach for the evaluation of vitamin B_{12} status. Holotranscobalamin is the form of vitamin B_{12} taken up by cells to meet metabolic demand.

Assays for sensitive functional markers of B_{12} status, such as the determination homocysteine and methylmalonic acid in serum, have also now been automated. These can be used to detect subtle disturbances in vitamin B_{12}-dependent metabolic pathways that stem from vitamin B_{12} insufficiency several years before deficiency-induced pathologies materialize. Unfortunately, it is not possible to predict which asymptomatic patients flagged as vitamin B_{12} insufficient will later develop a deficient state of clinical significance. In a 1.0- to 3.9-year follow-up study of 432 individuals not treated with vitamin B_{12} after an initial observation of an elevated methylmalonic acid concentration (defined in the study as >280 nmol/L) there was a longitudinal variation in concentrations of 34%. As the study progressed a substantial increase in methylmalonic acid concentration was detected in only 16% of participants, whereas 44% showed a decrease.[30] The imperfect prognostic utility of methylmalonic acid is set against the readiness with which vitamin B_{12} insufficiency may be corrected and an increasing emphasis in healthcare on disease prevention rather than disease treatment.

Laboratory Markers of Vitamin B_{12} Status

A decline in vitamin B_{12} status is initially signaled by a decrease in the abundance of holotranscobalamin. As tissue stores of vitamin B_{12} are utilized to sustain metabolic demand, eventual depletion leads to impaired performance of vitamin B_{12}-dependent pathways and an elevation in the plasma concentration of homocysteine and methylmalonic acid. The plasma concentration of vitamin B_{12} (as measured by serum vitamin B_{12} assays) falls more slowly and may not precede the pathological presentation of an advanced deficient state—this is because although vitamin B_{12} that is bound to haptocorrin accounts for the majority of the vitamin that is found in plasma, no known biological role has yet been described. Haptocorrin-bound vitamin B_{12} is also cleared from the circulation slowly (over a period of several months) which can mask the more rapid clearance of holotranscobalamin.

Preanalytical Sample Preparation

Sample collection

Venous blood (4–5 mL) should be collected from a peripheral vein into the appropriate serum collection tubes. The tubes should then be inverted eight to 10 times without shaking to ensure proper anticoagulation.

Serum is ideally required, but heparin plasma can also be used. Red cell vitamin B_{12} determination is occasionally performed by research groups,[31] yet the diagnostic utility of this test remains untested and will not be discussed here.[32] Lipemic samples are known to interfere with the holotranscobalamin assay and should be avoided if possible. Hemolyzed samples should be processed and interpreted with caution—grossly hemolyzed samples should be rejected. Vitamin B_{12} is frequently measured in tandem with folate. In some cases sodium ascorbate 5 mg/mL is added to samples to stabilize the folates and extend sample storage times,[33] which necessitates introduction of a separate vitamin B_{12} sample tube since ascorbate interferes with cobalamin analysis.

Serum preparation

The red cells should be removed after centrifugation for 10 min. at $1000–2000 \times g$ as soon as possible after collection. Samples must be fibrin free and without bubbles. The concentration of serum vitamin B_{12}, holotranscobalamin, and methylmalonic acid can reliably be determined if samples are stored at 2–8°C for up to 7 days. If it is not possible to analysis promptly the serum should be stored frozen at −20°C for up to 30 days or colder if longer storage duration is intended. Avoid more than three freeze/thaw cycles. If serum separating tubes are used, the analyst must visually inspect samples prior analysis to ensure that the gel has effectively separated cells from serum.

For homocysteine analysis, blood should be collected into serum tubes (serum) or tubes containing EDTA or lithium heparin (plasma). Alternatively, samples can be collected into purpose-manufactured homocysteine tubes (Kabevette Vacuum HCY 837V 3.5). The Kabevette tubes prolong the stability of unspun samples for up to 72 h.[34] Analysis on citrated plasma can be performed (homocysteine is included as part of some patient thrombophilia screens—for which citrated plasma is the preferred sample type) in which case the result should be multiplied by 1.1 to take into account the volume of citrate in the tube. Underfilled/overfilled tubes cannot be processed. Plasma should be separated from the red blood cells as soon as possible after centrifugation (ideally within 60 min and kept on ice prior to separation). Preanalytical sample handling is of critical importance. Delayed removal of red blood cells can result in artificially high concentrations of homocysteine. Factors affecting albumin concentrations will alter homocysteine because it is protein bound and venipuncture should not be performed after venous stasis or following the subject resting in supine position.

Competitive-binding luminescence-based serum B_{12} assays

Highly automated competitive-binding luminescence-based assays (CBLA) became widely available from the beginning of the 1990s. It is this technique that still predominates today. CBLA are rapid and well suited to meeting the current high demand for vitamin B_{12} analysis. Analytical limitations associated with this approach are known and include a propensity to generate erroneous results in some patients with proven pernicious anemia.[35]

Key steps of CBLA assays include:

(A) *Release of endogenous binding proteins and conversation to cyanocobalamin.* In serum, vitamin B_{12} is bound to either transcobalamin or haptocorrin. Vitamin B_{12} is released from the binding proteins using alkaline hydrolysis (NaOH at pH 12–13) in the presence of potassium cyanide to convert cobalamin to the more stable cyanocobalamin. Dithiothreitol is used to stop rebinding of released vitamin B_{12}.

(B) *Binding of vitamin B_{12} to the kit binder.* The pH of the sample is readjusted to be optimal for the competitive binding of the free cyanocobalamin by the kit-binding agent. The serum-derived cyanocobalamin competes with labeled cobalamin, which is complexed to a chemiluminescent or fluorescent substrate or enzyme, for limited binding sites on porcine intrinsic factor. Specificity for cobalamin is ensured by blocking potentially contaminating corrinoid binders (R binders) using an excess of blocking cobinamide.

(C) *Separation of bound and unbound vitamin B_{12}.* Following competitive binding, bound and unbound vitamin B_{12} is separated using a number of electro- or physicochemical and immunological methods. The *Roche Elecsys* utilizes an electrochemiluminescence measuring cell in which the bound B_{12}-ruthenium-intrinsic factor complex, attached to paramagnetic particles by biotin-streptavidin, is magnetically captured onto the surface of an electrode. The *Abbott Architect* uses polymer microparticle beads with an iron core, coated with porcine intrinsic factor to bind vitamin B_{12}. The bound vitamin B_{12} is then immobilized by positioning the assay reaction vessel in front of a magnet which pulls the paramagnetic microparticles onto the side wall of the reaction vessel. The reaction vessel contents are then aspirated and the reaction vessel refilled with buffer (in total the aspiration and refill step is repeated three times). Some assays use murine antiintrinsic factor antibody-enzyme conjugates as part of the signal generation.

(D) *Signal generation.* The bound vitamin B_{12} fraction is detected by addition of a chemiluminescent, fluorescent, or colorimetric enzyme substrate, which results in generation of fluorescence or light emission. There are two types of signal: flash, which is pH or electrically induced, and plateau, which is sustained. The initial rate of reaction or the area under the curve is used to calculate the result.

Electrochemiluminescence Immunoassay

In the *Roche Elecsys* platforms a voltage is applied to the electrode on which the bound vitamin B_{12}-ruthenium-intrinsic factor complexes have been immobilized by magnetic attraction. This generates electrochemical luminescence that is measured by the photomultiplier, the relative light units (RLUs) being inversely proportional to the sample vitamin B_{12} concentration.

Chemiluminescence

The VITROS vitamin B_{12} assay signal is measured by the use of a luminogenic substrate and an electron transfer agent-horseradish peroxide bound conjugate catalyzed the oxidation of the luminogenic substrate producing light. The amount of light present is indirectly related to the concentration of vitamin B_{12} present in the sample.

Enzyme-Linked Fluorescence

The *Siemens Centaur* and the *Abbott Architect* use acridinium esters bound to vitamin B_{12}-intrinsic factor complex coupled to paramagnetic particles; photons are emitted in response to pH change. The Siemens *Immulite 2000/2500* uses adamantyl dioxetane phosphate as an alternative substrate, which is cleaved by alkaline phosphatase-labeled B_{12}-intrinsic factor complex, resulting in generation of a plateau chemiluminescent signal. *Beckman Coulter Access* employs alkaline phosphatase/dioxatane phosphate (Lumi-Phos) for signal generation.

Limit of detection

The assay limit of detection is defined as the concentration of analyte at 2SD of 20 replicates above the zero standard, and for vitamin B_{12} assays is normally in the region of 22 pmol/L (30 ng/L). The functional sensitivity limit of serum vitamin B_{12} assays (concentration at which the coefficient of variation of the assay is <20%) is closer to 37 pmol/L (50 ng/L).

Comparability of serum vitamin B_{12} CBLA

Serum vitamin B_{12} assays are calibrated independently by manufacturers with traceability to an internally manufactured standard material rather than an international certified reference material. This confers poor agreement between commercially available assays for serum vitamin B_{12} as evidenced by UK NEQAS survey returns (Fig 12.4). The use of assay calibrants that are traceable to metrological standards assists with assay harmonization and the meaningful comparison of population data. The World Health Organization International Standard for serum B_{12}, 03/178, was ratified in 2007,[36] as a consensus of vitamin B_{12} protein-binding assays. However, poor alignment of the seven main analytical platforms continues.

Standards, Accuracy, and Precision of Cobalamin Assays

There is no reference method for serum vitamin B_{12} measurement. UK NEQAS Haematinics began an external quality assurance scheme for the determination of serum vitamin B_{12} in 1978 http://birminghamquality.org.uk/eqa-programmes/hic/. Work toward standardization began in the 1980s when seven laboratories participated

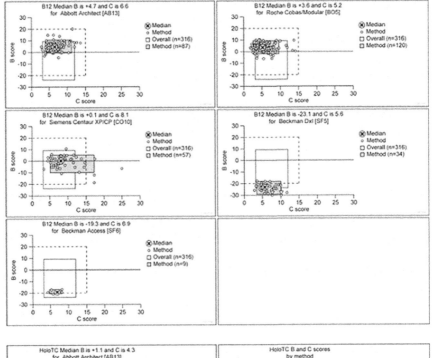

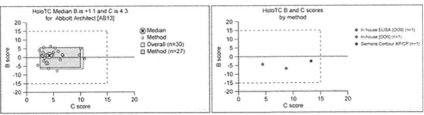

FIG. 12.4

(A) Data from the UK NEQAS Haematinics Scheme showing performance calculated over a rolling window of 6 months (18 External Quality Assurance specimens circulated) by five analytical methods used for the analysis of vitamin B$_{12}$ in serum. Methods clockwise from top left (UK NEQAS method abbreviation): Abbott Architect [AB13[; Roche Cobas/Modular [BO5]; Beckman Dxl [SF5]; Beckman Access [SF6]; Siemens Centaur [CO10]. (B) Data from the UK NEQAS Haematinics Scheme showing performance calculated over a rolling window of 6 months (18 External Quality Assurance specimens circulated) by three analytical methods used for the analysis of "Active B12" in serum. The B score is the average bias of all Specimen % biases [(result−target)/target]×100% during the rolling 6-month window. The C score is the SD of the B score and shows consistency of bias over the same rolling time period. The *blue box* indicates the 5th–95th centiles for each method. The *unfilled box* indicates the overall 5th–95th centiles irrespective of method. The *dotted box* indicates limits of acceptable performance defined as ±20% B score and 20% C score. All analyses were performed during 2017.

With permission from Birmingham Quality, University Hospitals Birmingham NHS Foundation Trust.

in a collaborative study using *E. gracilis* as the test organism in the turbidimetric Euglena vitamin B_{12} assay, for the analysis of an ampouled preparation of serum labeled 81/563. In 1985 the National Biological Standards Board established the preparation as the British Standard for Human Serum Vitamin B_{12} with an assigned potency of 320 pg/mL.[26] The standard was reclassified as the 1st WHO IS, (81/563) in 1992 but subsequently found to be positive for antihepatitis C virus (HCV) and HCV RNA. Twenty-three laboratories in seven countries analyzed pooled human serum from seven donors donated by UK NEQAS Haematinics and assigned a potency of 480 pg/mL to the 2nd WHO IS for serum vitamin B_{12} (03/178) that was ratified in 2007.[36] The simultaneous revaluation of the 1st WHO IS for human serum vitamin B_{12} (81/563) using contemporary methods was 332 pg/mL (103.75% of the value assigned in 1985).

The 1st and 2nd WHO IS for vitamin B_{12} have been little used. Poor uptake has partly been attributed to the lack of a primary reference measurement procedure which limits the degree of metrological traceability of the reference material. A second barrier is that in serum there are multiple forms of vitamin B_{12} at a range of proportions which prevents calibration in SI units.

Internal Adjustment Calibration

Most automated assay systems use calibration curves that are stored on barcode systems with each reagent lot. The barcode also contains the mathematical formulae for shifting or adjusting the observed responses to the master curve when the instrument requires routine calibration.

Reference Range

The lower limit of reference ranges for the interpretation of serum vitamin B_{12} assay is typically set in the region of 148 pmol/L (200 ng/L).[37] Although there is no consensus and cutoffs range from as low as 125 pmol/L to as high as 200 pmol/L.[38] As presented in Table 12.2 the assigned cutoff point determines the sensitivity and specificity of the serum vitamin B_{12} assay.

Table 12.2 Laboratory Test for the Evaluation of Vitamin B_{12} Status

Diagnostic Test	Deficiency Indicator	Comments
Full blood count	Macrocytosis	Late indicator of deficiency 20% of deficient patients present with no hematological deficit
Blood film	Oval macrocytes, hypersegmented neutrophils (>5% with >5 lobes); Howell-Jolly bodies	Hypersegmented neutrophils are not always present

Continued

Table 12.2 Laboratory Test for the Evaluation of Vitamin B$_{12}$ Status—Cont'd

Diagnostic Test	Deficiency Indicator	Comments
Reticulocyte count	Absolute count low pretreatment	Reticulocyte count increases by day 6 posttherapy and confirms vitamin B$_{12}$ deficit Coexisting iron deficiency may prevent reticulocyte increase
Serum vitamin B$_{12}$	Typical lower limit of reference range set at ~148 pmol/L Selectivity and specificity of 250 pmol/L cutoff is 85% and 45%, respectively[8, 38] Selectivity and specificity of 125 pmol/L cutoff is 35% and 95%, respectively[8, 38]	Evidence of metabolic vitamin B$_{12}$ deficiency (elevated homocysteine and/or methylmalonic acid) may be seen with serum vitamin B$_{12}$ concentrations >300 pmol/L. Conversely, with no clinical signs or symptoms and normal homocysteine and methylmalonic acid reflects poor specificity of total vitamin B$_{12}$ assay B$_{12}$ levels may be borderline low due to severe folate deficiency
Serum holotrans-cobalamin	Typical lower limit of reference range set at <25 pmol/L Selectivity and specificity of a 60 pmol/L cutoff is ~90% and ~55%, respectively[8, 38] Selectivity and specificity of a 30 pmol/L cutoff is ~65% and ~90%, respectively[8, 38]	If holotranscobalamin 25–70 pmol/L then suggest measure methylmalonic acid (or homocysteine if methylmalonic acid not available) Holotranscobalamin concentration in serum is influenced by recent dietary intake
Plasma or serum homocysteine	Typical upper limit of reference range set at <15 µmol/L. Age, sex, and renal function specific cutoffs rarely applied but desirable. Examples include: ≤10 µmol/L, children <15 year; ≤13 µmol/L, females (during pregnancy <10 µmol/L); ≤15 µmol/L, males aged 15–65 year; ≤15 µmol/L, females; ≤17 µmol/L males aged 65–74 year; ≤20 µmol/L all >74 year	Homocysteine concentration is heavily influenced by folate status. Vitamins B$_6$ and B$_2$ also influence concentration Homocysteine concentration heavily influenced by renal function
Serum methylmalonic acid	Poor consensus on upper limit of reference range. >280 nmol/L commonly adopted. Other frequently used cutoffs (and associated terminology) include >450 nmol/L (probable deficiency) and >750 nmol/L (definite deficiency)	Methylmalonic acid concentration heavily influenced by renal function and age

Holotranscobalamin

The novelty of holotranscobalamin as a laboratory marker of vitamin B_{12} status rests on the fact that cellular uptake of the vitamin is reliant on a receptor-mediated endocytosis process involving transcobalamin, a plasma protein that carries vitamin B_{12} and a cell surface receptor that specifically binds cobalamin-saturated transcobalamin. Only a minor fraction of circulating cobalamin measured by serum vitamin B_{12} assays is bound to transcobalamin. It is this fraction that is known as holotranscobalamin. Assays for holotranscobalamin are marketed as "active B12." Active B12 assays are increasingly superseding serum B_{12} assays in Australia, Austria, Canada, Germany, Holland, Nordic countries, Switzerland, and the United Kingdom.

Holotranscobalamin "active B_{12}" ELISA

An expert group published a method in 2002 that combined an ELISA method for the determination of transcobalamin with a procedure for the removal of transcobalamin not carrying vitamin B_{12}.[39]

Holotranscobalamin "active B_{12}" radioimmunoassay

The expert group collaborated with the development of commercial assays for holotranscobalamin, with the first of two iterations of radioimmunoassay from axis-shield released at the beginning of 2001 (the commercial radioimmunoassay was withdrawn in 2007). The now obsolete method used magnetic microspheres coated with monoclonal antibody to holotranscobalamin and achieved separation from haptocorrin by a magnetic separator. A $^{57}CoB_{12}$ tracer together with a reducing and a denaturing agent was added to destroy the holotranscobalamin linkage. When the vitamin B_{12} binder containing intrinsic factor was added, the free vitamin B_{12} and tracer competed for binding. The unbound tracer was removed by centrifugation and the bound fraction was measured using a gamma counter. The measured radioactivity reflected the competition between tracer and vitamin B_{12} bound to transcobalamin (i.e., holotranscobalamin). The concentration of vitamin B_{12} in the sample was calculated from a calibration curve using recombinant human holotranscobalamin. The assay required 0.4 mL of sample volume, had a coefficient of variation of <10%, with a limit of detection of 10 pmoL/L. The assay time was 4 h.

Holotranscobalamin "active B_{12}" immunoassay

An automated random-access analyzer method for holotranscobalamin (marketed as "active B12") on the *Abbott AxSYM* analyzer followed (worldwide launch excluding United States 27th June 2006)[40] which in turn was superseded by an assay on the *Abbott Architect*, a two-step quantitative immunoassay using chemiluminescent microparticle immunoassay technology (September 27, 2011). An ELISA assay from Axis-Shield was the next to launch (April 13, 2012) and more recently Siemens released the Centaur active-B12 (AB12) assay for use outside of the United States (January 22, 2016) and Roche Diagnostics released the ELECSYS assay in October 2017. The BioHit active B12 ELISA and IBL International active B12 (holotranscobalamin) ELISA are also available.

Calibration

All commercially available assays for holotranscobalamin use distinct calibrators that are traceable back to common frozen primary reference calibrators which are held by Axis-Shield. The reference calibrators were prepared gravimetrically from a stock solution of recombinant human holotranscobalamin and adjusted using a panel of serum samples of "known" holotranscobalamin concentration to correct for the difference in ionic strength and pH between the calibrator buffer and serum samples (the "known" values having been assigned by the radioimmunoassay (RIA) assay which was the predicate device for subsequent assays). Each batch of calibrators that is released to the market is matched against this material. Now that a value for holotranscobalamin has been added to IS 03/178 Axis-Shield are considering how this material may be used to maintain assay calibration and further aid assay alignment. However, this is rather circular since the calibrators used to assign the holotranscobalamin value to IS 03/178 overwhelmingly originate from Axis-Shield.

The *Abbott Architect* holotranscobalamin assay is a two-step quantitative immunoassay that uses chemiluminescent microparticle immunoassay technology. During the first step, sample and antiholotranscobalamin-coated paramagnetic microparticles are combined. Holotranscobalamin present in the sample binds to the antiholotranscobalamin-coated microparticles. After washing, the second assay step involves the addition of antitranscobalamin acridinium-labeled conjugate to create a reaction mixture. Following another wash cycle, pretrigger and trigger solutions are added to the reaction mixture. The resulting chemiluminescent reaction is measured as RLUs. There is a direct relationship between the amount of holotranscobalamin in the sample and the RLUs detected by the Architect I System optics. The assay is designed to have an assay range of <5.0 to 128 pmoL/L with a percent cross-reactivity of ≤10% from vitamin B$_{12}$-binding proteins apotranscobalamin and haptocorrin.

The assay from Siemens is performed using the ADVIA Centaur® XP and ADVIA Centaur® XPT Immunoassay Systems with an acridinium ester tracer signal generator and two murine antitranscobalamin conjugate antibodies. The assay was relaunched in 2017 as newly traceable to the WHO International Standard for vitamin B$_{12}$, serum folate, and holotranscobalamin (National Institute Biological Standards and Control, NIBSC code 03/178). In the literature from the manufacturer supporting the launch of the assay, the relationship of the active B12 assay (y) to the WHO IS NIBSC 03/178 (x) was determined by dilution recovery (observed range of 9.78–101.76 pmol/L) and is described using a weighted linear regression and a Pearson coefficient. ADVIA Centaur AB12 (y) = 0.90 (x) − 0.12 pmol/L, r = 1.00 The assay requires 50 μL of serum and has a reported lower limit of detection of 1.19 pmol/L, lower limit of quantification of 5.00 pmol/L, and a range up to 146 pmol/L. Percent cross-reactivity of ≤10% from vitamin B$_{12}$-binding proteins apotranscobalamin and haptocorrin.

The assay from Roche takes 18 min to perform and involves a first incubation step in which 18 μL of sample, a biotinylated monoclonal antiholotranscobalamin antibody, and a monoclonal antitranscobalamin antibody labeled with ruthenium complex react to form a sandwich complex. During a second incubation, after the addition of streptavidin-coated microparticles, the complex becomes bound to a solid phase via interaction of biotin and streptavidin. The reaction mixture is aspirated into the measuring cell where microparticles are magnetically captured onto the surface of an electrode. Unbound substances are then removed with ProCell II M. Application of a voltage to the electrode induces chemiluminescent emissions which are measured by a photomultiplier. Results are determined via a calibration curve which is generated by a two-point calibration curve.

Comparison of Active B$_{12}$ Assays From Abbott and Siemens

For 111 serum samples in the range of 11.39–116.96 pmol/L, the relationship of the ADVIA Centaur AB12 assay (y) and the Abbott Architect assay (x) has been described using Passing-Bablok regression and a Pearson coefficient using the ADVIA Centaur XP system.

ADVIA Centaur AB12 (y) = 1.26 (x) − 4.62 pmol/L (intercept), r = 0.94.

Comparison of Active B$_{12}$ Assays From Abbott and Roche

Using the Abbott Architect ci8200 and Roche Cobas e411, 120 serum samples were analyzed for active B12. Results covered the range of <25–>100 pmol/L. Method comparison was performed using weight Deming regression and Bland-Altman plots. Regent lots were also evaluated against WHO 03/178. The Roche Cobas method showed a small constant bias of 9 pmol/L against the Abbott Architect assay with a proportional bias of up to 23 pmol/L at the clinical decision point. Precision for the Roche method was also evaluated according to CLSI-EP5 A3 and demonstrated an intrarun precision of <4% and interrun precision of <6%.[41]

Limitations of Active B$_{12}$ Assays

Two groups have reported rare variants in the transcobalamin gene that interfere with the "active B$_{12}$" assay.[42, 43] The minor allele rs35838082 (p.R215W) is rare in Caucasians with a minor allele frequency (MAF) of <0.01 but more common in South Asians (MAF ~0.02) and those of African origin (MAF ~0.25). Holotranscobalamin results for these patients are erroneously low (~5 pmoL/L) despite all other laboratory markers reported as normal and an absence of clinical deficiency.[42, 43]

Holotranscobalamin External Quality Assurance

UK NEQAS Haematinics began offering an external quality assurance scheme in 2011 http://birminghamquality.org.uk/eqa-programmes/hic/.

Reference Range

In a study of 137 ostensibly healthy blood donors the serum holotranscobalamin concentration range was 40–150 pmol/L. In an audit of 4175 patients from a large London teaching hospital 5% of patients had a holotranscobalamin concentration < 25 pmol/L (classified as deficient) and 24% had a concentration between 25 and 50 pmol/L. In the latter group, 31% of patients were found to have a methylmalonic acid concentration > 280 nmol/L and were classified as deficient.[9]

Some 10% of circulating transcobalamin is saturated with a reference range of 5%–20%; and typically 15%–50% of vitamin B$_{12}$ is bound to transcobalamin. In subjects who were vitamin B$_{12}$ deficient, holotranscobalamin was 2–34 pmol/L and the transcobalamin saturation was 0.4%–3%, well below the reference interval, providing a clear cutoff from normal sera. Nexø's method combines a sensitive ELISA[44] for holotranscobalamin with a simple procedure for removal of the unsaturated transcobalamin or apotranscobalamin.

Methylmalonic Acid

The concentration of methylmalonic acid in serum reflects the availability and utilization of adenosylcobalamin in mitochondria. The interpretation of methylmalonic acid is considered the proxy gold standard and the most representative marker of metabolic vitamin B$_{12}$ insufficiency. There are, however, a number of confounding factors that should be considered when interpreting laboratory methylmalonic acid results. These key factors include age, renal function, and sex.[45] Examples of upper limits of the range are 280 nmol/L (<65 y) and 360 nmol/L for patients over the age of 65 years.

Methylmalonic Acid Measurement

Methylmalonic acid was first isolated from human urine in 1957.[46] Preliminary investigation on the excretion of methylmalonic acid by patients with severe megaloblastic anemia caused by vitamin B$_{12}$ deficiency followed in 1962.[47] At the same time, in the first example of methylmalonic acid being measured directly in urine by gas chromatography, a survey which included normal controls; patients with clinical evidence of vitamin B$_{12}$ deficiency; patients with low serum vitamin B$_{12}$ concentrations and no clinical signs; patients with related conditions with normal serum vitamin B$_{12}$ concentrations; and patients with other severe deficiencies was also reported.[48]

Early methods for methylmalonic acid determination in human urine included semiquantitative assays using paper[49] and thin layer chromatography.[50] Giorgio and Plaut used a spectrophotometric method although this was hampered by poor specificity.[51] Modifications using three wavelengths[52] and solvent extraction followed from other authors.[53]

A different approach was to determine methylmalonic acid indirectly by conversion to propionate but these were found to overestimate methylmalonic acid abundance.[54,55] Following Cox and White's first direct measurement,[47] improved gas chromatography methods for direct measurement in blood and urine were developed.[56,57] The first gas chromatographic method coupled to mass spectrometry was also reported bringing with it improved sensitivity and specificity although at the time the equipment was expensive and not widely available.[58] Methods using HPLC were also developed during the middle of the 1980s[59,60] but at the time were only sensitive enough for urine.

The early techniques were very manual and not sensitive enough to support routine determination in clinical laboratories. Measuring the concentration of methylmalonic acid in urine also had the disadvantage of requiring correction for creatinine excretion. Various methods are now available for the determination in serum and this is the preferred matrix.

Automated liquid chromatography tandem mass spectrometry with electrospray ionization-based method

An automated method capable of processing >200 samples daily has been available in the author's laboratory since 2012. Analysis is carried out using a Gerstel Multipurpose Sampler coupled to a LC-ESI-MS/MS. For sample preparation, 200 μL of serum or plasma is placed into a 2 mL glass screw top autosampler vial and the vial capped using magnetically transportable PolyMag caps (GESRTEL, Germany, http://www.gerstelus.com). The sample is then placed on the vial tray of the multipurpose sampler (MPS). Extraction of methylmalonic acid from serum or plasma is achieved through protein precipitation. Eight hundred microliters of acetonitrile containing 0.5% acetic acid and the internal standard MMA-d3 at a concentration of 165 nmol/L is added to the sample. The vial is then moved using the magnetic transport to the *CF*-100 centrifuge whereby the contents are thoroughly vortexed for 30 s to assist in the protein precipitation. The vial is then centrifuged at 3000 rpm for 1 min to separate the proteins from the supernatant in preparation for injection. Sample analysis is fully automated by means of an external injection valve and loop fitted onto the MPS, 20 μL of extract is injected. Separation is achieved by means of a Merck PEEK ZIC®-HILIC 2.1 I.D. × 100 mm; 3.5 μm particle size column (http://www.nestgrp.com). The chromatographic mobile phase consists of acetonitrile and 100 nM ammonium acetate adjusted to pH 4.5 with formic acid in water. An isocratic elution followed by a gradient column reconditioning is performed. LC run time is 8 min; column temperature is maintained at 30°C. The limit of quantification in a 200 μL sample is 60 nmol/L.

Gas chromatography mass spectrometry-based method

Plasma or serum methylmalonic acid is extracted, purified and, using *tert*-butyldimethylsilyl derivatives of methylmalonic acid, measured by GC-MS. A deuterated stable isotope of methylmalonic acid is used as an internal standard. The use of dicyclohexyl, another derivative of methylmalonic acid, is described by Rasmussen.[61]

High performance liquid chromatography-based method

Methylmalonic acid is extracted from 250 μL serum/EDTA plasma into ethyl acetate along with the internal standard: ethylmalonic acid (EMA).[62] To achieve sufficient assay sensitivity it is necessary to generate fluorescent derivatives of these analytes with dicyclohexylcarbodiimide prior to their separation by reversed-phase HPLC with an acidified mix of acetonitrile, tetrahydrofuran, and water containing dibutylamine. Using this method, the limit of detection and quantification for the measurement of MMA in a 250 μL sample is 15.4 nmol/L and 20.3 nmol/L, respectively.

Total Homocysteine

The total concentration of homocysteine in serum reflects the availability and utilization of methylcobalamin in the cytosol. Plasma homocysteine is elevated in both vitamin B_{12} and folate deficiency and in individuals with a common genetic polymorphism of methylenetetrahydrofolate reductase, C677T polymorphism. Folate status is a far more powerful determinant of serum plasma concentrations than vitamin B_{12} (~4–5-fold); therefore utility for the assessment of vitamin B_{12} status is improved in territories where supplementation of food with folic acid is mandated. Inborn errors in metabolism that manifest with greatly elevated elevations in serum homocysteine include those found in the *CBS* gene that encodes cystathionine-β-synthase (CBS), (EC 4.2.1.22). CBS catalyzes the first step of the transsulfuration pathway from homocysteine to cystathionine using the cofactor vitamin B_6 (pyridoxal-phosphate). Deficiency in vitamin B_6 may therefore also cause mild elevations in serum homocysteine.

Homocysteine Measurement

During the 1950s many tests were developed for the determination of methionine and cysteine, yet the lack of a specific test for homocysteine made it difficult to follow the pathway of methionine metabolism in complete detail. This led to the development of the first assay for homocysteine by Pasieka and Morgan in which homocysteine was detected in biological systems using paper chromatograms which were first sprayed with ninhydrin, heated, and then treated with mercuric nitrate solution to produce a cherry red spot surrounded by deep blue—a color specific for homocysteine.[63]

There was a proliferation in methods for the determination of homocysteine from the 1980s once elevated concentrations in serum were identified as a risk factor for thrombotic disease. For quantification, plasma homocysteine requires protein precipitation and reduction of disulfide bonds. The S–H group of homocysteine is derivatized using a thiol-specific reagent and the resulting adduct is detected. A variety of methods have evolved including ion-exchange amino acid analyzers,[64] radioenzymatic determination, capillary gas chromatography[65] stable isotope dilution combined with capillary GC-MS,[66] liquid chromatography electrospray tandem mass spectrometry,[67] and HPLC methods using fluorochromophore detection.

Automated enzyme immunoassays are commonly utilized today.[68] For a full overview of contemporary analytical methods, please refer to Chapter 11.

Identifying the Etiology of Vitamin B_{12} Deficiency
Radioisotopic absorption test—Schilling test

The radiolabeled cyanocobalamin and bovine intrinsic factor reagents required for the Schilling test have been withdrawn. The test was useful for identifying the cause of vitamin B_{12} deficiency and of up to four parts (part I, basic test; part II, with intrinsic factor; part III, following course of antibiotics; part IV, pancreatic enzymes taken for 3 days) in which the urinary excretion of radiolabeled vitamin B_{12} with and without intrinsic factor was established.[69,70] If during part I of the investigation <5%

of the labelled vitamin B_{12} was excreted and during part II excretion was normal or near to normal then malabsorption because of a result of lack of intrinsic factor (e.g., pernicious anemia) was indicated. If both parts I and II were abnormal then malabsorption not resulting from intrinsic factor deficiency was indicated, e.g., Crohn's disease. If Part III was normal after antibiotics then abnormal bacterial growth was likely. If part IV was normal after pancreatic enzyme replacement then pancreatic insufficiency was likely.

Nonisotopic absorption tests—CobaSorb

In healthy subjects, three to 4 h after an oral dose of vitamin B_{12} is ingested, the serum concentration of holotranscobalamin increases and reflects freshly absorbed vitamin B_{12}. A sensitive vitamin B_{12} absorption test developed by Nexø et al., and named the CobaSorb test, that relies on the holotranscobalamin assays to identify which patients may benefit from oral courses of vitamin B_{12} rather than the more commonly used replacement by IM injection is available.[71] The CobaSorb test involves measuring holotranscobalamin before and 2 days after daily intake of three times 9 μg B_{12} (cyanocobalamin form). The authors assigned a cutoff of >22% and >10 pmol/L to demonstrate active absorption with the caveat that it should not be used if the baseline B_{12} level is >65 pmol/L—in this situation the C-CobaSorb test is suggested.[72]

Plasma-intrinsic factor-antibodies

Two types of plasma-intrinsic factor-antibodies have been detected in the plasma of >60% of patients with pernicious anemia[15,73] with type I blocking the binding of B_{12} to intrinsic factor and type II stopping the attachment of intrinsic factor or intrinsic factor-B_{12} complex to ileal receptors. Assay methods have been reviewed.[73]

Achlorhydria

Achlorhydria as a cause of cobalamin malabsorption is indicated by the presence of raised gastrin concentrations.[74] Gastric parietal cell antibodies are detected in ~90% of patients with pernicious anemia; however, since these antibodies are also found in ~15% of elderly subjects, detection has limited utility.

Transcobalamin measurement

Haptocorrin binds 80% or more of vitamin B_{12} in serum and circulates in a predominately saturated firm referred to as holohaptocorrin. Conversely transcobalamin circulates in a predominately unsaturated form. Notable increases in the abundance of haptocorrin are seen in chronic myeloid leukemia, primary myelofibrosis, and other myeloproliferative neoplasms. Primary liver cancer (fibrolamellar hepatoma) is also associated with synthesis of large quantities of an abnormal form of haptocorrin. In laboratories using the serum vitamin B_{12} test (which is unable to discriminate between haptocorrin and transcobalamin bound vitamin B_{12}) high total values are detected. Conversely, low vitamin B_{12} concentrations without evidence of vitamin B_{12} deficiency may indicate a congenital decrease in haptocorrin.[75,76]

Congenital transcobalamin deficiency[77] leads to fulminating pancytopenia and megaloblastosis within 2 months of birth. Laboratory investigation using serum vitamin B_{12} assays will show normal total levels.

A rapid method to quantify total transcobalamin, which can be implemented using routine platforms using commercial "active B12" tests has recently been described.[78] In addition, apotranscobalamin can be assessed by subtracting endogenous holotranscobalamin concentration which can be measured in the same run, securing the same calibration level for all three parameters (holotranscobalamin, apotranscobalamin, and total transcobalamin). Using this approach the total transcobalamin 95% reference range was calculated to be 500–1276 pmol/L. The apotranscobalamin 95% reference interval was calculated to be 448–1184 pmol/L.[78]

Best Practice
First-Line Assessment of Vitamin B_{12} Status

The serum vitamin B_{12} and holotranscobalamin assays have both been highly automated using competitive binding luminescence technologies. Large numbers of samples can therefore be quickly screened at an affordable cost. Automated analytical solutions for homocysteine (immunoassays and LC-MS/MS) and methylmalonic acid (LC-MS/MS) are also available but have not been widely used for first-line assessment

Utility of Serum B_{12} Assays

The measurement of vitamin B_{12} in serum has been the most commonly used approach for the evaluation of vitamin B_{12} status since the 1960s. Cellular status is inferred through the abundance of vitamin B_{12} in the circulation and comparison against a predefined reference range. The long-standing application of this test is a consequence of the wide availability of highly automated laboratory platforms that perform serum vitamin B_{12} determinations at a low cost.

Screening of patients with macrocytic anemia

The serum vitamin B_{12} assay is an effective tool for the screening of vitamin B_{12} status in patients with macrocytic anemia. A serum vitamin B_{12} concentration below the lower limit of the laboratory reference range is diagnostic of a vitamin B_{12}-deficient state.[1, 13] In these patients the availability of vitamin B_{12} has been diminished to a point at which efficient DNA synthesis can no longer be sustained leading to defective cell nucleus maturation.[79–81] An increase in the MCV of up to 130 fL may be observed. If on investigation a serum vitamin B_{12} concentration within or above the reference range is found, then folate deficiency should be considered since it is the interplay between folate and B_{12} that is responsible for the megaloblastic anemia seen as a consequence of both vitamin deficiencies (Chapter 11).

Utility as a screening test in mixed patient populations

Patients with suspected vitamin B_{12} deficiency often do not have anemia but present with common neurologic symptoms such as symmetric paresthesias, numbness, and gait problems.[14, 82] The severity of megaloblastic anemia is not correlated with neurologic dysfunction.[14, 82] Serum vitamin B_{12} assays provide an estimate of total vitamin B_{12} abundance, but can give no indication of cellular utilization; for this reason, possible vitamin B_{12} deficiency cannot confidently be excluded when results fall in the indeterminate range of 125–250 pmol/L. This limitation is illustrated by the evaluation of a serum vitamin B_{12} assay for the detection of vitamin B_{12} deficiency against a methylmalonic acid cutoff of 750 nmol/L (a concentration indicative of "definite" B_{12} deficiency) in which sensitivities of ~35% and ~85% were found at 125 and 250 pmol/L, respectively. The corresponding assay specificity was ~95% and ~45%, respectively.[83] The lower serum vitamin B_{12} cutoff for diagnosing B_{12} deficiency is typically set at ~148 pmol/L (200 ng/L).[37] Elevations in serum methylmalonic acid and homocysteine that signify disturbances in metabolic networks consistent with vitamin B_{12} deficiency occur as high as 300 pmol/L.[84] Further investigation using a second-line test is necessary for serum B_{12} results that fall within the indeterminate range.

Application during pregnancy

Vitamin B_{12} utilization increases during pregnancy from 2.4 μg/d to 6.0 μg/d. Failure to diagnosis deficient states is associated with adverse outcomes.[85–87] Laboratories typically interpret results against inappropriate reference range cutoffs derived from nonpregnant population, including men. Hemodilution and a decrease in abundance of haptocorrin can lead to a ~50% decline in serum vitamin B_{12} at term.[88] Since serum vitamin B_{12} assays are unable to discriminate between haptocorrin- and transcobalamin-bound (holotranscobalamin) B_{12} interpretation of results is challenging.

Interpretation of results > 1000 pmol/L

Serum vitamin B_{12} concentrations >1000 pmol/L are not uncommon. With the exception of ongoing vitamin B_{12} replacement regimes, an unexpectedly high concentration of vitamin B_{12} in serum can frequently be traced to changes in the abundance of vitamin B_{12}-binding proteins. An increase in the abundance of haptocorrin is a feature of some malignant diseases. An increased concentration of serum vitamin B_{12} can also be caused by the presence of auto antibodies against transcobalamin which does not appear to be related to the development of vitamin B_{12} deficiency.[89] High vitamin B_{12} concentrations may also be a consequence of immunoglobulin-complexed B_{12} resulting in assay interference.

Analytical challenges associated with serum B_{12} assays based on competitive binding luminescence technologies

Automated assays for the measurement of serum B_{12} are based on competitive binding luminescence technologies. In the presence of high-titer antiintrinsic factor antibodies in serum from patients with pernicious anemia, assays based on competitive binding luminescence technologies generate spurious results.[35] A study illustrated failure rates of serum vitamin B_{12} assays in the analysis of samples from patients previously diagnosed unequivocally with pernicious anemia. Six of 23 (26%) patients were missed by the *Beckman Coulter Access assay*, which used the UniCel DxI

800 Immunoassay System, 5 of 23 (22%) by the *Roche Elecsys Systems Modular Analytics E170*, and 8 of 23 (35%) by the *Siemens Advia Centaur assay.*[35]

Utility of Holotranscobalamin Assays

Holotranscobalamin is a test that is currently applied in several different modes. Some laboratories use the test as a sole status indicator[90]; others as a first-line screening test in conjunction with a second-line test[9]; and others as a second-line test in conjunction with a serum B_{12} assay.[91] Holotranscobalamin has been shown to be unaffected by assay interference from high-titer intrinsic factor antibody concentrations.[92] In addition, holotranscobalamin is not subject to the ~50% fall in total vitamin B_{12} concentrations seen in normal pregnancy.[88]

Utility as a screening test in mixed patient populations

A holotranscobalamin concentration <25 pmol/L strongly suggests vitamin B_{12} deficiency. In common with the serum vitamin B_{12} assay there is an indeterminate range of results. For holotranscobalamin results that fall in the range 25–70 pmol/L it is not possible to confidently exclude vitamin B_{12} deficiency. Further investigation using a second-line test is necessary. An evaluation of the holotranscobalamin assay for the detection of serum B_{12} deficiency against a methylmalonic acid cutoff of 750 nmol/L showed sensitivities of ~65% and ~90% at 30 and 60 pmol/L, respectively. The corresponding assay specificity was ~90% and ~55%, respectively.[83]

Comparison of Holotranscobalamin and Serum B_{12} Assays

At this time a definitive direct comparison between holotranscobalamin and serum vitamin B_{12} assays is not possible. This is because no single "gold standard" laboratory marker of vitamin B_{12} status has been identified. The best data available is in the form of receiver operator characteristic curves produced by a study in which subjects were classified as vitamin B_{12} deficient if they had a serum methylmalonic acid concentration >750 nmol/L. This study showed holotranscobalamin to be a moderately more reliable marker of vitamin B_{12} status than serum vitamin B_{12} with superior sensitivity and specificity.[83]

Holotranscobalamin concentrations do not correlate well with total serum vitamin B_{12}. Holotranscobalamin <25 pmol/L occurs in ~5% of laboratory requests in General Patient hospital population and are indeterminate (25–50 pmol/L) for ~25% of requests.[9] However, elevations in metabolic markers of vitamin B_{12} status are known to occur at holotranscobalamin levels 50–70 pmoL/L at a frequency of 50–54 pmol/L, 17%; 55–59 pmol/L, 21%; 60–64 pmol/L, 15%; and 65–70 pmol/L, 14%.[9, 93]

Second-Line Testing

It is possible to address the proportion of results which are considered to be indeterminate by the holotranscobalamin and serum vitamin B_{12} assays through further secondary testing using methylmalonic acid and homocysteine.

Utility of Homocysteine Assays

In territories where the addition of folic acid to all enriched cereal-grain foods is not mandated, the utility of an elevation in plasma homocysteine concentration to identify vitamin B_{12} deficiency is limited by a codependency on 5-methyltetrahydrofolate abundance. 5-Methyltetrahydrofolate is a more powerful nutritional determinant of homocysteine than methylcobalamin. Other, more modest, nutritional determinants of homocysteine include vitamin B_6 and vitamin B_2.[94]

It is preferable to apply reference ranges for the interpretation of homocysteine that are age and sex specific, and which consider renal function. Examples of reference ranges suitable for use in the clinical setting are ≤10 μmol/L, children <15 years.; ≤13 μmol/L, females (during pregnancy <10 μmoL/L); ≤15 μmol/L, males aged 15–65 years.; ≤15 μmol/L, females; ≤17 μmol/L males aged 65–74 years.; ≤20 μmol/L all >74 years.

Utility of Methylmalonic Acid

The determination of methylmalonic acid in serum is considered the proxy gold standard marker vitamin B_{12} status. A consensus view on reference range cutoffs has yet to be achieved, with some laboratories citing a serum concentration >280 nmol/L as indicative of vitamin B_{12} insufficiency in patients <65 years with normal renal function.[9] Other laboratories adopt cutoff values as high as 750 nmol/L.[1] Interpretation of this useful marker is more challenging in the elderly and those with impaired renal function[44] since to some extent the elevated concentrations seen in these patients are independent of vitamin B_{12} status. Measuring methylmalonic acid concentrations in these patients before and several days post a pharmacological dose of vitamin B_{12} can be useful, with a drop of >200 nmol/L indicative of vitamin B_{12} deficiency.

Multiassay Approaches to the Laboratory Diagnosis of Vitamin B_{12} Status

Second-line screening of vitamin B_{12} status

Mathematical models that combine multiple markers of vitamin B_{12} status into a single diagnostic indicator have been developed.[95] The most comprehensive model consists of a "four-variable" analysis to calculate a combined indicator of vitamin B12 status (cB12), with cB12 expressed as cB12 = log10[(holotranscobalamin · s erum B12)/(methylmalonic acid homocysteine)] (age factor).[95] Using the formula the four individual markers are transformed into a single variable dependent on age and is known as the "Fedosov's wellness score." cB12 is interpreted as *high-normal, normal, low-normal, deficient,* and *severely deficient.* "Fedosov's wellness score" has been used to evaluate B_{12} status in nonanemic healthy Swiss senior citizens.[96] This approach has also been taken in a preliminary study to evaluate the genetic epidemiology of static and functional biomarkers of vitamin B_{12} status in older adults.[97]

Recommended Approaches for the Laboratory Assessment of Vitamin B$_{12}$ Status

At this time the application of a "four-variable" analysis consisting of serum vitamin B$_{12}$, holotranscobalamin, homocysteine, and methylmalonic acid to generate Fedosov's wellness score for the evaluation of vitamin B$_{12}$ status is not possible in most laboratories, barriers include cost and access to all four assays to generate the required data. To improve accessibility to this approach Fedosov has adapted his model and validated "three-variable" or "two-variable" variants. Analysis of these approaches has shown that the new models estimate vitamin B$_{12}$ status within acceptable margins of error when compared to the "four-variable" analysis.[98]

Recommended "Three-Variable" Analysis: Serum Vitamin B$_{12}$, Holotranscobalamin, and Methylmalonic Acid

Using models developed that calculate the Fedosov's wellness score based on the application of three assays being used in combination with a correction for folate on the model through its modulation of homocysteine[98] the best simulation of the "four-variable" analysis: serum B12, holotranscobalamin, and methylmalonic acid. Homocysteine should be omitted.

The next best "three-variable" analysis is achieved by omitting either serum B12 or holotranscobalamin from the "four-variable" analysis and combining the chosen assay with methylmalonic acid and homocysteine.[98]

Recommended "Two-Variable" Analysis: Holotranscobalamin and Methylmalonic Acid

The smallest "two-variable" analysis error (when compared to the "four-variable" analysis) is achieved when holotranscobalamin is combined with methylmalonic acid.[98] It is this approach that the author recommends for laboratory diagnosis of vitamin B$_{12}$ status since an acceptable deviation from the "four-variable" analysis is achieved without the expense of performing additional tests.

Example Laboratory Assessment Algorithm

An example laboratory assessment algorithm that has been used extensively in the author's laboratory since 2012 that utilizes holotranscobalamin (first-line assay) with MMA (second-line assay) is shown in Fig. 12.5. In laboratories where holotranscobalamin is not available the recommended approach is serum vitamin B$_{12}$ combined with methylmalonic acid (Fig. 12.5). Note that in the interest of economy in this example a sequential selection algorithm is followed, i.e., whether a second-line assay is performed is dependent on the outcome of the first-line assay rather than both tests being performed on all samples. In the author's laboratory ~5% of samples from a mixed patient population have a holotranscobalamin <25 pmol/L and are classified as deficient; an indeterminate concentration of 25–70 pmol/L

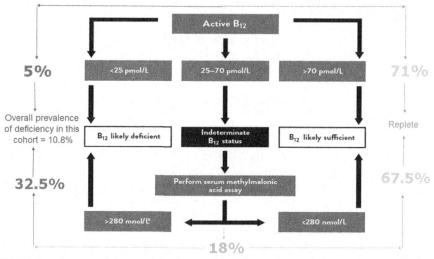

FIG. 12.5

An example laboratory assessment algorithm for vitamin B_{12} status assessment in use in the author's laboratory since 2012 that uses holotranscobalamin and methylmalonic acid. Proportion of patients defined as deficient, requiring second-line testing, and replete represents experience over a 3-month period and the analysis of 4175 consecutive patients referred to Guy's and St. Thomas' NHS Foundation Trust (January to March 2012). *Cutoff applied to patients <65 years old with good renal function (a cutoff of >360 nmol/L applied to patients >65 years).

is measured in ~25% of samples leading to second-line MMA analysis. All other samples are classified as replete. A methylmalonic acid concentration >280 nmol/L (used as the cutoff to define "deficient") is subsequently present in the indeterminate samples as follows: holotranscobalamin 25–29 pmol/L, elevated methylmalonic acid in 41% of samples; 30–34 pmol/L, 32%; 35–39 pmol/L, 33%; 40–44 pmol/L, 30%; 45–49 pmol/L, 26%; 50–54 pmol/L, 17%; 55–59 pmol/L, 21%; 60–64 pmol/L, 15% and 65–70 pmol/L, 14%. Total prevalence of B_{12} deficiency in the mixed patient population is 10.8%.[9, 93]

Acknowledgments

I remain indebted to Dr. Malcolm Hamilton and Mrs. Sheena Blackmore, for their willingness to share their significant experience and expertise in the diagnosis of vitamin B_{12} status. Thanks to Dr. Agata Sobczyńska-Malefora, Renata Gorska, David Card, Denise Oblein, and the Nutristasis Unit at Guy's and St. Thomas' Hospital NHS Foundation Trust for ongoing expertise in the development and application of assays to establish vitamin status.

References

1. Bor M, Nexø E. Vitamin B12–cobalamin. In: Herrmann W, Obeid R, editors. *Vitamins in the prevention of human diseases*. Berlin: De Gruyter; 2011. p. 187–271.
2. Fang H, Kang J, Zhang D. Microbial production of vitamin B12: a review and future perspectives. *Microb Cell Factories* 2017;**16**:15.
3. Hunt A, Harrington D, Robinson S. The diagnosis, investigation and management of vitamin B_{12} deficiency. *BMJ* 2014;**349**:5226.
4. Green R. Vitamin B12 deficiency from the perspective of a practicing hematologist. *Blood* 2017;**129**:2603–11.
5. Brito A, Grapov D, Fahrmann J, Harvey D, Green R, Miller JW, et al. The human serum metabolome of vitamin B_{12} deficiency and repletion, and associations with neurological function in elderly adults. *J Nutr* 2017; https://doi.org/10.3945/jn.117.248278.
6. Heyssel R, Bozian R, Darby W, Bell MC. Vitamin B_{12} turnover in man. The assimilation of vitamin B_{12} from natural foodstuff by man and estimates of minimal daily dietary requirements. *Am J Clin Nutr* 1966;**18**:176–84.
7. Pfeiffer C, Johnson C, Jain R, Yetley EA, Picciano MF, Rader JI, et al. Trends in blood folate and vitamin B-12 concentrations in the United States, 1988-2004. *Am J Clin Nutr* 2007;**86**:718–27.
8. Clarke R, Grimley Evans J, Schneede J, Nexø E, Bates C, Fletcher A, et al. Vitamin B_{12} and folate deficiency in later life. *Age Ageing* 2004;**33**:34–41.
9. Sobczyńska-Malefora A, Gorska R, Pelisser M, Ruwona P, Witchlow B, Harrington DJ, et al. An audit of holotranscobalamin ('Active' B12) and methylmalonic acid assays for the assessment of vitamin B12 status: application in a mixed patient population. *Clin Biochem* 2014;**47**:82–6.
10. Fyle J, Madsen M, Højrup P, Christensen EI, Tanner SM, de la Chapelle A, et al. The functional cobalamin (vitamin B12)-intrinsic factor receptor is a novel complex of cubilin and amnionless. *Blood* 2004;**103**:1573–9.
11. Hom B, Olesen H. Plasma clearance of 57 cobalt-labelled vitamin B12 bound in vitro and in vivo to transcobalamin I and II. *Scand J Clin Lab Invest* 1969;**23**:201–11.
12. Nexø E, Gimsing P. Turnover in humans of iodine- and cobalamin-labeled transcobalamin I and of iodine-labeled albumin. *Scand J Clin Lab Invest* 1975;**35**:391–8.
13. Devalia V, Hamilton M, Molloy A. British committee for standards in haematology. Guidelines for the diagnosis and treatment of cobalamin and folate disorders. *Br J Haematol* 2014;**166**:496–513.
14. Lindenbaum J, Healton EB, Savage DG, Brust JC, Garrett TJ, Podell ER, et al. Neuropsychiatric disorders caused by cobalamin deficiency in the absence of anemia or macrocytosis. *N Engl J Med* 1988;**318**:1720–8.
15. Waters H, Smith C, Howarth J, Dawson DW, Delamore IW, et al. A new enzyme immunoassay for the detection of total type I and type II intrinsic factor antibody. *J Clin Pathol* 1989;**42**:307–12.
16. Oosterhuis WP, Niessen RW, Bossuyt PM, Sanders GT, Sturk A. Diagnostic value of the mean corpuscular volume in the detection of vitamin B12 deficiency. *Scand J Clin Lab Invest* 2000;**60**:9–18.
17. Harrington DJ. Investigation of megaloblastic anaemia: cobalamin, folate and metabolite status. In: Bain BJ, Bates I, Laffan DM, editors. *Dacie and lewis practical haematology*. 12th ed. London: Elsevier; 2016. p. 187–213.
18. British Columbia Medical Association. Cobalamin (vitamin B12) deficiency—investigation and management. In: *Guidelines and protocols*. 2013. 1 May.

19. McDowell LR. *Vitamin history, the early years.* 1st ed. Florida: University of Florida; 2013272.
20. Hill JA, Pratt JM, RJP W. The chemistry of vitamin B_{12}. Part I. The valency and spectrum of the coenzyme. *J Chem Soc* 1964;5149–53.
21. Boxer GE, Rickards JC. Chemical determination of vitamin B_{12}; I. Determination of 5,6-dimethylbenzimidazole by colorimetric and fluorometric methods. *Arch Biochem* 1950;**29**:75–84.
22. Boxer GE, Rickards JC. Chemical determination of vitamin B12. II. The quantitative isolation and colorimetric determination of millimicrogram quantities of cyanide. *Arch Biochem* 1951;**30**:372–81.
23. Hutner SH, Provasoli L, Stockstad EL, Hoffmann CE, Belt M, Franklin AL, et al. Assay of anti-pernicious anemia factor with Euglena. *Proc Soc Exp Biol Med* 1949;**70**:118–20.
24. Ross GI. Vitamin B_{12} assay in body fluids. *Nature* 1950;**166**:270–1.
25. Sauberlich HE, Dowdy RP, Skala JH. *Laboratory tests for the assessment of nutritional status.* Cleveland, Ohio: CRC Press; 197460–70.
26. Curtis AD, Mussett M, Kennedy D. British standard for human serum vitamin B12. *Clin Lab Haematol* 1986;**8**:135–47.
27. Amos RJ. Investigation of megaloblastic anaemia. In: Lewis SM, Bain BJ, Bates I, editors. *Dacie and lewis practical haematology.* 9th ed. London: Churchill Livingstone; 2001. p. 129–48.
28. Barakat RM, Ekins RP. Assay of vitamin B_{12} in blood. A simple method. *Lancet* 1961;**2**:25–6.
29. Raven JL, Robson MB, Morgan JO, Hoffbrand AV. Comparison of three methods for measuring vitamin B_{12} in serum: radioisotopic, *Euglena gracilis* and *Lactobacillus leichmannii. Br J Haematol* 1972;**22**:21–31.
30. Hvas AM, Ellegaard J, Nexø E. Increased plasma methylmalonic acid level does not predict clinical manifestations of vitamin B_{12} deficiency. *Arch Intern Med* 2001;**161**:1534–41.
31. Valente E, Scott JM, Ueland PM, Cunningham C, Casey M, Molloy AM. Diagnostic accuracy of holotranscobalamin, methylmalonic acid, serum cobalamin, and other indicators of tissue vitamin B12 status in the elderly. *Clin Chem* 2011;**57**:856–63.
32. Carmel R. Holotranscobalamin: not ready for prime time. *Clin Chem* 2012;**58**:644–5.
33. Blackmore S, Pfeiffer C, Hamilton M, et al. Recoveries of folate species from serum pools sent to participants of the UK NEQAS haematinics scheme in february and march 2004. *Clin Chim Acta* 2005;**355**:S459. UKNP 1.2. abstract.
34. Sobczyńska-Malefora A, Critcher M, Harrington DJ. *The suitability of Kabevette sample collection tubes for measurement of total plasma homocysteine using the Abbott Architect chemiluminescence method (poster).* In: *10th conference on one carbon metabolism, vitamins b and homocysteine. Nancy, France, from July 7th–11th*; 2015.
35. Carmel R, Agrawal YP. Failures of cobalamin assays in pernicious anemia. *N Engl J Med* 2012;**367**:385–6.
36. Thorpe S, Heath A, Blackmore S, Lee A, Hamilton M, O'Broin S, et al. International standard for serum B12 and serum folate: international collaborative study to evaluate a batch of lyophilised serum for B12 and folate content. *Clin Chem Lab Med* 2007;**45**:380–6.
37. Snow CF. Laboratory diagnosis of vitamin B12 and folate deficiency: a guide for the primary care physician. *Arch Intern Med* 1999;**159**:1289–98.
38. Harrington DJ. Laboratory assessment of vitamin B_{12} status. *J Clin Pathol* 2017;**70**:168–73.
39. Nexø E, Christensen AL, Hvas AM, Petersen TE, Fedosov SN. Quantitation of holotranscobalamin, a marker of vitamin B_{12} deficiency. *Clin Chem* 2002;**48**:561–2.

40. Brady J, Wilson L, McGregor L, Valente E, Orning L. Active B12: a rapid, automated assay for holotranscobalamin on the Abbott AxSYM analyzer. *Clin Chem* 2008;**54**:567–73.
41. Heil SG, Rodenburg P, Findeisen P, De Rijke YB. *Multicenter evaluation of the Roche Elecsys® active B12 (holotranscobalamin) electro-chemiluminesence immunoassay abstracts voorjaarscongres NVKC.* April 201811–3.
42. Sobczyńska-Malefora A, Pangilinan FJ, Plant GT, Velkova A, Harrington DJ, Molloy AM, et al. Association of a transcobalamin II genetic variant with falsely low results for the holotranscobalamin immunoassay. *Eur J Clin Investig* 2016;**46**:434–9.
43. Keller P, Rufener J, Schild C, Fedosov SN, Nissen PH, Nexø E. False low holotranscobalamin levels in a patient with a novel TCN2 mutation. 2016. *Clin Chem Lab Med* 2016;**54**:1739–43.
44. Nexø E, Christensen AL, Petersen TE, Fedosov SN. Measurement of transcobalamin by ELISA. *Clin Chem* 2000;**46**:1643–9.
45. Vogiatzoglou A, Oulhaj A, Smith AD, Nurk E, Drevon CA, Ueland PM, et al. Determinants of plasma methylmalonic acid in a large population: Implications for assessment of vitamin B$_{12}$ status. *Clin Chem* 2009;**55**:2198–206.
46. Thomas K, Stalder E. Isohenmg von methylmalonsaure aus normalen menschlichen ham. *Chem Ber* 1957;**90**:970–4.
47. White AM. Preliminary investigations on the excretion of methylmalonic acid by patients with severe megaloblastic anaemia. *Biochem J* 1962;**84**:41.
48. Cox EV, White AM. Methylmalonic acid excretion: an index of vitamin B$_{12}$ deficiency. *Lancet* 1962;**7261**:853–6.
49. Barness LA, Young D, Mellman WJ, Kahn SB, Williams WJ. Methylmalonate excretion in a patient with pernicious anemia. *N Engl J Med* 1963;**268**:144–6.
50. Bashir HV, Hinterberger H, Jones BP. Methylmalonic acid excretion in vitamin B$_{12}$ deficiency. *Br J Haematol* 1966;**12**:704–11.
51. Giorgio AJ, Plaut GWE. A method for the colorimetric determination of urinary methylmalonic acid in pernicious anemia. *J Lab Clin Med* 1965;**66**:667–76.
52. Dale RA. The assay of methylmalonic acid in urine. *Clin Chim Acta* 1972;**41**:141–7.
53. Westwood A, Taylor W, Davies G. Colorimetric method for determination of urinary methylmalonic acid. *Ann Clin Biochem* 1979;**16**:161–4.
54. Millar KR, Lorentz PP. A gas chromatographic method for the determination of methylmalonic acid in urine. *J Chromatogr* 1974;**101**:177–81.
55. Barton EP, Elliot JM. A technique for measurement of methylmalonic acid in cattle urine. *J Dairy Sci* 1977;**60**:1816–9.
56. Schiller CM, Summer GK. Gas-chromatographic determination of methylmalonic acid in urine and serum. *Clin Chem* 1974;**20**:444–6.
57. McMurray CH, Blanchflower WJ, Rice DA, McLoughlin M. Sensitive and specific gas chromatographic method for the determination of methylmalonic acid in the plasma and urine of ruminants. *J Chromatogr* 1986;**378**:201–7.
58. Marcell PD, Stabler SP, Podell ER, Allen RH. Quantitation of methylmalonic acid and other dicarboxylic acids in normal serum and urine using capillary gas chromatography-mass spectrometry. *Anal Biochem* 1985;**150**:58–66.
59. Bennett MJ, Brady CE. Simpler liquid-chromatographic screening for organic acid disorders. *Clin Chem* 1984;**30**:542–6.
60. Morgan DK, Danielson ND. Determination of methylmalonic acid after diazonium derivatization by high-performance liquid chromatography. *Anal Chim Acta* 1985;**170**:301–10.

61. Rasmussen K. Solid phase sample extraction for rapid determination of methylmalonic acid in serum and urine by a stable isotope dilution method. *Clin Chem* 1989;**35**:260–4.

62. Babidge P, Babidge W. Determination of methylmalonic acid by high-performance liquid chromatography. *Anal Biochem* 1994;**216**:424–6.

63. Pasieka AE, Morgan JF. The detection of homocysteine in biological systems. *Biochim Biophys Acta* 1955;**18**:236–40.

64. Rasmussen K, Moller J. Methodologies of testing. In: Carmel R, Jacobsen D, editors. *Homocysteine in health and disease.* Cambridge: Cambridge University Press; 2001. p. 9–20.

65. Kataoka H, Takagi K, Makita M. Determination of total plasma homocysteine and related aminothiols by gas chromatography with flame photometric detection. *J Chromatogr B Biomed Appl* 1995;**664**:421–5.

66. Stabler S, Marcell P, Podell E, Allen RH. Quantitation of total homocysteine, total cysteine and methionine in normal serum and urine using capillary gas chromatography-mass spectrometry. *Anal Biochem* 1987;**162**:185–96.

67. Magera MJ, Lacey JM, Casetta B, Rinaldo P. Method for the determination of total homocysteine in plasma and urine by stable isotope dilution and electrospray tandem mass spectrometry. *Clin Chem* 1999;**41**:1517–22.

68. Ueland PM, Refsum H, Stabler SP, Malinow MR, Andersson A, Allen RH. Total homocysteine in plasma or serum: methods and clinical applications. *Clin Chem* 1993;**39**:1764–79.

69. Schilling RF. Intrinsic factor studies. II. The effect of gastric juice on the urinary excretion of radioactivity after the oral administration of radioactive vitamin B_{12}. *J Lab Clin Med* 1953;**42**:860–6.

70. International Committee for Standardization in Haematology. Recommended method for the measurement of vitamin B_{12} absorption. *J Nucl Med* 1981;**22**:1091–3.

71. Bhat DS, Thuse NV, Lubree HG, Joglekar CV, Naik SS, Ramdas LV, et al. Increases in plasma holotranscobalamin can be used to assess vitamin B-12 absorption in individuals with low plasma vitamin B_{12}. *J Nutr* 2009;**139**:2119–23.

72. Hardlei TF, Mørkbak AL, Bor MV, Bailey LB, Hvas AM, Nexø E. Assessment of vitamin B_{12} absorption based on the accumulation of orally administered cyanocobalamin on transcobalamin. *Clin Chem* 2010;**56**:432–6.

73. Shackleton P, Fish D, Dawson D. Intrinsic factor antibody tests. *J Clin Pathol* 1989;**42**:210–2.

74. Slingerland DW, Cardarelli JA, Burrows BA, Miller A. The utility of serum gastrin levels in assessing the significance of low serum B12 levels. *Arch Intern Med* 1984;**144**:1167–8.

75. Jacob E, Herbert V. Measurement of unsaturated 'granulocyte-related' (TCI and TCIII) and 'liver-related' (TCII) binders by instant batch separation using a microfine precipitate of silica (QUSO G32). *J Lab Clin Med* 1975;**88**:505–12.

76. Briddon A. Homocysteine in the context of cobalamin metabolism and deficiency states. *Amino Acids* 2003;**24**:1–12.

77. Whitehead V. Acquired and inherited disorders of cobalamin and folate in children. *Br J Haematol* 2006;**134**:125–36.

78. Griffioen PH, van Dam-Nolen DHK, Lindemans J, Heil SG. Measurement of total transcobalamin employing a commercially available assay for active B12. *Clin Biochem* 2017;**50**:1030–3.

79. Harmening DM. *Megaloblastic anemia: clinical haematology and fundamentals of haemostasis.* F A Davis Company; 2001112–9.

80. Provan D, Singer C, Baglin T, et al. Red cell disorders. In: *Oxford handbook of clinical haematology*. 3rd ed. Oxford University Press; 2010. p. 46–7.

81. Hamilton MS, Blackmore S. Investigation of megaloblastic anaemia cobalamin, folate and metabolite status. In: Bain BJ, Bates I, Laffan MA, editors. *Dacie and lewis practical haematology*. 11th ed. Churchill Livingstone; 2012. p. 202–24.

82. Healton EB, Savage DG, Brust JC, Garrett TJ, Lindenbaum J. Neurologic aspects of cobalamin deficiency. *Medicine (Baltimore)* 1991;**70**:229–45.

83. Clarke R, Sherliker P, Hin H, Nexø E, Hvas AM, Schneede J, et al. Detection of vitamin B12 deficiency in older people by measuring vitamin B12 or the active fraction of vitamin B12, holotranscobalamin. *Clin Chem* 2007;**53**:963–70.

84. Herrmann W, Obeid R, Schorr H, Geisel J. The usefulness of holotranscobalamin in predicting vitamin B12 status in different clinical settings. *Curr Drug Metab* 2005;**6**:47–53.

85. Murphy MM, Molloy AM, Ueland PM, Fernandez-Ballart JD, Schneede J, Arija V, et al. Longitudinal study of the effect of pregnancy on maternal and fetal cobalamin status in healthy women and their offspring. *J Nutr* 2007;**137**:1863–7.

86. Vanderjagt DJ, Ujah IA, Ikeh EI, Bryant J, Pam V, Hilgart A, et al. Assessment of the vitamin B_{12} status of pregnant women in Nigeria using plasma holotranscobalamin. *ISRN. Obstet Gynecol* 2011;365894.

87. Sobczyńska-Malefora A, Ramachandran R, Cregeen D, Green E, Bennett P, Harrington DJ...Lemond H. An infant and mother with severe B12 deficiency—vitamin B12 status assessment should be determined in pregnant women with anaemia. *Eur J Clin Nutr* 2017;**71**:1013–5.

88. Morkbak AL, Hvas AM, Milman N, Nexø E. Holotranscobalamin remains unchanged during pregnancy. Longitudinal changes of cobalamins and their binding proteins during pregnancy and post partum. *Haematologica* 2007;**92**:1711–2.

89. Chanarin I. *The megaloblastic anaemias*. 3rd ed. London: Blackwell Scientific Publications; 1990.

90. Heil SG, de Jonge R, de Rotte MC, van Wijnen M, Heiner-Fokkema RM, Kobold AC, et al. Screening for metabolic vitamin B12 deficiency by holotranscobalamin in patients suspected of vitamin B12 deficiency: a multicentre study. *Ann Clin Biochem* 2012;**49**:184–9.

91. Miller JW, Garrod MG, Rockwood AL, Kushnir MM, Allen LH, Haan MN, et al. Measurement of total vitamin B12 and holotranscobalamin, singly and in combination, in screening for metabolic vitamin B12 deficiency. *Clin Chem* 2006;**52**:278–85.

92. Hamilton M, Blackmore S, Lee A, et al. Ability of holotranscobalamin assay (active B12) to detect severe cobalamin deficiency evidenced by high methylmalonic acid in the presence of high titre intrinsic factor antibody and false normal B12 results. *Br J Haematol* 2010;**149**(Supp 1):54. 26Abs.

93. Sobczyńska-Malefora A, Critcher MS, Harrington DJ. The application of holotranscobalamin and methylmalonic acid in hospital patients and total vitamin B_{12} in primary care patients to assess low vitamin B_{12} status. *J Hematol Thromb* 2015;**1**:8–16.

94. Refsum H, Smith AD, Ueland PM, Nexø E, Clarke R, McPartlin J, et al. Facts and recommendations about total homocysteine determinations: an expert opinion. *Clin Chem* 2004;**50**:3–32.

95. Fedosov SN. Metabolic signs of vitamin B(12) deficiency in humans: computational model and its implications for diagnostics. *Metab Clin Exp* 2010;**59**:1124–38.

96. Risch M, Meier DW, Sakem B, Medina Escobar P, Risch C, Nydegger U, et al. Vitamin B12 and folate levels in healthy Swiss senior citizens: a prospective study evaluating reference intervals and decision limits. *BMC Geriatr* 2015;**15**:82.

97. Elliott M, Harrington DJ, Andrew T, Sobczyńska-Malefora A, Ahmadi KR. *Genetic epidemiology of static and functional biomarkers of vitamin B$_{12}$ status in older adults.* In: *11th hyperhomocysteinaemia conference, Aarhus, Denmark*; 2017. May. Abstract O34.
98. Fedosov SN, Brito A, Miller JW, Green R, Allen LH. Combined indicator of vitamin B12 status modification for missing biomarkers and folate status and recommendations for revised cut-points. *Clin Chem Lab Med* 2015;**53**:1215–25.

Methods for assessment of vitamin C

13

David J. Card

Nutristasis Unit, Viapath, St. Thomas' Hospital, London, United Kingdom

Chapter Outline

Structure and Function

The term "vitamin C" refers to L-ascorbic acid (the reduced form of vitamin C), which is the predominant biologically active form, and its oxidation product dehydroascorbic acid (DHAA), which is rapidly converted back to L-ascorbic acid in vivo thus increasing its bioavailability (Figs. 13.1 and 13.2). Ascorbic acid has two enantiomers (L- and D-) and their equivalent epimers (L-isoascorbic acid and D-isoascorbic acid). Only L-ascorbic acid occurs endogenously while D-ascorbic acid and L-isoascorbic acid have no vitamin C activity.[1]

The majority of mammals are able to synthesize vitamin C. However, humans and some other mammals such as primates and guinea pigs are dependent on dietary sources, mainly through fresh fruit and vegetables. Because of the abundance of dietary sources of vitamin C, humans have evolved a loss of function in L-gulonolactone oxidase, the enzyme responsible for vitamin C biosynthesis.[2] Other animals synthesize vitamin C from glucose, hence the evolutionary advantage in terms of improved energy utilization.

Laboratory Assessment of Vitamin Status. https://doi.org/10.1016/B978-0-12-813050-6.00013-9

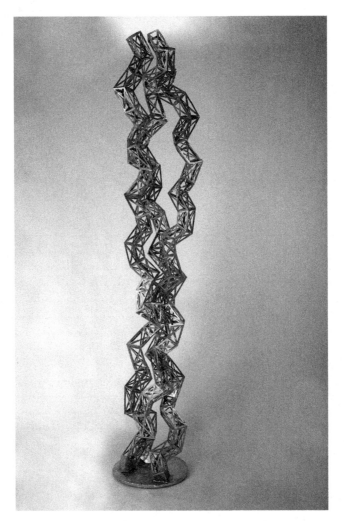

FIG. 13.1

Unravelling collagen.

From Voss-Andreae J. Stainless steel. Stamford, CT: Downtown Abstractions; 2015.
www.julianvossandreae.com.

Vitamin C has specific deficiency symptoms which are collectively referred to as scurvy. Many of these symptoms can be attributed to the role of vitamin C in collagen biosynthesis as impairment of this process results in destabilization of connective tissues. Three of the enzymes that require vitamin C as a cofactor are essential to the formation of functional collagen: prolyl-4-hydroxylase, prolyl-3-hydroxylase, and lysyl hydroxylase, which catalyze cross-link formation between prolyl and lysyl residues on its alpha-helices. The five other vitamin C-dependent enzymes are required for carnitine

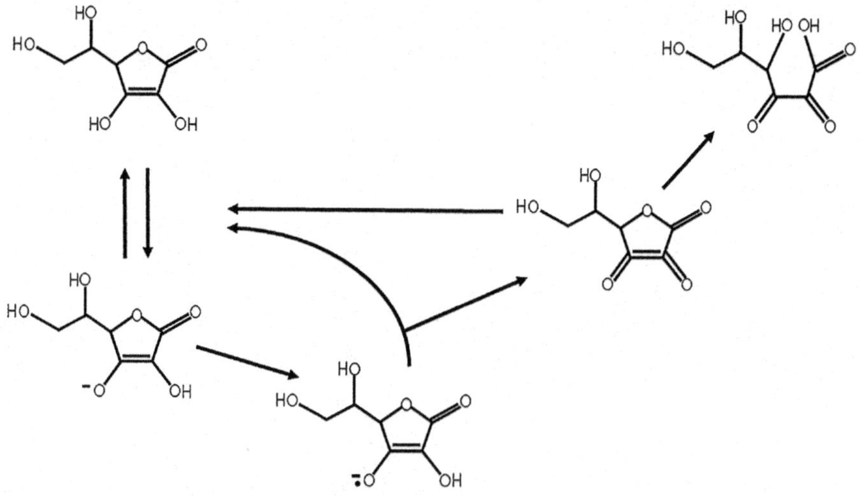

FIG. 13.2

Structure of L-ascorbic acid and its oxidation products.

synthesis, noradrenalin synthesis, peptide amidation, and tyrosine metabolism. In addition to its role as a cofactor, vitamin C is a potent antioxidant capable of absorbing reactive oxidation species that would otherwise damage tissues. In particular, lipid oxidation is prevented through direct antioxidant action and through the reduction of other antioxidants such as the tocopheroxyl radical (Chapter 4) and glutathione.[3]

Dietary Intake

Foods rich in vitamin C are fresh fruits, especially citrus fruits and green vegetables. Vitamin C is also present at high concentrations in potatoes and certain meats such as liver, kidney, and heart (although not in meats derived from muscle). For healthy individuals bioavailability is high, with ascorbic acid absorption from foods shown to be similar to that of the equivalent intake from a pharmaceutical preparation.[4]

Recommendations for dietary vitamin C intake vary between different national agencies, with values that range from 40 to 110 mg/day. The World Health Organization recommends an intake of 45 mg/day, which is based on 50% tissue saturation in 97.5% of adult males. On the dose response curve published by Levine et al., where steady-state plasma ascorbic acid concentrations were achieved at 200 mg/day, this represents an intake at the base of the steep slope.[5]

Absorption, Transport, Kinetics, and Storage

Ascorbic acid and DHAA are absorbed through the entire length of the intestinal lumen via enterocytes. Because of their size and charge (at physiological pH) they do not diffuse easily across lipid bilayers.[6] Estimates based on analysis in animal

Table 13.1 Vitamin C Content of Human Tissues

Tissue	Mg/100g Wet Tissue	Mg/Total Tissue	% of Total Pool
Pituitary gland	40–50	0.22–0.28	0.01
Adrenal glands	30–40	3–4	0.17
Eye lens	25–31		
Brain	13–15	185–215	10.3
Liver	10–16	158–253	8.8
Spleen	10–15	16–24	0.9
Kidneys	5–15	15–24	0.8
Heart muscle	5–15	16–48	0.9
Lungs	7	52–155	2.9
Skeletal muscle	3	1200	66.6
Testes	3	1.0	0.06
Thyroid	2	0.55	0.03
Leukocytes	35	63	3.5
Plasma	0.4–1.0	93	5.2

From Omaye S, Schaus E, Kutnink M, Hawkes W. Measurement of vitamin C in blood components by high-performance liquid chromatography. Implication in assessing vitamin C status. Ann N Y Acad Sci 1987;**498**:*389–401.*

tissues, which have been extended to their equivalent in humans, suggest ascorbic acid in plasma makes up ~5% of the total body pool, with the majority of ascorbic acid present in skeletal muscle (~67%), brain (~10%), and liver (~9%),[7] (Table 13.1). Tissues with a higher requirement for vitamin C reflect their higher metabolic turnover, particularly through collagen synthesis and absorption of reactive oxygen species. The differential distribution of vitamin C around the body relies on active transport mechanisms, which are discussed in detail elsewhere.[6] In the kidney, ascorbic acid and DHAA are reabsorbed in the renal tubules via epithelial cells. Previously leukocytes were assumed to be involved in the transport of vitamin C, with plasma levels reflecting metabolic turnover.[7] However, no evidence for this mechanism has since been found and so it is more likely that leukocyte concentrations correspond to other tissues.

Much of the useful experimental data on vitamin C requirements in humans originates from experiments which would not be considered ethical in the modern era. Hodges et al. investigated a human experimental model of scurvy in the 1960s using prison volunteers. The subjects were fed a vitamin C-deficient diet until symptoms of scurvy appeared, in order to deplete the body pool of vitamin C. After 100 days, they were given ^{14}C labeled L-ascorbic acid at daily doses ranging from 4 to 64 mg. They estimated that symptoms of scurvy appear when the total body pool of vitamin C has decreased to ~300 mg, and that the basic requirement for intake is 10 mg/day with 30 mg/day adequate to eliminate the risk of insufficiency.[8, 9] Since then, further studies have attempted to quantify body pool size and optimal vitamin C requirements using factors such as dose-function relationships, availability in food, steady-state concentrations, and urinary excretion.[10] Estimating optimal intake from such a range of complex factors introduces error, discrepancies have been found between estimates of body pool size and concentrations found in tissues.[11] Clearly optimal intake and estimates of body pool size would be easier to estimate were a suitable functional marker of vitamin C status available. Further to these issues, the kinetics of vitamin C homeostasis are known to be altered during deficiency when compared to periods of adequate status, further complicating models used to estimate optimal status.[12, 13] More recently the total body pool of vitamin C has been estimated to be 900 mg with a RDI set at 40 mg/day for men and 30 mg/day for women.[14]

Gene knockout (Gulo(−/−)) mice have been introduced as an animal model for study of vitamin C insufficiency/deficiency providing interesting data on the possible outcomes of chronic suboptimal status.[15] However, because of concomitant adaptations that have coincided with the loss of function of L-gulonolactone oxidase, gene knockout mice may not be the best model. For example, the loss of the ability to catabolize uric acid, possibly indicating that uric acid has taken over some of the functions of ascorbic acid.[16] Given this possibility it may be the case that gene knock-in guinea pigs would be a more appropriate animal model of vitamin C deficiency.

Clinical Application of Measurement

Dr. James Lind famously identified a cure for scurvy while working as a Royal Navy physician in the 18th century.[17] The disease was widespread and mortality was high among sailors because they undertook extended sea voyages with insufficient access to foods containing vitamin C. Lind's discovery meant that many sailors were saved and represented one of the earliest examples of the identification of a specific link between diet and health. Over 150 years later, Albert Szent-Gyorgyi and Norman Haworth received their Nobel prizes for the discovery and characterization of the compound responsible. Today, the relationship between vitamin C status and human health is still not fully understood, particularly in terms of optimal intake and the effects of long-term suboptimal status. Methodologies developed in order to assess vitamin C status have so far had a limited impact on the investigation of optimal status.

The laboratory assessment of vitamin C status is hampered by the poor stability of vitamin C and the lack of available markers that accurately reflect tissue stores or functional impairment. Measurement of L-ascorbic acid in plasma provides a basic method capable of confirming clinically overt scurvy and identifying possible suboptimal status. However, plasma concentrations are subject to rapid changes in response to intake[18] and are influenced by acute phase response.[19] People that smoke cigarettes have been shown to have significantly lower plasma vitamin C concentrations (as well as other antioxidants), which has been attributable to reaction with oxidants in cigarette smoke. Various functional markers of vitamin C status have been identified,[20] but so far none have been developed that demonstrate any diagnostic validity or research utility in humans. High doses of vitamin C have been shown to be nontoxic; therefore laboratory analysis in the clinical setting has focused on the confirmation of clinical deficiency. The relationship between plasma concentrations of vitamin C and dietary vitamin C intake has previously been assessed by systematic review and meta-analysis. There was found to be a moderate positive correlation between plasma L-ascorbic acid concentrations and intake, and the conclusion that vitamin C measurement is affected by multiple confounding factors.[21]

Hypoascorbemia, classically manifesting as scurvy, results in an array of symptoms including osteoporosis, hypercholesterolemia, atherosclerosis, anemia, hemolysis, microvascular disease, capillary hyperperfusion, and hemorrhage. The symptoms of scurvy are largely the result of impaired collagen synthesis. Collagen is the most abundant protein in humans and forms the major component of connective tissue. Vitamin C acts as a cofactor for two prolyl hydroxylase and one lysyl hydroxylase enzymes that are responsible for the posttranslational hydroxylation of proline and lysine residues located on the three alpha helices of collagen. These residues are then capable of forming peptide cross-links which confer structural stability to the collagen molecule. Inhibition of this process because of insufficient vitamin C downregulates procollagen secretion[22] and results in destabilized collagen in tissues leading to increased vascular permeability and lysis. This in turn leads to hemolysis, hemarthrosis, purpura, bleeding, and anemia.[23, 24] Clotting screens (consisting of prothrmbin time (PT), activated partial prothrombin time (APTT), and fibrinogen) are unable to identify patients with vitamin C deficiency. However, patients with malnutrition may have other concomitant nutritional deficiencies, e.g., a raised PT because of vitamin K deficiency (Chapter 5), or a raised mean corpuscular volume (MCV) because of folate (Chapter 11) or vitamin B_{12} deficiency (Chapter 12). Other hematological tests would typically indicate a normocytic anemia with a normal platelet count. Ascorbic acid facilitates the absorption of inorganic iron (nonheme iron) and thus poor status can also contribute to iron deficiency anemia. Diagnosis is primarily based on clinical symptoms with a plasma sample taken prior to ascorbate treatment for confirmation by determination of L-ascorbic acid. Plasma L-ascorbic acid concentrations in patients with scurvy are typically close to or below the detection limit of the assay, several times lower than the lower limit of the healthy reference range. Improvement of symptoms in response to treatment with vitamin C provides an additional indication of diagnosis, which occurs within days of treatment.[25]

Suboptimal vitamin C status is difficult to assess from a single L-ascorbic acid measurement. Long-term plasma ascorbic acid concentrations were inversely associated with increased all-cause mortality, cardiovascular disease, and ischemic heart disease in 45- to 79-year-old men and women.[26] Measurement of leukocyte vitamin C concentrations gives improved assessment of body stores but is technically challenging.[7] There are currently no functional assays of vitamin C status available to indicate suboptimal status at the tissue level. Generally, patients who are not showing signs of scurvy are not routinely monitored for impaired status; however, there is one patient group recently identified that could benefit from monitoring and/or routine supplementation. Patients with graft versus host disease appear to be at high risk of deficiency, in particular, patients with mucous membrane involvement are likely to have impaired intake due to the reduced consumption of citrus fruits which irritate the oral mucositis. They may also have increased requirements due to the enhanced immune response to the graft.[27]

Analytical Methods
Historical Methods

Early methods for vitamin C determination were bioassays that used vitamin C-deficient guinea pigs.[28] Later, chemical methods appeared such as that of Harris and Ray (1935) based on dye titration[29] which used 2,6-di-chlorophenolindophenol. Early colorimetric assays used the 2,4-dinitrophenyl-hydrazine derivative of DHAA with 85% H_2SO_4.[30] Generally, these methods were prone to interference from other reducing substances, were not 100% specific to vitamin C, and could not differentiate between ascorbic acid and DHAA. Later spectrophotometric methods using color change reactions, e.g., perinaphthindanetrione hydrate and its derivative 2-nitroperinaphthindanetrione hydrate reacting with ascorbic acid, avoided some of these problems.[31] Enzymatic assays were later developed that used ascorbic acid oxidase or peroxidase to convert ascorbic acid to DHAA.[1] These methods tended to be hampered by poor sensitivity and specificity and were more suited to measurement in pharmaceutical tablets and fruit juices. Fluorometric assays require ascorbic acid to be oxidized to DHAA and then condensed with *o*-phenylenediamine forming a fluorescent quinoxaline derivative.[1] By performing the same measurement without addition of an oxidizing agent the ascorbic acid concentration could be determined. Again this method had various drawbacks as it was prone to interferences and lacked sensitivity. Gas-liquid chromatographic methods require complete removal of water from the sample by lyophilization and then derivatization of ascorbic acid to a trimethylsilyl ether in order to improve its volatility. In standards, derivatization is rapid (~5 min) but may take up to 48 h in biological samples, detection was either by flame ionization or beta ionization. Few laboratories implemented these methods, probably because of their complexity.

These methods were all superseded by HPLC methods that could reliably measure vitamin C with high sensitivity, precision, and specificity. High performance liquid chromatography methods generally use ion-pair reversed-phase chromatography or ion exchange chromatography with UV or EC detection and are discussed later in this chapter. Most assays used today measure L-ascorbic acid. The free radical intermediate semidehydroascorbic acid (also called ascorbate free radical) can also be measured but it is highly unstable (decay rate constant $= 2.8 \times 10^5 \, m^{-1} s^{-1}$)[1] making it unsuitable for measurement in routine clinical practice. Older methods measured total ascorbate, i.e., the sum of ascorbic acid plus DHAA, which can negate the problem of interconversion; however, ascorbic acid as the biologically active form has become established as the superior marker of status.

Functional Markers

Currently, no functional markers for assessment of vitamin C status have been conclusively demonstrated to reflect status in humans. This has left the assessment of optimal intake open to speculation. Ideally a functional marker would not be subject to the same acute changes in response to intake that takes place with plasma ascorbate determination and would better indicate clinically asymptomatic, suboptimal status. The marker would need to be specific for vitamin C status and relatively straightforward to measure.

Various candidate functional markers of vitamin C status have been postulated, their utility was reviewed by Benzie in 1999 who identified three classes of marker: molecular markers including enzymes that have enhanced activity dependent on vitamin C; biochemical markers that require vitamin C for their synthesis, e.g., carnitine- and collagen-derived hydroxyproline; and physiological markers such as DNA oxidation products and blood pressure.[20]

Bates et al. carried out a series of experiments in the 1990s and 2000s to investigate the utility of collagen cross-link ratios in urinary collagen metabolites, which showed some promise as functional markers. The two key cross-links were pyridinoline and deoxypyridinoline. A vitamin C-restricted diet was shown to give altered collagen cross-link ratios despite no symptoms of scurvy in the animals tested. Having demonstrated the relationship in animals, human studies were conducted comparing British and Gambian boys.[32] The study showed a significantly higher cross-link ratio (4.36) in British boys when compared to their Gambian counterparts (3.83) which correlated with a significantly lower vitamin C intake. The Gambian cohort also demonstrated a seasonal decrease in collagen cross-link ratios associated with poor vitamin C intake in the rainy season when the availability of foods containing vitamin C was very poor. However, in a seven-week controlled intervention of 100 mg/d ascorbic acid the collagen cross-link ratios failed to respond. In cases of scurvy, symptoms associated with impaired collagen synthesis respond to supplementation with vitamin C in just two to three weeks.[24] It is possible that the cross-link ratios of collagen represent a much longer term marker of vitamin C status, particularly as collagen has a very long half-life in certain tissues. Genetic factors also cannot be ruled out which were alluded to in cited studies into human genetic abnormalities (Ehler-Danlos type VI, Ullrich-Turner and Bruck syndromes). Their evidence suggested that genetic conditions in humans, where vitamin C requirements are increased, correlate with alterations in collagen cross-link ratios and that these alterations can be corrected by increasing their vitamin C intake. Urinary collagen cross-links are reasonably straightforward to measure by reversed-phase high performance liquid chromatography (HPLC) and fluorescence (FL) detection[33] and kits are available that simplify the method further.[34] Therefore it seems feasible that collagen cross-links could, with further research, provide the marker needed to assess optimal intake.

Assessment of vitamin C status, to some extent, falls into the category of assessment of antioxidant status and more generally, oxidative stress. For markers of vitamin C status, as with markers of oxidative stress in general, the usefulness of the results produced suffers because of the multifactorial nature of the biochemical processes occurring. This includes external oxidant forces, e.g., smoking, infection, and so on, and other contributors to antioxidant status such as vitamin E (Chapter 4) and the body's repair mechanisms. A recently established example is the determination of DNA methylation which, to some extent, represents the contribution of vitamin C status, but to what extent remains to be established. Rossi et al. observed that there are variations in the stability of ascorbic acid between samples, which they postulated could be used as a marker of redox potential and antioxidant capacity. Various other markers of antioxidant status and oxidative stress have also been evaluated, and in-depth discussion of these markers is beyond the scope of this book chapter.

Matrix Variations

Whereas plasma vitamin C measurement is commonly employed in the routine diagnostic setting, those who wish to probe deeper in to the understanding of vitamin C may need to make measurements in other matrices. Plasma concentrations are thought to represent a transient reservoir of vitamin C, in which the vitamin is transported to tissues. Leukocyte concentrations are thought to be more equivalent to tissues as measurements represent intracellular concentrations. They are also thought to be superior to red blood cells as they possess the same organelles as tissues, whereas red blood cells lack a nucleus and mitochondria.[7] It has been demonstrated in animal models that leukocyte ascorbic acid concentrations correlate with liver stores ($r=0.683$) and total body pool ($r=0.923$).[13] When compared with plasma, leukocytes have been estimated to contain much higher concentrations of ascorbic acid, 0.4–1.0 mg/100 g wet tissue in plasma compared to 35 mg/100 g in leukocytes, which are less susceptible to changes in intake and fall more slowly during depletion.[7] Methods for measurement of leukocyte ascorbic acid concentrations determine mixed cell populations: mononuclear, polymorphonuclear, and platelets. The relative concentrations of vitamin C in these populations vary as does the relative proportion of each cell type. Other factors which can influence concentrations are physiological stress, exercise, and smoking.[7]

Other matrices have been less fruitful in terms of published research. Ascorbic acid can be measured in human urine and responds significantly to changes in intake up to 715 mg/day.[35] Previously the buffy coat was investigated as a possible matrix for vitamin C analysis.[36] Because the buffy coat contains the white cells and platelets it is likely to provide a better indication of tissue stores than plasma while being easier to harvest and more plentiful than leukocyte preparations. Despite this, buffy coat vitamin C determination has not become established as a marker, presumably this is because the matrix is too difficult to harvest for routine analysis and lacks the specificity required for the research setting.

One possible, as yet unexplored matrix for vitamin C analysis is sweat—this is interesting as potentially sweat could provide access to vitamin C originating from tissues. Early studies prior to the development of methods specific for ascorbic acid and DHAA detected reducing substances in sweat which were attributed to vitamin C,[37] today the presence of vitamin C in sweat has yet to be confirmed using reliable methods. Examples of other organic molecules found in sweat are ethanol, cortisol, urea, and lactate.[38] Since compounds found in the sweat must be able to pass through lipid bilayers it is likely any presence of vitamin C in sweat is a result of active transport and because there is no transporter specific to vitamin C it is likely to be the product of other, more dominant processes.

Sample Stability

The primary factor influencing the accurate determination of ascorbic acid and DHAA is stability. Assays have for several decades demonstrated high sensitivity, reproducibility, and recovery in plasma samples post addition of a stabilization agent when stored at subzero centigrade temperatures. However, factors that influence accuracy prior to these steps are to some extent unknown, as rates of decay may vary between samples, an effect that was observed by Rossi et al. and was attributed to variations in biological redox potential.[39] Ascorbic acid decays by oxidation while DHAA is hydrolyzed. Their rates of decay are influenced by various factors such as their concentration in the sample, temperature, light, pH, dissolved oxygen, and the presence of oxidizing or reducing agents.[1]

Various stability studies have been carried out, but because of variations in the techniques employed, care should be taken when comparing results. It is recommended that laboratories performing vitamin C analysis should assess stability locally. For the purposes of research, collection and immediate processing of samples should be tightly controlled. In the clinical environment, sample collection and transport should also be well controlled wherever possible. However, sample collection may not always be carried out as the laboratory would wish, especially if receiving referred samples from organizations outside of their own. Therefore a good working knowledge of the relative stability of the analytes will help in the assessment of decay, which should be weighed up against the consequences of a patient rebleed if the original sample is deemed unsuitable for analysis and rejected. This is particularly relevant if the patient has been supplemented with vitamin C after the sample was taken as plasma ascorbic acid concentrations will normalize rapidly.

Serum is not a suitable matrix for vitamin C analysis, as DHAA can be converted back to ascorbic acid giving erroneous results.[40] The consensus is that heparinized plasma provides the most stable anticoagulant.[39] Vitamin C is at its most unstable prior to addition of a stabilization agent, this time period should be minimized and the plasma sample should be stored and transported frozen and protected from light. Rossi et al. showed that a delay in the addition of metaphosphoric acid (MPA) to heparinized samples at room temperature resulted in a 90% yield in plasma over 3 h and an 85% yield over 4 h. Results from Karlsen et al. were comparable; they demonstrated that heparinized whole-blood samples left at room temperature yielded ~95% ascorbic acid after 3 h and ~70% after 6 h. They also showed that an unacidified plasma sample left at room temperature yielded 20% ascorbic acid after 24 h and 0% ascorbic acid after 72 h, while an unacidified plasma sample left at 4°C yielded 70% ascorbic acid after 24 h and 40% ascorbic acid after 72 h.[41]

Methods commonly employ the addition of an equal volume of freshly prepared MPA (10% *w/v*) to precipitate proteins and stabilize the analytes, followed by immediate storage at −70°C. Other additives have previously been evaluated by Kand' Ar et al.: perchloric acid (1.0 mol/L), TCA (10%), sulfosalicylic acid (10%), TCA (10%) plus oxalic acid (10 mmol/L) mixture, acetonitrile plus hydrochloric acid (0.1 mol/L), acetonitrile plus acetic acid (0.1 mol/L), ethanol plus hydrochloric acid (0.1 mol/L), and ethanol plus hydrochloric acid (0.1 mol/L). It was concluded that MPA provided the best stability without introducing any assay interferences.[18] Karlsen et al. observed that addition of MPA improved yields greatly; a sample left at room temperature yielded 90% ascorbic acid after 24 h and 70% ascorbic acid after 72 h while a sample left at 4°C yielded 95% ascorbic acid after both 24 h and 72 h.[39] They also saw no degradation of ascorbic acid until 80 days while preserved in MPA and stored at −70°C, after this time a 6.8% decline was seen during the first year, which reached 14% after 2 years. Similarly Margolis et al. estimated that MPA preserved plasma ascorbic acid stored at −70°C declines by 4%–7% per year. Additional insight was provided by Kand'ár et al. who showed that ascorbic acid samples extracted with MPA were stable on a cooled autosampler for at least 10 h and that MPA preserved sample vials purged with N_2 gas gave a 2.9% improvement in yield over 10 h stored at −80°C.[42] Reducing agents have also been employed with great effect. Rossi et al. added dithioerythritol (DTE), (~7.5 mmol/L) to blood samples resulting in yields of ~100% after 24 h at 4°C and 90 days at −25°C compared to 80% and 0% when no DTE was added. Kand'ár et al. described a similar effect using 1,4-dithiothreitol (DTT) where ascorbic acid in plasma samples treated with and stored at −80°C are stable for at least 5 years.[42] This does mean, however, that only total ascorbic acid can be measured as DHAA is converted to ascorbic acid. It is debatable whether DHAA would significantly contribute to ascorbic acid concentrations using this technique but it would not be advisable to use this procedure for determination of ascorbic acid concentrations in the clinical setting because of the unknown contribution of DHAA in patient samples.

Best Practice

Currently, the most reliable way to assess vitamin C status is to measure L-ascorbic acid in plasma or leukocytes.[18] As leukocytes are thought to provide a more accurate indication of cellular stores[7] they are likely to be a more appropriate matrix for use in the research setting. Plasma is the preferable matrix in the clinical diagnostic laboratory because of the ease of collection and measurement; however, sample stability remains a significant challenge. Measurement of ascorbic acid in plasma is a clinically useful marker which can be used to confirm scurvy when a preascorbic acid treatment sample is obtained. Samples taken posttreatment are of no diagnostic utility as the L-ascorbic acid concentration in the plasma rises rapidly. A commonly employed cutoff for deficiency is 11 μmol/L,[43] which in practical terms tends to be a more useful cutoff than reference intervals generated from healthy subjects.

High performance liquid chromatography with UV detection provides a cost effective and convenient method, and if DHAA quantitation is desirable a dual wavelength UV detector can be employed. The peak UV absorptions of L-ascorbic acid are 265 nm (neutral solution) and 245 nm (acidic solution). Methods employing wavelengths ranging from 220 to 268 nm have been published,[1] where use of lower wavelengths may give improved selectivity at the expense of sensitivity. High performance liquid chromatography with UV detection provides ample sensitivity, where an on-column sensitivity of 50 pmol should be achievable,[1] enabling the analyst to differentiate scurvy at the all-important cutoff of 11 μmol/L. Sensitivity is improved by injecting a larger volume of sample but higher volumes may disrupt the chromatography, a typical injection volume of 20–50 μL should be adequate.[44] The samples in the autosampler should ideally be protected from light and refrigerated to 4°C. Isocratic HPLC systems provide improved chromatographic stability over gradient systems where disturbances in the column equilibrium can affect the reproducibility of the chromatography. Chromatographic run times are typically <10 min per injection, Karlsen et al. achieved more rapid run times by using a reversed-phase monolithic HPLC column that allowed higher throughput analysis required for large-scale studies.[41]

Because of its acidic nature, L-ascorbic acid ionizes easily and is effectively separated by ion chromatography (either ion-pair reversed-phase chromatography or ion exchange chromatography).[45] Ion chromatography utilizes a positively charged stationary phase which has affinity for the negatively charged, deprotonated L-ascorbic acid. A mobile phase is required to consist of an ion pair reagent such as dodecyltrimethyl ammonium chloride combined with sodium dihydrogen phosphate and sodium-EDTA that will coat the octadecyl moieties of the reversed-phase HPLC column and generate a positively charged surface. It should also contain a suitable solvent, e.g., acetonitrile or methanol that will allow optimization of the chromatography and a suitable acid, e.g., acetic acid that will enable optimization of pH balance for effective ion chromatography. The HPLC columns employed are typically reversed-phase C18, 5 μm, and 250 × 4.6 mm.

Sample preparation involves mixing an equal volume of preserved sample supernatant with an internal standard solution prior to injection on to the HPLC system. Internal standards are usually 3,4-dihydroxybenzylamine hydrobromide or isoascorbic acid. Isoascorbic acid tends to be less used in modern methods probably due to its presence in some biological samples.[46, 47] Some methods have no internal standard[48] which, if the autosampler injects a reliably consistent volume of sample, can perform as well as methods that utilize internal standards. If no internal standard is used a loop-fill injection mode is preferable as it is more reliable; however, all methods which do not utilize an internal standard are potentially subject to error if the autosampler malfunctions and injection volume is not consistent, an error that will not be noticeable in samples of unknown concentration. Lyophilized calibrators and quality control materials are available from commercial suppliers, e.g., Chromsystems, making the assay significantly simpler to perform and maintain.

A good example of the method for leukocyte ascorbate determination is provided by Omaye et al.[7] A higher volume of blood is required than for plasma determination, typically 15 mL, which should be obtained using plastic equipment in order to minimize cell damage. Whole blood is centrifuged immediately at $90 \times g$ for 15 min, prior to removal of the upper layer consisting of platelet-rich plasma. An aliquot of the platelet-rich plasma should then be removed for determination of the leukocyte fractions (mononuclear and polymorphonuclear leukocytes) as well as red blood cells and platelets. The remainder of the platelet-rich plasma is centrifuged at $800 \times g$ for 10 min resulting in a "platelet pellet" left after the remaining supernatant is decanted. The platelets are gently resuspended in a salt solution and centrifuged again twice to remove wash away all the plasma. Finally the pellet is resuspended in 50 μL of an aqueous solution containing 5% MPA and 0.54 mM disodium EDTA. The sample is then stored at −70°C prior to analysis.

Standardization

A standard reference material (reference: 970) for measurement of ascorbic acid in frozen human serum was previously provided by National Institute for Standards and Technology (NIST) in 2003,[46] but has since been discontinued. The metrology of the material was based on the gravimetric addition of ascorbic acid to depleted plasma, stabilized with MPA and stored under nitrogen in sealed glass ampoules at −70°C. It is possible that to some extent this material was affected by issues of stability. Margolis et al. validated the production of the material indicating that it was stable for up to 6 years in these conditions, contrary to results obtained by other investigators using similar conditions.[41] It is also possible that to some extent the ascorbic acid concentration was depleted during transit. Results from the interlaboratory comparison carried out by NIST using this material gave coefficient of variations (CVs) of 22% (10.1 mmol/L) and 19% (30.6 mmol/L), ($n = 17$). Recently the external quality assurance (EQA) provider Instand, set up an assurance program for ascorbic acid in lyophilized human plasma.[47] The use of lyophilized plasma appears to have solved many of the issues of stability, enabling samples to be transported in ambient temperatures. Similarly use of this type of material enables distribution of commercial calibrators and quality control (QC) materials.[34] The initial rounds of the EQA scheme have so far given interlaboratory CVs ranging from 9.4 to 17.1% (mean CV = 11.8%, mean concentration = 70.5 μmol/L, mean number of returns from participants = 43), (Fig. 13.3). Typical published intralaboratory CVs for measurement of ascorbic acid in human plasma are relatively low at <5%[7, 48] reflecting the relatively high concentration of the analyte. Using the Horwitz curve it can be estimated that an interlaboratory CV of approximately 10% should be expected[49] much lower than the values obtained by NIST, which may reflect the atypical stability issues associated with ascorbic acid determination.

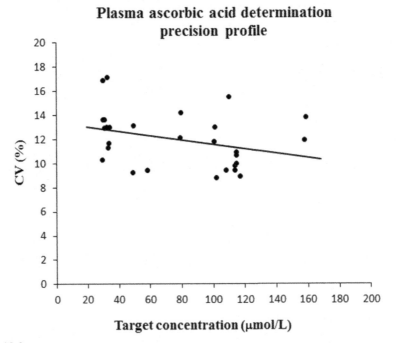

FIG. 13.3

Inter-laboratory precision profile for determination of ascorbic acid in lyophilized human plasma, 2014–2017.

Acknowledgments

I would like to thank Dr. Chris Bates for his advice on functional markers of vitamin C status. Thanks also to Julian Voss-Andreae for allowing me to reproduce the image of his collagen sculpture. For more artistic creations inspired by proteins and other biological molecules, please visit http://www.julianvossandreae.com. Many Thanks to Dr. Jason Heikenfeld, Department of Electrical Engineering and Computer Systems, University of Cincinnati, for his advice on the presence of vitamin C in sweat.

References

1. Washko P, Welch R, Dhariwal K, Wang Y, Levine M. Ascorbic acid and dehydroascorbic acid analyses in biological samples. *Anal Biochem* 1992;**204**:1–14.
2. Stone I. Homo sapiens ascorbicus, a biochemically corrected robust human mutant. *Med Hypotheses* 1979;**5**:711–21.
3. Du J, Cullen JJ, Buettner GR. Ascorbic acid: chemistry, biology and the treatment of cancer. *Biochim Biophys Acta* 2012;**1826**:443–57.
4. Ball GFM. *Vitamins, their role in the human body*. Blackwell Publishing Ltd; 2004.

5. Levine M, Conry-Cantilena C, Wang Y, Welch RW, Washko P, Dhariwal K, Park J, Lazarev A, Graumlich JF, King J, Cantilena L. Vitamin C pharmacokinetics in healthy volunteers: evidence for a recommended dietary allowance. *Proc Natl Acad Sci* 1996;**93**:3704–9.

6. Wilson J. Regulation of vitamin C transport. *Annu Rev Nutr* 2005;**25**:105–25.

7. Omaye S, Schaus E, Kutnink M, Hawkes W. Measurement of vitamin C in blood components by high-performance liquid chromatography. Implication in assessing vitamin C status. *Ann N Y Acad Sci* 1987;**498**:389–401.

8. Hodges R, Baker E, Hood J, Sauberlich H, March S. Experimental scurvy in man. *Am J Clin Nutr* 1969;**22**:535–48.

9. Baker E, Hodges R, Hood J, Sauberlich H, March S. Metabolism of ascorbic-1-14C acid in experimental human scurvy. *Am J Clin Nutr* 1969;**22**:549–58.

10. Kallner A, Hartmann D, Hornig D. Steady-state turnover and body pool of ascorbic acid in man. *Am J Clin Nutr* 1979;**32**:530–9.

11. Ginter E. What is truly the maximum body pool size of ascorbic acid in man? *Am J Clin Nutr* 1980;**33**:538–9.

12. Bluck L, Izzard A, Bates C. Measurement of ascorbic acid kinetics in man using stable isotopes and gas chromatography/mass spectrometry. *J Mass Spectrom* 1996;**31**:741–8.

13. Lindblad M, Tveden-Nyborg P, Lykkesfeldt J. Regulation of vitamin C homeostasis during deficiency. *Nutrients* 2013;**5**:2860–79.

14. Olson JA, Hodges R. Recommended dietary intakes (RDI) of vitamin C in humans. *Am J Clin Nutr* 1987;**45**:693–703.

15. Yu R, Schellhorn H. Recent applications of engineered animal antioxidant deficiency models in human nutrition and chronic disease. *J Nutr* 2013;**143**:1–11.

16. Proctor P. Similar functions of uric acid and ascorbate in man? *Nature* 1970;**228**:868.

17. Bartholomew M. James Lind and scurvy: a revaluation. *J Marit Res* 2002;**4**(1):1–14.

18. Jacob RA. Assessment of human vitamin C status. *J Nutr* 1990;**120**:1480–5.

19. Block G, Jensen C, Dalvi T, Norkus E, Hudes M, Crawford P, Holland N, Fung E, Schumacher L, Harmatz P. Vitamin C treatment reduces elevated C-reactive protein. *Free Radic Biol Med* 2009;**46**:70–7.

20. Benzie I. Vitamin C: prospective functional markers for defining optimal nutritional status. *Proc Nutr Soc* 1999;**58**:469–76.

21. Dehghan M, Akhtar-Danesh N, McMillan C, Thabane L. Is plasma vitamin C an appropriate biomarker of vitamin C intake? A systematic review and meta-analysis. *Nutr J* 2007;**13**(6):41.

22. Schwarz R, Kleinman P, Owens N. Ascorbate can act as an inducer of the collagen pathway because most steps are tightly coupled. *Ann N Y Acad Sci* 1987;**498**:172–85.

23. Peterkofsky B. Ascorbate requirement for hydroxylation and secretion of procollagen: relationship to inhibition of collagen synthesis in scurvy. *Am J Clin Nutr* 1991;**54**:1135–40.

24. Fleming J, Martin B, Card D, Mellerio J. Pain, purpura and curly hairs. *Clin Exp Dermatol* 2013;**38**:940–2.

25. Pimentel L. Scurvy: historical review and current diagnostic approach. *Am J Emerg Med* 2003;**21**:328–32.

26. Khaw K, Bingham S, Welch A, Luben R, Wareham N, Oakes S, Day N. Relation between plasma ascorbic acid and mortality in men and women in EPIC-Norfolk prospective study: a prospective population study. *Lancet* 2001;**357**:657–63.

27. Kletzel M, Powers K, Hayes M. Scurvy: a new problem for patients with chronic GVHD involving mucous membranes; an easy problem to resolve. *Pediatr Transplant* 2014;**18**:524–6.

28. Coward K, Kassner E. The determination of vitamin C by means of its influence on the body weight of Guinea-pigs. *Biochem J* 1936;**30**:1719–27.

29. Harris D, Ray M. Diagnosis of vitamin C subnutrition by urine analysis: with a note on the antiscorbutic value of human milk. *Lancet* 1935;**225**:71–7.

30. Roe J, Kuether C. A color reaction for dehydroascorbic acid useful in the determination of vitamin C. *Science* 1942;**95**:77.

31. Ridi M, Moubasher R, Hassan Z. Spectrophotometric assay of ascorbic acid with perinaphthindanetrione hydrate. *Science* 1950;**112**:751–2.

32. Munday K, Fulford A, Bates C. Vitamin C status and collagen cross-link ratios in Gambian children. *Br J Nutr* 2005;**93**:501–7.

33. Kraenzlin M, Kraenzlin C, Meier C, Giunta C, Steinmann B. Automated HPLC assay for urinary collagen cross-links: effect of age, menopause, and metabolic bone diseases. *Clin Chem* 2008;**54**:1546–53.

34 Chromsystems. Instruction manual for the HPLC analysis of vitamin C in plasma/serum. Chromsystems Instruments & Chemicals GmbH, Germany Order number: 65065n.d..

35. Fukuwatari T, Shibata K. Urinary water-soluble vitamins and their metabolite contents as nutritional markers for evaluating vitamin intakes in young Japanese women. *J Nutr Sci Vitaminol (Tokyo)* 2008;**54**:223–9.

36. Loh H. The relationship between dietary ascorbic acid intake and buffy coat and plasma ascorbic acid concentrations at different ages. *Int J Vitam Nutr Res* 1972;**42**:80.

37. Cornbleet T. Vitamin C content of sweat. *Arch Dermatol Syphilol* 1936;**34**:253–4.

38. Sonner Z. The microfluidics of the eccrine sweat gland, including biomarker partitioning, transport, and biosensing implications. *Biomicrofluidics* 2015;**9**:031301.

39. Rossi B, Tittone F, Palleschi S. Setup and validation of a convenient sampling procedure to promptly and effectively stabilize vitamin C in blood and plasma specimens stored at routine temperatures. *Anal Bioanal Chem* 2016;**408**:4723–31.

40. Margolis S, Duewer D. Measurement of ascorbic acid in human plasma and serum: stability, intralaboratory repeatability, and interlaboratory reproducibility. *Clin Chem* 1996;**42**:1257–62.

41. Karlsen A, Blomhoff R, Gundersen T. High-throughput analysis of vitamin C in human plasma with the use of HPLC with monolithic column and UV-detection. *J Chromatogr B Anal Technol Biomed Life Sci* 2005;**824**:132–8.

42. Kand'ár R, Záková P. Determination of ascorbic acid in human plasma with a view to stability using HPLC with UV detection. *J Sep Sci* 2008;**31**:3503–8.

43. Smith J, Hodges R. Serum levels of vitamin C in relation to dietary and supplemental intake of vitamin C in smokers and nonsmokers. *Ann N Y Acad Sci* 1987;**498**:144–52.

44. Farber C, Kanengiser S, Stahl R, Liebes L, Silber R. A specific high-performance liquid chromatography assay for dehydroascorbic acid shows an increased content in CLL lymphocytes. *Anal Biochem* 1983;**134**:355–60.

45 Merck Index, 12th ed., Merck and Co. Incn.d.

46. Margolis S, Vangel M, Duewer D. Certification of standard reference material 970, ascorbic acid in serum, and analysis of associated interlaboratory bias in the measurement process. *Clin Chem* 2003;**49**:463–9.

47 Gesellschaft zur Förderung der Qualitätssicherung in medizinischen Laboratorien e.V.Ubierstr.2040223 Düsseldorf, www.instand-ev.de.

48. Ross M. Determination of ascorbic acid and uric acid in plasma by high-performance liquid chromatography. *J Chromatogr B Biomed Appl* 1994;**657**:197–200.

49. Horwitz W. Evaluation of analytical methods used for regulation of foods and drugs. *Anal Chem* 1982;**54**:67–76.

Index

Note: Page numbers followed by *f* indicate figures and *t* indicate tables.